Physical Methods
in Modern Chemical Analysis

Volume 3

PHYSICAL METHODS IN MODERN CHEMICAL ANALYSIS

Edited by

THEODORE KUWANA
Department of Chemistry
The Ohio State University
Columbus, Ohio

Volume 3

 1983

ACADEMIC PRESS
A Subsidiary of Harcourt Brace Jovanovich, Publishers
New York London
Paris San Diego San Francisco São Paulo Sydney Tokyo Toronto

ACADEMIC PRESS, INC.
111 Fifth Avenue, New York, New York 10003

United Kingdom Edition published by
ACADEMIC PRESS, INC. (LONDON) LTD.
24/28 Oval Road, London NW1 7DX

Library of Congress Cataloging in Publication Data
Main entry under title:

Physical methods in modern chemical analysis

Includes bibliographies and indexes.
1. Chemistry, analytic. I. Kuwana, Theodore.
QD75.2.P49   543   77-92242
ISBN 0-12-430803-1 (v.3)

PRINTED IN THE UNITED STATES OF AMERICA

83 84 85 86   9 8 7 6 5 4 3 2 1

Contents

X-Ray Spectrometry

Donald E. Leyden

Transform Techniques in Chemistry

Alan G. Marshall

Electrochemical Characterization of Chemical Systems

Larry R. Faulkner

Global Optimization Strategy for Gas-Chromatographic Separations

R. J. Laub

List of Contributors

Numbers in parentheses indicate the pages on which the authors' contributions begin.

Larry R. Faulkner (137), Department of Chemistry, University of Illinois at Urbana-Champaign, Urbana, Illinois 61801

R. J. Laub (249), Department of Chemistry, San Diego State University, San Diego, California 92182

Donald E. Leyden (1), Department of Chemistry, Colorado State University, Fort Collins, Colorado 80523

Alan G. Marshall (57), Departments of Chemistry and Biochemistry, The Ohio State University, Columbus, Ohio 43210

Preface

Chemists today are faced with chemical analyses that are becoming increasingly complex. Principal component analysis has been replaced with identification and determination of trace species in complex mixtures. Furthermore, the trace analyses frequently involve levels below the microgram range, extending to the nanogram and femtogram ranges. The problem of identification is also nontrivial, extending from organic and inorganic compounds in various matrices to complex biological macromolecules. New tools often associated with sophisticated instrumentation are also constantly being introduced. Surface analysis is a good example of an area for which recent years have seen the advent of many new methods, and the abbreviations ESCA, SIMS, XPS, LEEDS, etc., are now common in the literature. These methods have made it possible to analyze and characterize less than monolayers on solid surfaces. Thus the demand upon a practicing chemist is to have a working knowledge of a wide variety of physical methods of chemical analysis, both old and new: the new ones as they are developed and applied, and the old ones as they are better understood and extended. It is the aim of "Physical Methods in Modern Chemical Analysis" to present a description of selected methodologies at a level appropriate to those who wish to expand their working knowledge of today's methods and for those who wish to update their background. It should also be useful to graduate students in obtaining a basic overview of a wide variety of techniques at a greater depth than that available from textbooks on instrumental methods.

"Physical Methods in Modern Chemical Analysis" will contain chapters written by outstanding specialists who have an intimate working knowledge of their subject. The chapters will contain descriptions of the fundamental principles, the instrumentation or necessary equipment, and applications that demonstrate the scope of the methodology.

It is hoped that these volumes continue the standard exemplified by the earlier volumes, "Physical Methods in Chemical Analysis," edited by Walter Berl in the 1950s and 1960s.

The patience and assistance of my wife, Jane, during the editing process are gratefully acknowledged.

Contents of Other Volumes

X-Ray Spectrometry

Donald E. Leyden

Department of Chemistry
Colorado State University
Fort Collins, Colorado

I. Introduction

X-ray spectrometry (x-ray fluorescence) is an emission spectroscopic technique of important usefulness for the qualitative and quantitative determination of elements of atomic number greater than ten. The technique depends upon the characteristic emission of x radiation, generally in the 1–60-keV energy range (12–0.2-Å wavelength range), following excitation of atomic energy levels by an external energy source such as other x rays, electron beams, or accelerated charged particles. In most sample matrices, x-ray fluorescence can detect elements at concentrations of about one

1

microgram per gram of sample (1 ppm), or in total amounts of a few micrograms. X-ray spectrometry has proved to be of great value in metallurgical and geochemical analyses and has found many other industrial applications.

X rays were discovered by Roentgen (thus, Roentgen rays in German) in 1895. Moseley laid the foundation for the relationships between atomic structure and the x-ray emission spectra of the elements. He recognized the potential for quantitative analysis using x rays. The chain of events involving the development of the Coolidge x-ray tube (1913), the development of the Soller collimator for x rays (1924), and the improvement of gas x-ray detectors by Geiger and Muller (1928) eventually led to the design of a prototype commercial spectrometer by Friedman and Birks (1948). Thus a sensitive, selective, nondestructive (in most cases) analytical spectroscopic tool was introduced. Today there are many analytical determinations using x-ray spectrometry. A large percentage of analyses performed in the mining and metals industry are done with the aid of x-ray spectrometry. The introduction of energy dispersive x-ray spectrometers created a renewed interest in the technique and has been partially responsible for additional applications such as multielement determinations in environmental samples.

II. Principles of X-Ray Spectrometry

Spectroscopic techniques depend generally upon the interaction of electromagnetic radiation with matter. Figure 1 shows the commonly useful range of the electromagnetic spectrum. Low energy radiation interacts with low energy transitions in atoms and molecules. For example, a transition in the energy content of the vibrational modes of chemical bonds requires energy in the infrared region of $\sim 5 \times 10^{-5}$ keV. Transitions in the electronic energy levels in the valence or bonding electrons of atoms, ions, or molecules re-

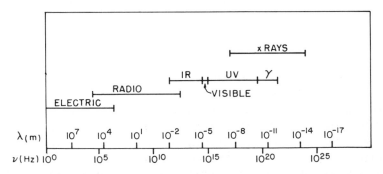

Fig. 1 Wavelengths and frequencies of electromagnetic radiation.

quires energy in the visible–ultraviolet region of $\sim 5 \times 10^{-3}$ keV. However, sufficient excitation of an atom or ion to result in the emission of x-radiation requires energy in the x-ray region of the electromagnetic spectrum of ~ 5 keV. Thus x-ray spectrometry utilizes radiation energy that lies at the high energy side of radiation that can be used with relative safety. Although modern commercial instruments have been designed with many safety features, the user should be aware of proper procedures to ensure safety [1, 2], as well as local and national codes for installation and inspection.

A.   Absorption of X Radiation

When x-ray photons impinge upon a collection of atoms, there are two types of interaction that will be of concern in this chapter: scattering of the x-ray photons and absorption of the x-ray photon energy. Scattering occurs when the x-ray photon interacts with electrons in the outer orbitals of the atom. These electrons will likely require much less energy to be ejected from the atom than the x-ray photon energy. Thus when the x-ray photon energy quantum "collides" with these low energy electrons, two results may be obtained as shown in Fig. 2. Elastic (Rayleigh) scattering may occur, which causes no change in the energy (wavelength) of the incident radiation. Rayleigh scattering is the principal source of the background observed in

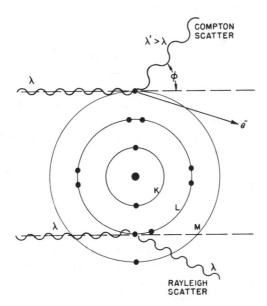

Fig. 2   Illustration of Compton scattering and Rayleigh scattering of x radiation.

x-ray spectra. The background will thus show a scattering spectrum characteristic of the spectrum of the incident x-ray source.

If the collision is inelastic (Compton), scattering of the incident radiation will occur with loss of energy or a shift of the scattered radiation to longer wavelength. This shift is predicted from theory to be

$$\lambda'_A - \lambda_A = \Delta\lambda_A = \frac{10^8 h}{M_e C} (1 - \cos\phi) \qquad (1)$$

where h is Planck's constant (6.6×10^{-27} erg sec), C is the velocity of electromagnetic radiation ($\sim 3 \times 10^{10}$ cm/sec), M_e is the electron mass (9.1×10^{-28} g), and 10^8 is the conversion for Å/cm. Because most commercial instruments have the source–sample–detector angle near 90°, ϕ becomes a supplemental angle near 90°, resulting in $\cos\phi = 0$. This results in $\Delta\lambda \cong 0.024$ Å, which is known as the Compton wavelength. All incident radiation may be scattered. If the incident radiation is a sharp, nearly monochromatic beam, a scattering peak is observed at a wavelength ~ 0.024 Å longer than the incident beam; this is the Compton peak. Rayleigh scattering will lead to a scattering peak unmodified in wavelength. The Compton-to-Rayleigh scattering peak intensity ratio increases as the incident x-ray energy increases and as the atomic number of the sample decreases. This latter dependence proves useful in correction of emission intensity for matrix variation.

With a few exceptions, scattering is a nuisance in x-ray spectrometry because of the contribution to background. A more important interaction is the one responsible for the characteristic emission of x rays from atoms in the sample. Chemists are familiar with the absorption spectral peak shape in most spectroscopic methods. As illustrated in Fig. 3a, the electronic transition is between two quantized energy levels. The transition occurs only where $\Delta E = h\nu$, where ν is the frequency of the incident radiation. Therefore, when a monochromator is scanned, an absorption peak occurs at this frequency broadened because of low energy processes such as changes in the vibrational energy levels. However, if a similar experiment were performed with incident x radiation, a sudden jump or edge in the absorption would be seen rather than a peak, as illustrated in Fig. 3b. To understand this shape for the energy dependence of absorption one must consider the details of the process.

If the energy of the hypothetical monochromatic incident x-ray beam is less than that required to photoeject an electron from an orbital lower in energy than the valence or bonding orbital, the absorption of incident radiation will be primarily a result of the scattering processes. As the energy of the incident beam is increased, an energy is eventually reached at which an inner orbital electron is photoejected from the atom as illustrated in Fig. 4. The energy required to photoeject an electron decreases as the principal

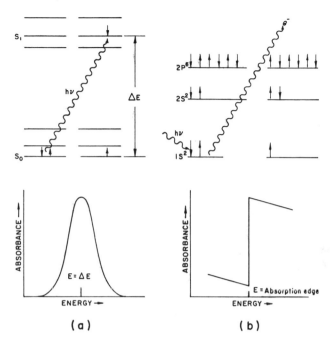

Fig. 3 Excitation of electronic energy levels: (a) transition between two quantized energy levels and (b) photoejection of electrons by x radiation.

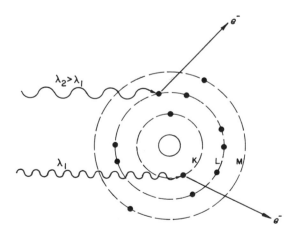

Fig. 4 Illustration of photoejection of K electrons by higher energy radiation and L electrons with lower energy radiation.

quantum number of the affected orbital increases. Thus 7.1-keV x rays are required to photoeject 1s electrons from copper, whereas x rays of only about 0.8 keV are required for 2s or 2p electrons. In the case of magnesium, the values are 1.3 and 0.06 keV, respectively. This photoejection is an example of the well-known photoelectric effect. In summary, the total absorption of x rays by a pure element is a linear combination of all the photoelectric and scattering processes. The net result is expressed in a way very similar to Lambert's law:

$$I = I_0 \exp -(\mu/\rho)\rho \cdot t \qquad (2)$$

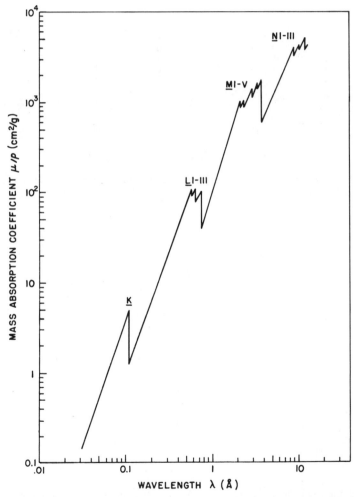

Fig. 5 X-ray absorption curve for uranium as a function of wavelength. [After Bertin [3], with permission of Plenum Press.]

where μ/ρ is the mass absorption coefficient (in square centimeters per gram), ρ is the density (in grams per cubic centimeter), and t is the thickness of the absorbing sample. As in the Beer–Lambert law, this observation does not require a knowledge of which processes are controlling the attenuation of the x rays, but does assume a monochromatic beam. An example of the effect of wavelength is shown in Fig. 5. However, it is only the photoelectric absorption mechanism that leaves a vacancy in the inner orbitals, which may result in characteristic x-ray emission.

Before discussion of the x-ray emission process, it should be mentioned that x-ray absorption measurements also represent an analytical method, but of considerably more limited application than emission measurements. The primary reason for the limitation lies in the fact that an absorption jump occurs rather than a peak as a function of wavelength or energy as described above. This means that the absorption is not selective. For example, if one wishes to measure Cr in steel by x-ray absorption, the Fe in the steel will have a mass absorption coefficient larger than Cr at any wavelength above the Fe absorption edge. Thus to get a measurement for Cr one must use two measurements; one wavelength just below an absorption edge for Cr and one just above that value. The difference in the two values would be mainly due to the photoelectric effect of Cr. This creates a more complicated set of experiments than measuring the Cr emission intensity directly. Thus x-ray absorption as an analytical tool has been limited primarily to routine process control applications.

B. Emission of X Radiation

The absorption of x radiation as a result of the photoelectric effect results in an excited electronic state in the target atom. A simple partial illustration of the electronic transitions for copper is given in Fig. 6. Figure 6a shows the absorption spectrum for copper for the photoejection of a 1s (K) electron. This photoejection has created a vacancy in the 1s orbital of the copper orbital. This is in fact an ionization, but from the core electrons rather than the valence electrons as is more commonly the case. In fact, the energy of the 1s electrons is so shielded from the state of the valence electrons that the absorption edge energy (K_{ab}) and the energy of the emitted x radiation are essentially independent of the oxidation state and bonding of the atoms. Such effects can be observed with atoms of low atomic number, but only with difficulty, and such shifts and spectral fine structure are not important in general applications of x-ray spectrometry. Once the vacancy has been created by the photoelectric effect, the result is a high energy state that may relax in part by transitions of electrons from higher energy orbitals, filling the vacancy in the inner orbital, but creating a lower energy state having an

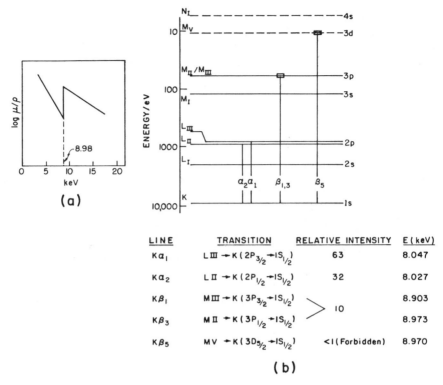

Fig. 6   Transition diagram for copper: (a) absorption curve and (b) transitions.

orbital vacancy. Eventually an electron is picked up from the environs to regain electroneutrality. The transitions must follow quantum mechanical selection rules: $\Delta n > 0$, $\Delta l = \pm l$, and $\Delta j = \pm l$ or 0, where n is the principal quantum number, l is the angular quantum number, and $j = l + s$ is the vector sum of l and s, the spin quantum number for spin–orbit coupling. The transitions that follow the selection rules are termed allowed (diagram) lines, those that do not are called forbidden, and those that result in atoms that have two or more vacancies in inner orbitals at the time of the emission are called satellite (nondiagram) lines. One of the unfortunate aspects of x-ray spectrometry is that a confusing, unsystematic, and unconventional scheme of notation for the x-ray spectral lines appears in most text and reference books. It cannot be the purpose of this chapter to change that notation. The notation more or less conventional to most texts will be used here. The reader may wish to read further to be clear on the notation [3, 4]. Figure 6b shows the transitions for the K lines of copper. The number of K lines observed for an element depends in part upon the number of filled orbitals.

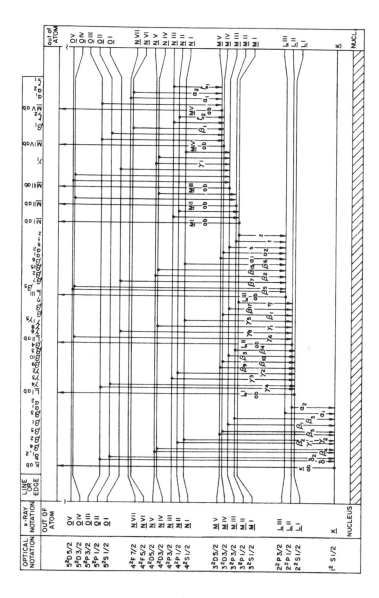

Fig. 7 General transition diagram for x-ray emission.

For example, the weak, forbidden K_{β_5} line for copper is observed because there are no $4p_{1/2,\,3/2}$ electrons to provide the K_{β_2} line of nearly the same energy, which would thus obscure the much weaker K_{β_5}. A much more complete transition diagram is shown in Fig. 7.

Note that if the atom is irradiated by an x ray of sufficient energy to photo-eject a K electron ($h\nu > K_{ab}$) from an element, then both the K and L (etc.) series of lines may be observed as the L orbital vacancies are created by the process of emission of the K_α lines. However, if the incident x ray is only greater in energy than the L absorption edge energy ($K_{ab} > h\nu > L_{I}, L_{II}, L_{III}$), only the L and M (etc.) series will be observed.

In summary, atoms may be excited by the photoelectric absorption of x rays, leading to the creation of a vacancy in an inner orbital of the atom. X-ray absorption (or apparent absorption) resulting from scattering does not lead to such an excited state. Transitions that obey the quantum mechanical spectroscopic selection rules yield strong lines and those that are forbidden yield only very weak lines, usually not observed. The energy and relative intensities of these lines create an x-ray emission spectrum characteristic of each element.

C. Relationships between X-Ray Spectrometry and Other Techniques

It is appropriate to mention the relationships between x-ray spectrometry and other spectroscopic methods at this point in this chapter. First, to do so will put the method qualitatively in perspective with other techniques. Second, it will provide an opportunity to discuss some important side points of x-ray emission by atoms.

Figure 8 shows in more detail the events that may occur in an excited atom; in this case following the photoejection of a K ($1s_{1/2}$) electron from the target atom. When the electron is ejected, it escapes the sample surface with a kinetic energy equal to the difference between the incident photon energy $h\nu$ (assuming a monochromatic beam) and the binding energy ϕ_K of the ejected electron. Less fundamental but important terms such as the instrument surface work function will not be considered here. If the electrons are collimated, passed into a magnetic energy analyzer, and detected as a function of kinetics, a spectrum of intensity versus electron kinetic energy is obtained. If the incident x ray is so low in energy that outer orbital (not valence) electrons are ejected, the binding energy ϕ_n becomes substantially dependent upon the oxidation state and bonding of the target atom. This technique is known as Electron Spectroscopy for Chemical Analysis (ESCA) and was pioneered by Siegbahn and co-workers [5]. ESCA provides an excellent tool for the investigation of surface chemistry. The technique is

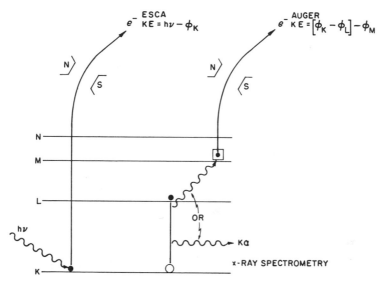

Fig. 8   Illustration of ESCA and Auger processes as related to x-ray spectrometry.

limited to surfaces because the escape depth of the photoejected electron is 100 Å or less, or to gases at very low pressures.

Once the photoejection is accomplished, an excited state exists. The relaxation mechanism leading to x-ray emission as represented by the $K_{\alpha_{1,2}}$ line shown in Fig. 8 has been described. However, there is a competing mechanism: when the L electron undergoes the L $\rightarrow$ K ($2p_{1/2,\,3/2} \rightarrow 1s_{1/2}$) transition, the energy released may result in the ejection of a second electron, rather than an x-ray photon. This secondary electron emission is known as the Auger process [6]. As in the case of the ESCA electron, the Auger electron will escape from a few atomic layers beneath the surface with a kinetic energy that may be analyzed. The input energy is the difference between the binding energies of the K and L electrons ($\phi_K - \phi_L$), which is essentially the same as the energy of the $K_{\alpha_{1,2}}$ x ray for the target atom. Thus by measurement of the kinetic energy of the Auger electron, the valence and bonding-state-dependent binding energy of the M electron may be determined.

It is important to recognize that the Auger effect is in competition with the x-ray emission process. As indicated in Fig. 8, either an x-ray photon or an Auger electron is emitted from an atom in the excited state, i.e., with an inner orbital vacancy. Electrons in atoms of low atomic number have low binding energies and the x-ray photons emitted from these atoms couple well to give the internal absorption for ejection of a secondary electron. The ratio of excited atoms that relax by x-ray emission to the total atoms excited is called the fluorescent yield (ω) and increases with an increase in the atomic number of the target atom. A plot of fluorescence yield versus atomic number

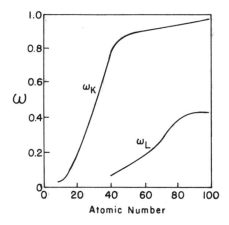

Fig. 9 Fluorescence yield as a function of atomic number.

is shown in Fig. 9. Note that the value ω_K rises from about 8% for phosphorus to approximately 90% for lead for the K lines, whereas ω_L remains much lower. The low values of the fluorescent yield are a major contributor to the relatively low sensitivity of x-ray spectrometry for elements of low atomic number. This is a result of events within the atom and cannot be altered by instrument design.

Two other techniques that are related to conventional x-ray spectrometry are electron-probe microanalysis [7] and scanning electron microscopy. These two methods are closely related. The electron-probe instruments are designed for x-ray analysis and operate with a high electron beam current and low resolution ($\sim 1~\mu m$). Scanning electron microscopes are designed for the highest possible resolution (~ 100 Å) and image display and operate at lower beam currents. In both cases a focused electron beam may be rastered across the sample. X radiation resulting from the ejection of inner orbital electrons by the high energy electron beam is detected and an elemental map of the surface may be constructed. In modern instruments this elemental map may be displayed in color, with each color or tone representing a different element. Quantitation with these techniques presents many difficulties. However, the ease of qualitative identification of elements on surfaces makes them extremely powerful tools.

III. Instrumentation for X-Ray Spectrometry

X-ray spectrometers may be considered similar to any other type in block diagram from. The principal components are an energy source, a sample holder, an optical system, a dispersion device, a detector, and some form of data readout. Some of these components are substantially different from those found in infrared, visible, and ultraviolet spectrometers because of the high energy of the x radiation. Safety features to avoid accidental exposure of the operator to x rays are also present. However, a casual glance at a modern energy dispersive x-ray spectrometer will reveal no more of the nature of the technique than any other spectrometer.

The fact that there are two major types of instrumentation complicates the discussion of instrument design in a limited chapter. The older, but important, wavelength dispersive instruments have been discussed in detail in many books and monographs [8, 9]. The discussion of energy dispersive x-ray spectrometers has been much more limited. A discussion of both types of instrumentation will be given here. It is beyond the scope of this chapter to discuss the several types of portable instruments that use filters, dual detectors, and other arrangements to perform selected functions [10, 11].

A. *Wavelength Dispersive Instrumentation*

Wavelength dispersive x-ray spectrometers are one of two important types of x-ray spectrometers available commercially. The term is meant to reflect the fact that these instruments contain a physical device for the angular dispersion of the radiation emitted by the sample so that specific characteristic lines may be selected for intensity measurement. The sequence of spectroscopic events can be used as an organizational basis for the components.

1. *Sources of Excitation*

In Section II.A, the principles of excitation of atoms prior to x-ray emission described the use of x-ray photons as an energy source. As it happens, there are two other important energy sources that may be used to photoeject electrons from the inner orbitals of atoms. These are accelerated electrons and charged particles such as protons. With the exception of electron-probe microanalysis, scanning electron microscopy, and some limited applications to portable or on-line spectrometers, electron beams as energy sources are not widely used in x-ray spectrometry. Principal reasons for this are that to use electron beams, the sample must be in an inconvenient high vacuum and considerable difficulties are encountered from charging effects if the sample is not a conductor. Proton or other ion beams may be obtained by utilizing a Van de Graaff generator. X-ray spectrometry is done using these elaborate devices [12].

Wavelength dispersive x-ray spectrometers require a high power efficient excitation to perform to maximum capacity. Because these instruments are used for precise quantitative measurements, and very often in off-line process control of expensive operations, stability and reliability of the source are extremely important. The solution to these needs has been found in the x-ray tube, a direct descendant of the one designed by Coolidge [13]. A schematic representation is shown in Fig. 10. All components are in a high vacuum. A filament (a) is heated by a filament voltage of a few volts. The heated filament thermally ejects electrons. As the filament voltage is increased by adjustment of a highly regulated circuit, the filament current and thus its temperature increases. The result is a higher rate of production of electrons, which leads to a higher x-ray intensity. This current is usually measured in milliamperes; the tube current I ("mA") is an important instrumental parameter.

A potential of several kilovolts is applied between the filament (cathode) and an anode (b) that serves as a target for the electrons that are accelerated to the anode; the "kV" or tube potential V (in kilovolts) is also an important instrumental parameter. The anode is prepared from or plated with a pure element such as W, Mo, Cr, Ag, Au, Rh, or other metal. Because the x-ray tubes used in wavelength dispersive x-ray spectrometers operate at 2–3 kW provisions are made for cooling of the anode by flowing water. The electrons strike the anode with a maximum kinetic energy equal to the applied tube potential. As the electrons penetrate the surface of the target material, two processes that produce x rays may take place. If the kinetic energy of the electron is equal to or greater than the absorption edge energy corresponding to the ejection of an inner orbital electron of the atom of the target, the electron will be ejected by collision between electrons in the beam and those

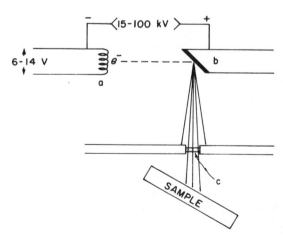

Fig. 10   Schematic representation of a Coolidge x-ray tube: (a) filament, (b) target anode, and (c) beryllium window.

in the atomic orbitals. Once an orbital vacancy has been created, the target atoms will emit characteristic line spectra as discussed earlier. As the electrons in the beam enter the target, several mechanisms of inelastic scattering may lead to deceleration of the electrons. The loss in kinetic energy is manifest as x-ray photons. The maximum amount of energy released in one unit is the kinetic energy of the electron, which in turn is the x-ray tube operating voltage. The Duane–Hunt law, Eq. (3), expresses the wavelength equivalent as the shortest wavelength x ray that can be obtained from an x-ray tube operated at a potential of V (in kilovolts):

$$\lambda_{min} \text{ (angstroms)} = 12.4/V \tag{3}$$

A continuum distribution of wavelengths longer than λ_{min} is produced as shown in Fig. 11. The characteristic line spectrum of the target atom will be superimposed upon the continuum, provided the tube V is sufficiently large. Again referring to Fig. 10, x rays produced at the anode are transmitted by a beryllium window (c) and impinge upon the sample.

Selection of V and I as well as the anode target element for a particular x-ray spectrometric determination is an important part of the procedure. Several considerations of detail must be made. However, the principal factors are to select a target material such that an intense characteristic line of the target lies at a wavelength just shorter than the absorption edge of the sample element, so that the latter is efficiently excited for x-ray emission.

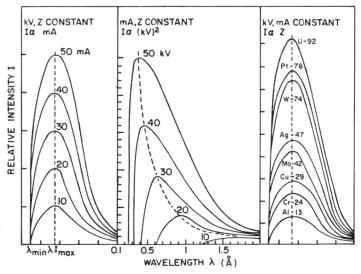

Fig. 11 Effect of current, potential, and target element on the continuum spectrum. [After Bertin [3], with permission of Plenum Press.]

The operating V is selected to ensure that this anode characteristic line is excited and to position the maximum intensity of the continuum at a wavelength efficient for excitation of the sample. The tube current is adjusted so that the intensity of the measured x-ray emission lies in a workable range. Obviously, many compromises are usually required.

2. Optical Geometry and Dispersion

Figure 12 shows a schematic representation of the optical geometry of an x-ray spectrometer with inverted optics; that is, a geometry in which the sample is located above the x-ray tube. The x radiation from the x-ray tube impinges upon the sample from below. All atoms in the sample that have absorption edge energies below the maximum x-ray energy provided from the tube are excited and emit x rays. Some of those x rays pass through a primary collimator as shown in Fig. 12. This collimator is usually a Soller [14] type, which consists of many layers of metal or metal-coated plastic

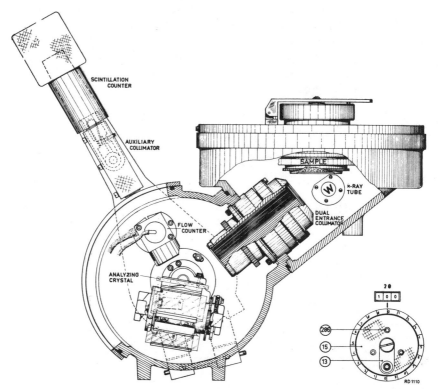

Fig. 12   X-ray optical system for Philips PW-1410 x-ray spectrometer. [With permission from Philips Electronic Instruments, Inc.]

foils separated by a millimeter or so. Only parallel x-ray beams pass these foils as divergent beams strike the foil and are absorbed or scattered. Once collimated, the ray of parallel x rays containing the radiation characteristic of all emitting elements in the sample impinges upon a crystal. This crystal may be flat or curved in one or two dimensions. This "analyzing crystal" is the heart of the dispersive nature of a wavelength dispersive x-ray spectrometer.

Figure 13 shows how the analyzing crystal serves to disperse the radiation of impinging upon it so that selection of a desired wavelength of x ray for measurement can be made. As the parallel, polychromatic beam of x rays impinges upon the crystal, it is diffracted from the different crystal lattice planes. Because those diffracted from different planes must travel different distances, phase shifts leading to destructive interference will occur unless the difference is equal to the wavelength of the radiation or an integer multiple of the wavelength. In this case no phase change occurs and the radiation is reinforced. The Bragg law,

$$n\lambda = 2d \sin \theta \qquad (4)$$

permits the calculation of the angle θ at which a wavelength λ will be selected if the analyzing crystal has a lattice spacing of d; both d and λ are in angstroms. Because of the mechanical arrangement of the goniometer, the device that adjusts the sample–crystal–detector angle, it is convenient to use 2θ rather than θ. The value of n can assume integer values $1, 2, 3, \ldots$, and the resulting values of $\lambda, \lambda/2, \lambda/3$ that solve Bragg's law are called first-

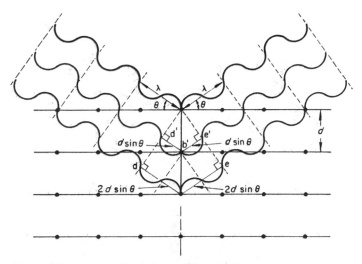

Fig. 13   Illustration of Bragg's law diffraction from an analyzing crystal.

TABLE I

Common Analyzing Crystals

Chemical name and common name[a]	Chemical formula	$2d$ (Å)
Lithium fluoride: LiF (220)	LiF	2.848
Lithium fluoride: LiF (200)	LiF	4.028
Sodium chloride: NaCl	NaCl	5.641
Germanium: Ge (111)	Ge	6.532
Pentaerythritol: PET (002)	$C(CH_2OH)_4$	8.742
Ammonium dihydrogen phosphate: ADP (101)	$NH_4H_2PO_4$	10.640

[a] Numbers in parentheses are the Miller indices to show the diffracting plane.

order, second-order, etc. lines; any of these present in the sample spectrum will reach the detector. Table I shows some common analyzing crystals and the d spacing for them. The reader may use information in Fig. 6b and Table I to determine that the first-order K_α line for copper will be at 44.97° 2θ if a LiF (200) analyzing crystal is used.

The schematic diagram shown in Fig. 12 is for a sequential spectrometer in which the angle 2θ is selected, either manually or by a computer, so that the intensities of x rays from elements in the sample are measured sequentially. The goniometer may be driven in synchronization with a recorder so that an x-ray spectrum of the sample is recorded. Alternatively, a series of analyzing crystals and detectors may be placed radially around the sample. In these "multichannel" instruments in which the sample–crystal–detector angle is fixed for each channel, several elements may be determined simultaneously. These simultaneous multichannel instruments are very expensive and serve best when speed in routine determinations is paramount.

3. Detectors and Associated Electronics

Two functions are served by the detector(s) and associated electronic circuitry in wavelength dispersive x-ray spectrometers. First and foremost is the detection of the x rays and amplification of the signal. Second, the rejection of any unwanted signal, such as higher or lower order x rays from diffraction by the analyzing crystal, and detector noise, is accomplished by the electronics. In modern sequential spectrometers it is common to find two detectors in tandem: a proportional and a scintillation detector. The proportional detector shown in Fig. 14 is a gas-filled or flowing-gas detector usually using P-10 gas, which is 90% Ar and 10% methane. This detector consists of a hollow metal cylinder with a wire from end to end down the major axis and insulated electrically from the cylinder. On opposite sides of

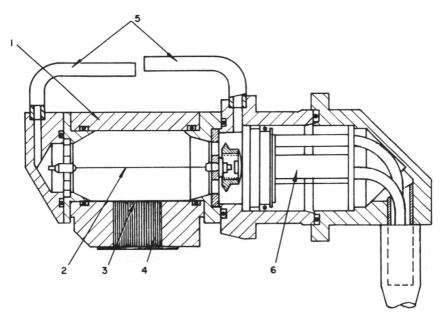

Fig. 14 Flow proportional detector: (1) cylindrical chamber, (2) anode wire, (3) window, (4) collimator, (5) flow gas connections, and (6) amplifier.

the cylinder, Mylar windows are mounted with a tight seal over openings in the cylinder. The filled detectors are sealed, whereas P-10 gas flows at a controlled flow rate and pressure in the flow detectors. A potential of several hundred volts is applied between the cylindrical case and the wire. When an x-ray photon enters the detector, electrons are photoejected from the valence orbitals of argon atoms. Because the energy required to do this is much less than the energy of x-ray photons, many $Ar^+ - e^-$ pairs are created per photon. The electrons are captured by the anodic wire and the Ar^+ ions pick up electrons at the cathodic wall of the detector. This is a fast process resulting in a short current pulse. Only a modest background current flows when there are no $Ar^+ - e^-$ pairs to act as conductors. The current pulse is easily converted into a voltage pulse and amplified. Because the magnitude of the current pulse is proportional to the number of $Ar^+ - e^-$ pairs produced, which in turn is proportional to the energy of the x-ray photon that entered the detector, the resulting voltage pulse is proportional to the x-ray photon energy. A few argon atoms may photoeject a core electron, then emit an Ar K_α x ray that escapes the detector unabsorbed. The result is a quantity of undetected energy equal to that of the Ar K_α photon. A pulse will result equivalent to the energy of the incident x rays minus the energy of the Ar K_α x ray (2.96 keV). When a scan of the pulse distributions is made, these escape peaks are seen at positions equivalent to 2.96 keV less energy than the main peak, especially for intense peaks.

Proportional counters are most efficient for detection of x-ray photons of energies less than ~8 keV and of wavelengths greater than ~1.5 Å. More energetic x rays tend to pass through the detector unabsorbed. It is common to find a scintillation detector located behind the proportional detector (Fig. 12). As shown in Fig. 15, these detectors consist of a scintillation crystal such as thallium-doped sodium iodide, NaI(Tl), which emits a burst of blue (410 nm) light when struck by an x-ray photon. The number of light photons produced is proportional to the incident x-ray photon energy. The scintillation crystal is mounted directly onto the optical window of a photomultiplier tube. After electronic processing, the scintillation burst is converted to an amplified voltage pulse.

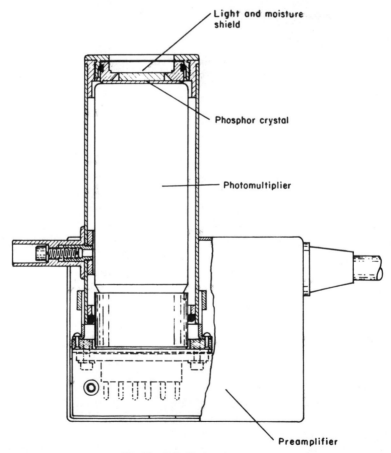

Fig. 15   Scintillation detector.

These two detectors may be operated separately and independently, or simultaneously. In simultaneous operation, the detector operating potential (or output gain) must be adjusted so that an x-ray photon of a given energy gives the same voltage pulse output from both detectors. Both detector types need time (~ 1 μsec) to recover. At incident photon rates greater than $\sim 30,000/\text{sec}$, some counts may be lost.

As mentioned earlier, the analyzing crystal disperses radiation from the sample so that the desired analyte line is selected for incidence upon the detector(s). However, the Bragg law is satisfied not only by first-order diffraction of the primary wavelength λ but also by second-order ($\lambda/2$), third-order ($\lambda/3$), and so on. These shorter wavelengths are those of higher energy x rays; if they are present in the spectrum of the sample they will result in higher pulse heights from the detector than those from the first-order wavelength λ. An important electronic component in wavelength dispersive x-ray spectrometers is the pulse-height discriminator. This device permits the selection of the pulse-height range that will be processed by the counting circuits. Any pulse below the lower setting and any above the upper limit will be rejected and not counted. In this way, low energy noise pulses and pulses with higher energy than first-order lines will be rejected.

The output of the pulse-height discriminator is inputted into a scaler timer. The timer may be preset and the number of counts accumulated during that time displayed. Alternatively, the scaler may be preset and the time required to accumulate the selected number of counts displayed.

4. *Variations in Modern Wavelength Dispersive Instrumentation*

Modern wavelength dispersive x-ray instrumentation consists of two important types. The less expensive versatile, sequential instruments are still the most common. There has been only limited improvement in the optical designs in recent years. The counting electronics have been greatly improved by the exclusive use of solid state components. The use of minicomputer control of the instrumental parameters such as tube current and tube potential, choice of collimator, selection of one of several crystals, selection of detector, adjustment of pulse-height-discriminator limits, goniometer settings, and the acquisition, storage, and processing of the x-ray intensity data have been the most significant advancements. In situations involving several different routine applications, a program stored on disk or casette may be loaded into the computer and the instrumental parameters adjusted entirely by the computer. Once a program is prepared for a certain job by a skilled spectroscopist, it can be executed by a less skilled technician with minimal chance of experimental error. Such software changes of the instrumental function are much faster and less complicated than the hardware changes required to change the function of multichannel, simultaneous instruments.

However, some of the latter instruments incorporate a scanning goniometer along with several fixed simultaneous channels. Even more recently, micro-computers such as the commercially available personal computers have been employed in control and acquisition functions of wavelength dispersive x-ray spectrometers.

B. Data Treatment in Wavelength Dispersive X-Ray Spectrometry

The output of the detector pulse-height selector of a wavelength dispersive x-ray spectrometer is displayed either as total counts in a fixed time (digital) or through a ratemeter (analog) that displays the count rate (counts per second) and may be output onto a recorder. The accumulated count display is useful in quantitative measurements and the ratemeter for qualitative applications.

1. Qualitative Analysis

Because each element emits a characteristic x-ray spectrum, elemental identification is readily performed using x-ray fluorescence. The ratemeter output on a strip recorder displays the instantaneous x-ray intensity. If the goniometer is driven at a constant rate, a record of intensity versus 2θ is made; an x-ray spectrum is recorded. Figure 16 shows an x-ray spectrum of steel recorded on a sequential x-ray spectrometer. Elemental identifica-tions are made by comparison of the peak 2θ positions found in tables for each analyzing crystal [15] with those in the spectrum. Care should be taken to use more than one peak, because accidental overlap may occur. For example, the arsenic K_α line is at 1.178 Å and the lead L_α line is at 1.179 Å. Only a careful match of the As K_α/K_β or Pb $L_\alpha/L_\beta/L_{\gamma 1}$ pattern with reason-ably close to theoretical intensity ratios can give a reliable elemental identi-fication. One must also be alert to the possibility of second-order lines, scattered x rays from the characteristic spectrum of the x-ray tube, and escape peaks corresponding to strong lines in the spectrum. Each of these can be misleading. However, with reasonable care qualitative identification can be made. Some recent instruments can avoid the slow scanning between possible peak positions by computer control of the goniometer so that a rapid slew of the goniometer is done between possible peak positions as calculated by the computer from peak wavelength data stored in memory.

2. Quantitative Analysis

In cases of quantitative applications of wavelength dispersive x-ray spectrometry, a spectrum is usually recorded to locate the precise peak 2θ

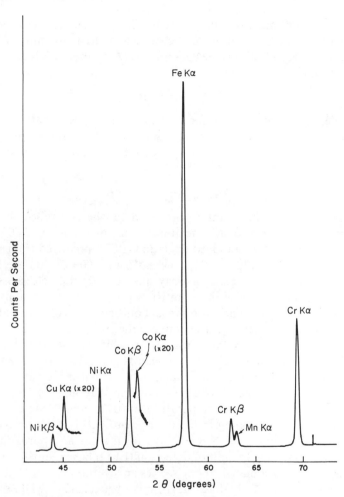

Fig. 16 Wavelength dispersive x-ray spectrum of AISI Type 347 stainless steel. Mo x-ray tube, 30 kV, 30 mA; P-10 flow proportional detector; LiF (200) analyzing crystal; fine collimation; 100 kcps full scale.

position, determine whether there are peaks interfering with the analyte line selected for use, and locate 2θ positions off the peak at which background readings may be taken. The older practice of reading peak heights from spectral scans for quantitative measurements has declined in use with the introduction of digital display and computer reading of the data.

Because the scatter of x-ray tube continuum radiation from the sample creates a substantial background, readings off the peak are taken as background. The scatter increases with decreasing atomic number of the matrix and is more significant when the analyte concentration is low. Counts are accumulated at the 2θ position of the peak and the net counts N for the analyte are the peak (P) minus the background (B) counts if both are acquired for equal time:

$$N = P - B \tag{5}$$

A convenient result of counting random emission events is that Poisson statistics are followed to a close approximation. Poisson statistics predict that the standard deviation in the number of counts is simply the square root of the number of counts:

$$\text{SD} = \sqrt{N} = \sqrt{P - B} \tag{6}$$

It is easily seen that the counting precision can be controlled to a great degree by the number of counts accumulated. For example, if 100 counts are collected, the relative standard deviation (RSD) predicted from Poisson statistics is $(10/100) \times 100 = 10\%$; whereas for 10,000 counts the RSD is 1%. It is common in many texts on x-ray spectrometry to find discussions of various lengths on the use of these statistics [16]. These discussions can be misleading, since the calculations represent only statistics of repeated counting of the same sample without any change in the instrument. In fact, repeated counting of the sample on modern instruments with calculation of the mean and the standard deviation about the mean give results very close to those calculated from N. However, any analyst knows that more enters into the determination of the precision of a measurement than just the measurement step. Factors such as sample preparation, instrumental parameter adjustment, and sample positioning in the instrument usually determine the precision of the determination. Nonetheless, if taken as an approximate but useful calculation, a lower limit of detection (LLD) based only upon counting statistics can be estimated. If one makes an arbitrary choice that a signal will be real (i.e., statistically different from the background) when it is three times the standard deviation of the background, then the lower limit of analyte concentration detectable is that which gives a net count N greater than $3 \times \sqrt{B}$. For example, if 100 counts were collected for the background, a detectable net count would be 30; the total counts in the peak would be $(N + B) > 130$. Given a standard, one can use this concept to estimate the lowest detectable amount. Recently, a more careful and statistically sound treatment of the calculation of lower limits of detection has been given by Currie [17]. These proposals, rather than the simple estimates given by the counting statistics, should be used. It is usually variations in sample preparation rather than instrumental variation that set the detection values. Under any circumstances, the LLD is to be reported and interpreted only as a ballpark guide to the instrument limitations and not as some precise and

TABLE II

Detection Limits[a] for Various X-Ray Spectrometric Configurations[b]

Element	Tube excited, energy dispersive	Secondary fluorescer, energy dispersive	Wavelength dispersive
Na	1000	2000	500
Al	100	200	50
Cl	180	100	4
Ca	90	5	2
Co	15	45	1
Br	50	10	5
Zr	60	25	5

[a] ppm, for 200 sec counting time.
[b] From Jenkins [22].

highly significant value, as is sometimes done. Detection limits for a variety of instrument types are given in Table II. Generally quantitative measurements can be made satisfactorily at levels above about three times the LLD.

Methods of calibration for x-ray quantitative determinations using wavelength dispersive x-ray spectrometry are similar to those of other analytical methods. Whenever possible a set of calibration standards is used to prepare an intensity versus concentration working curve. For some sample types, such as metals and geological materials, standards are commercially available from the U.S. National Bureau of Standards or commercial suppliers. In other areas, acquisition of standards may be the single most difficult task. Often, wet chemical analysis of a portion of the sample may be required. The remaining portion is then used as a secondary standard. In cases in which the matrix has little or no effect on the intensity of the analyte line, a simple plot of peak minus background intensity versus concentration gives excellent results. When the composition of the sample matrix or the particle size of powders varies widely, or certain specific interelement effects are present, corrections may be required for these effects. These are problems that arise from the sample and are not a function of instrument design; they are discussed in Section IV.

C. Energy Dispersive X-Ray Instrumentation

Recent developments in detector and computer technology have resulted in the significant advancement of instruments that are known as energy dispersive x-ray spectrometers. This nomenclature is somewhat confusing, since there are no dispersion devices. Instead, the sorting of wavelengths by a crystal in wavelength dispersive x-ray spectrometers is replaced by energy analyzers using fast electronic circuits and computer or multichannel analyzers. There are features common to both wavelength dispersive and energy dispersive spectrometers, and other features very different in the two. The most outstanding differences is that energy dispersive x-ray spectrometry represents a truly simultaneous multielement capability without the need for either mechanical movement or hardware changes.

The first difference observed when one views modern energy dispersive and wavelength dispersive x-ray spectrometers is the difference in the physical size of the systems. Figure 17 shows photographs of two such instruments. Because only one emission line at a time is observed in the wavelength dispersive mode, an intense x-ray source is needed to compensate for this inefficient process. This means that a high power (2–3 kW) x-ray generator weighing several hundred pounds is required. Water cooling of the x-ray tube is also necessary. The mechanical components of the goniometer are heavy and require space. None of these components is necessary in an energy dispersive x-ray spectrometer.

1. *Sources of Excitation*

One may consider the analyzing crystal as the heart of a wavelength dispersive x-ray spectrometer, as is the dispersive device in any dispersive spectrometer. The detector holds that honor in energy dispersive x-ray spectrometers. The nature of the detector is such that much lower intensities of source x rays are required, indeed can be tolerated, than in wavelength dispersive instruments. Therefore, a more varied list of x-ray sources may be considered.

Isotopic sources were among the first used in early energy dispersive x-ray spectrometers. Combined with the recognition that low intensity sources were needed, isotopes offered the advantage that no power was required. However, as the instrumentation developed, these sources proved to provide too little intensity from the quantities of isotope that were safe to use in the laboratory. However, they have found use in portable instruments; instruments so portable in fact, that x-ray spectrometers have operated on Mars [18]. Radioisotopes may decay by a variety of mechanisms to emit x rays or gamma rays that may in turn be used to excite a sample. As an illustration, ^{55}Fe decays by K capture of a 1s electron by the nucleus, resulting in the formation of ^{55}Mn with a vacancy in the K orbital. This excited state relaxes by emission of Mn K x rays at 5.9 keV. Table III shows other useful radioisotopes. In addition to the low intensity of the isotopic sources, they emit sharp, narrow bandwidths of radiation. Any absorption edge below that energy will not be excited. The excitation efficiency drops significantly when an absorption edge lies well below the source energy. The result of this is

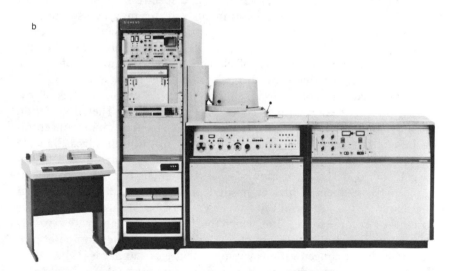

Fig. 17 Modern x-ray spectrometers. (a) Energy dispersive x-ray spectrometer. [With permission from United Scientific Corporation.] (b) Wavelength dispersive x-ray spectrometer. [With permission from Siemens Corporation.]

TABLE III
Isotopic Sources Used in Energy Dispersive X-Ray Spectrometry

Isotope	Energy of emitted radiation (keV)
$^{55}_{26}\text{Fe}$	5.9
$^{57}_{27}\text{Co}$	6.4, 14, 122, 136
$^{109}_{48}\text{Cd}$	22, 88
$^{135}_{53}\text{I}$	27, 35
$^{153}_{64}\text{Gd}$	42, 97, 103
$^{241}_{95}\text{Am}$	11–22, 26, 59.6

the need for several expensive isotopic sources to cover even commonly encountered energy ranges. This defeats the purpose of the simultaneous capability of energy dispersive x-ray spectrometry.

A source type that is similar in principle to isotopic sources is the secondary fluorescer. Secondary fluorescer sources are usually pure element metal foils that are placed between the sample and an x-ray tube. Instead of irradiating the sample directly, the x-ray tube irradiates the secondary source. This source then emits the characteristic line spectrum of the element of which it is composed. In many instruments a turret may be used to hold and position several fluorescers. Some instruments propose the use of compounded elemental mixtures in a binder to obtain an energy spread of the source lines. A sufficient number of elements would begin to synthesize a continuum readily produced by an x-ray tube. In general, secondary fluorescer sources have the same disadvantages as radioisotopic sources of low intensity and poor energy range coverage for excitation. They do offer an advantage when there is a need to determine an element of minor concentration in a matrix of elements that have higher absorption edges than the analytes. A secondary source may be selected that has an emission line of energy higher than the absorption edge of the analyte, but lower than those of any matrix element. Only the desired analyte will emit x rays. This is sometimes possible, but usually not with an x-ray tube. The use of a secondary fluorescer or isotopic source from which the excitation radiation is essentially monochromatic simplifies the use of fundamental parameter computer programs to estimate concentrations without the use of standards [19].

Disregarding nonroutine, noncommercial sources such as accelerated charged particles [12], the development of energy dispersive x-ray spectrometers has seen most of the commercial systems use x-ray tubes for sample excitation. The lack of a dispersion device in an energy dispersive x-ray spectrometer means that a much larger percentage of the x rays emitted by the sample impinge upon the detector. As in the case of proportional and scintillation detectors, those in energy dispersive spectrometers can process only about 30,000 x-ray photon counts per second.

Therefore a much lower source intensity is required. A low power x-ray tube is used to excite the sample directly, or a high power tube is used to excite a secondary fluorescer. Whereas modern wavelength dispersive instruments use 2–3-kW tubes, requiring water cooling, energy dispersive systems use 40–60-W air-cooled x-ray tubes in conjunction with filters located between the tube and sample. Figure 18 shows a typical optical arrangement for a modern energy dispersive spectrometer. In the particular instrument shown, two x-ray tubes may be mounted simultaneously. The transmission tube is an end-window x-ray tube in which a metal foil serves both as the window and as the anode. For low wattage tubes, this arrangement offers an advantage. The x-ray continuum and characteristic lines are generated as described earlier. However, rather than be collected at an angle from the surface as before and as is done in the side window tube also shown, x rays transmitted by this foil anode play a role in the tube characteristics. Remember that an element has a relatively low mass absorption coefficient

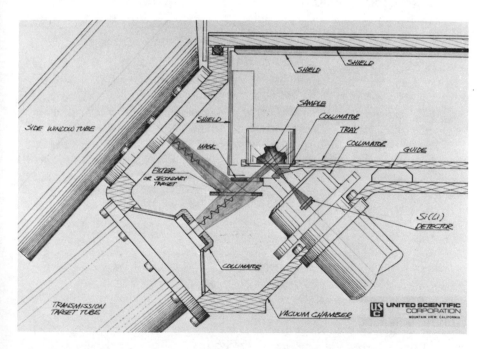

Fig. 18 Optics for an energy dispersive x-ray spectrometer. [With permission from United Scientific Corporation.]

for x rays emitted by that element. This is because the energy of all emitted x rays is below the corresponding absorption edge energy. However, continuum x rays of energy higher than the absorption edge will be efficiently absorbed and converted into x-ray lines characteristic of the target. Lower energy continuum radiation will be scattered. The result of these processes leads to a spectrum of x rays emitted by a transmission tube that is primarily the target characteristic lines. The disadvantage of the transmission tube is that lower intensities than desired are obtained because problems in cooling the foil target limit the current that can be used. A Coolidge (side window) x-ray tube may also be used in this position.

When an x-ray tube of either type is mounted in the position shown by the transmission tube in Fig. 18, a filter may be positioned between the tube and the sample. When a side window or conventional x-ray tube is used, a thin (0.01–0.2-mm) filter of the anode material can be used to accomplish the same functions as the transmission anode described above for the transmission tube, resulting in essentially characteristic lines of the tube anode element irradiating the sample. Because of this and the higher output of the side window tubes, they are most often used.

The second position for the x-ray tube in the design shown in Fig. 18 permits the irradiation of a secondary fluorescer. In this case the same unit that holds the tube filter may hold the fluorescer. A mask may be placed between the filter–fluorescer position and the sample to reduce background by reduction of the irradiated area. The distance between the tube and sample is only a few centimeters. As shown in Fig. 18, the detector is located only about 3 cm from the sample. Notice the only moving parts are the sample tray, which rotates to change samples, and the filter wheel, which permits a change of filters. The close coupling of all these optical components is very important in maintaining the best sensitivity. Thus the optical geometry in an energy dispersive x-ray spectrometer can be very simple, engineered and cast as fixed position units. Problems of crystal alignment and other variables associated with the moving parts of scanning wavelength dispersive systems are eliminated.

2. Detectors and Associated Electronics

There is no inherent, fundamental difference between the way that detectors function in energy dispersive x-ray spectrometry and in wavelength dispersive x-ray spectrometry. However, they do differ in the fact that the solid state detectors used in energy dispersive x-ray instruments have a much greater resolution than those used in wavelength dispersive instruments. In this case, by resolution one means the ability to distinguish between the energies of the x-ray photons that impinge upon the detector. It is this increased resolution brought about by the advances in solid state semiconductor detectors that led to great advances in the field of energy dispersive x rays from the period of about 1965. Figure 19 shows a Si(Li) solid state detector. This detector consists of a disk of highly purified silicon; carefully doped with lithium, this disk or wafer is essentially a semiconductor material. On the front surface of this wafer will be a thin area of inactive p-type silicon. A gold layer is plated onto this surface to serve as electrical contact point. On the back side of the wafer will be an n-type silicon layer, followed again by a gold-plated layer to provide the electrical contact. This back layer serves as a ground point for the detector. As an x-ray photon enters the detector, the photon loses its energy by the ionization of the silicon atoms. That is, they form silicon–electron pairs in much the same way as in the case of the proportional counter in which Ar^+-e^- pairs were formed. In the case of the silicon detector, the energy required to form each of these pairs is ~ 3.8 eV; therefore, the number of such pairs formed within the detector is equal to the energy of the entering x-ray photon divided by 3.8 eV. This is a considerably lower energy requirement per ion pair than we find with either the scintillation detector or the proportional detector. Therefore, the resolution of the silicon detector is much greater than that of either of the other two. A bias potential is applied across the silicon detector. Once these ion pairs are formed, the electrons migrate rapidly to the anode with reference to this bias potential. As a result, the charge generated by the formation of these ion pairs can be quickly collected and measured on a microsecond

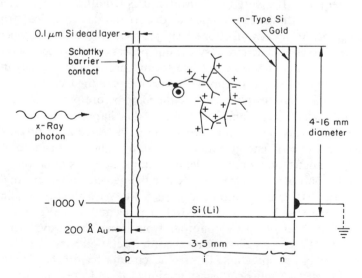

Fig. 19   Silicon solid state x-ray detector.

time scale. The resulting current from collecting the charge created within the detector is amplified directly on the detector housing using a field effect transistor (FET) amplifier. This results in a pulse, the amplitude of which is directly proportional to the energy of the x-ray photon that impinged upon the detector, being sent to the electronic processing system. The processing of these pulses will be described later. As with the previously discussed detector types, the solid state silicon detector has a varied efficiency of response as a function of energy. The detector is most efficient for x-ray photons of energies above ∼2 keV and below ∼12 keV. This is a fortuitous energy range since it covers the K emission lines for elements ranging from silicon to bromine in atomic number. Obviously, many elements with L lines lying in the range of 2 to 12 keV also exist and will be efficiently detected by the silicon detector. As with other detector types, losses in count rate may occur if the input count rate or the x-ray intensity is too high. Although somewhat complicated by the processing electronics, count rate losses for silicon detectors are observed at x-ray intensities leading to counts above ∼20,000–30,000 counts/sec. This is not significantly different from the proportional and scintillation counters discussed previously. Because it is the resolution of the silicon detector that is extremely critical in the application to x-ray spectrometry, it is important also to realize that resolution losses and energy gain shifts occur when count rates are too large, as well as actual count rate losses themselves. For these reasons, as well as the fact that all fluorescent x rays from the sample enter the detector, a low power x-ray source is required. If the fluorescent intensities from the sample are in fact too high, several things can happen to distort the data acquired from the x-ray detector. As already mentioned, simple count rate losses may occur. In these cases the electronic circuitry cannot process and properly count the x rays as they enter the detector. The second problem that exists in these detector types is known as pileup. This term refers to the fact that more than one photon can conceivably enter the detector before processing of a previous photon has been completed. When this happens it is possible to generate sum peaks. These sum peaks may be detected after processing as representing a photon energy equal to the sum of the two or more photons that enter the detector. For example, if a sample happens to be very high in iron, the K_α photons from iron may enter the detector at a very high rate. Rather than observe the photon energy as 6.4 keV for the iron K_α line, one would observe that line and in addition a peak at 12.8 keV, representing the sum of the energies of two photons from iron. Obviously, such a phenomenon would severely complicate the interpretation as well as the quantitation of an x-ray spectrum. Electronic circuits have been devised that prevent this summation or at least minimize it and are known as pulse-pileup rejection circuits. The purpose of such circuits is to detect and thereby reject the processing of the charge within the detector when two photons arrive. Obviously, the system cannot detect absolutely simultaneous arrival of photons, and therefore the count rate must be maintained at a reasonably low level. In addition to this pulse pileup rejection circuitry to prevent the formation of summation peaks, the development of pulsed x-ray tubes further refines the possibility of minimizing count rate losses and pulse pileup problems. The pulsed x-ray tube contains circuitry that allows the electron beam to be deflected from the anode target upon command. When a photon is detected as arriving at the detector by means of a special circuit, a signal is sent through the electronic circuitry to deflect the electron beam within the x-ray tube and thereby terminate momentarily the generation of x rays. This minimizes the probability of pileup occurring. As a result of this sophisticated electronic design, the probability of summation peaks, count rate losses, and high detector deadtime due to the pileup rejection process have been minimized. Therefore, the modern energy dispersive x-ray spectrometer will consist of a combination of sophisticated pileup rejection circuits working in concert with sophisticated x-ray tube design.

The resolution inherent in the silicon detector provides the advantage that x-ray photon energies can be discriminated without the use of an analyzing crystal. There are of course accompanying disadvantages. One of these is that in order for the detector to function properly and to have a lifetime greater than a few milliseconds, it must be maintained at liquid-nitrogen temperatures. This is done by mounting a detector within a stainless steel container that houses both the detector and the FET preamplifier circuit. The detector is cooled by attaching a cold finger inserted into a Dewar of liquid nitrogen. In most modern instruments there are protection circuits to shut down the instrument should the liquid-nitrogen container run dry. A silicon detector is not necessarily destroyed by warmup; however, if the bias potential is maintained with a warm detector, the lithium is rapidly diffused out of the silicon wafer and the detector is destroyed. A second and also important reason for maintaining the system at liquid-nitrogen temperatures is to lower the electronic noise. Detectors operated at room temperature do not provide the high quality resolution of approximately 150 eV that is available in commercial instruments today. Nonetheless, the inconvenience, cost, and sacrifice in portability demanded by the liquid-nitrogen Dewar is an important disadvantage. Recent work has shown that mercuric iodide (HgI_2) detectors offer promise in this area. Mercuric iodide detectors do not have the inherent resolution of silicon detectors; however, they do offer the advantage of being opertive with a bias potential at room temperature. At the time of the writing of this chapter, approximately 300 eV resolution is achievable with mercuric iodide detectors in the laboratory [20, 21]. These mercuric iodide detectors hold much promise for the future, inasmuch as portable energy dispersive x-ray systems not requiring liquid nitrogen may be feasible. These detectors will never have the resolution of laboratory-grade silicon-based energy dispersive x-ray spectrometers; however, many functions may be performed with such detectors. It is important therefore to watch for their development in the future.

The use of microcomputer-based multichannel analyzers is just being combined with these detectors and may soon offer highly portable analytical systems. Data reduction programs may be included in firmware (ROMs and PROMs) so that substantial analytical capabilities will be available in a small package. The availability of a small, 30-kV, 0.5-mA power supply–x-ray tube combination would make this an even more attractive possibility.

D. Data Treatment in Energy Dispersive X-Ray Spectrometry

As described earlier, the processing of counts of x-ray pulses arriving from detectors in wavelength dispersive x-ray spectrometers is relatively simple. A simple single-channel energy discriminator or pulse-height analyzer may be employed to reject pulses at energies both lower and higher than those desired. Remember, in wavelength dispersive systems one is interested in counting essentially one x-ray photon energy at a time. Energies have been dispersed or resolved by use of the analyzing crystal to disperse the radiation and the goniometer to set the Bragg law angle to the desired value. In the case of energy dispersive x-ray spectrometers, all photon energies emitted from the fluorescent sample are impinging upon the detector. It is the pulse-height resolution generated within the detector that allows the energy discrimination required for the energy dispersive x-ray spectrometer. The pulse heights then must be carefully processed and sorted in order to generate the x-ray spectrum. This process may be compared with scanning the 2θ angle with the goniometer to obtain an x-ray spectrum. In one case the process is achieved mechanically and in the second case electronically. In order to understand the mechanism of sorting these energies, we must discuss the basic principles of multichannel analysis.

At the risk of oversimplification, one can compare a multichannel analyzer with a mail sorting room. Pulses generated from within the detector and amplified by the FET preamplifier are transmitted, perhaps with some pulse shaping, to an analog-to-digital converter (ADC). The pulse height (maximum voltage) transmitted is directly proportional to the energy of the x-ray photon that impinged upon the detector. The function of the ADC is to convert these analog voltage pulses to a digital number that is proportional in value to the pulse height. That is, the magnitude of the analog voltage pulse height is represented by the magnitude of a binary number: an analog-to-digital conversion. Once this conversion has been made, usually in a straight binary form, the digital number is considered to be an address. In modern energy dispersive x-ray spectrometers, the boxes for these addresses lie in the memory of a minicomputer. Once the address for a particular pulse height has been determined, that address is located within memory and the content of that memory is increased by one. That is to say, a pulse of a certain energy has been detected, converted to an address, that address located within the computer and its contents increased by one count. Thus if we consider an address on a letter as being a certain post office box and if the mail is sorted into those boxes, each time a letter arrives it is located in that post office box number. This is not just the single-channel analyzer, but a multichannel analyzer that is capable of sorting energy into locations within a computer memory. The computer is capable of displaying the contents of memory as a function of energy or channel number simultaneously with the processing of the data. Therefore, as one watches a video display screen on an energy dispersive x-ray spectrometer, the increase in count content in each channel may be seen to grow in an apparently simultaneous manner. Again referring to the mechanical scanning wavelength dispersive x-ray spectrometer, one watches the chart paper as the goniometer scans the 2θ angle. In most cases, the computer can be directed by programming to display on the video screen a certain energy range, for example, from 0 to 15 keV. If we consider only the K lines, this would display the spectra of elements ranging up to atomic number ~ 38. The resolution of the spectrum would be such that we could resolve the K_α lines of all of the elements within that energy range. However, if one observes that there are considerable similarities in the energies of the K_α line for element Z and the K_β line for element $Z - 1$, it would not be surprising to find that there is considerable overlap of these peaks. The K_β line of element $Z - 1$ will be buried beneath the K_α lines of element Z. However, if the composition of the sample is such that these elements of adjacent atomic number are not present in high frequency, it is a simple task to identify the elements present within the sample. Quantitative measurements are performed with considerably more difficulty, but made easier by the fact that because a multichannel analyzer system is required a minicomputer is usually present in the system.

Little improvement is expected in the performance of the detectors themselves in terms of resolution. The electronic circuits have been greatly improved and further improvement may be expected in the future. In all probability, the most significant advance in energy dispersive x-ray systems will lie in the use of microcomputer systems in the analyzers to reduce the cost and increase the speed. A second area of improvement may lie in the use of other detector forms such as mercuric iodide. As mentioned before, mercuric iodide offers little promise for improvement in overall resolution, but does offer the possibility of portability.

1. *Qualitative Analysis*

As described earlier, qualitative analysis with sequential scanning, wave-length dispersive, x-ray spectrometers requires that the goniometer be scanned in synchronization with a strip chart recorder. In this way, peaks are observed corresponding to the 2θ angles for the various x-ray emission lines of the elements present in the sample. This is a time-consuming process. The recording must be taken and removed from the recorder, 2θ angles must be determined on the recording and then matched with 2θ angles appearing in tabulations for each crystal. That is, for a given crystal used in the spectrometer there will be the necessity of tabulating and comparing the 2θ angles for each line present in the spectrum. As can readily be seen this is a tedious process. However, in the case of the energy dispersive x-ray spectrometer, again the advantages of the computer and display combination may be realized. As a spectrum is acquired, the counts stored in the computer memory may be displayed on a cathode ray tube (CRT). Each vendor will have various kinds of cursor control, x-ray characteristic line spectra display, color or black and white display, and many other types of features. The main point is that it is a fast and simple job to superimpose theoretical x-ray line positions and relative intensities shown as histogram bars upon the experimental spectrum. Usually by rotating a knob on the control panel these characteristic elemental spectra may be scanned through the energy range until a fit occurs. It is important that a complete, and carefully evaluated fit be acquired. Remember that there is no difference in the energy of x rays observed for elements by energy dispersive and by wavelength dispersive x-ray spectrometry; therefore, accidental overlaps such as the arsenic-K_α lead-L_α lines will still exist. Nonetheless, the qualitative identification of elements present in the sample, both expected and unexpected, is one of the most powerful advantages of energy dispersive x-ray spectrometry.

2. *Quantitative Analysis*

Data acquired from the wavelength dispersive x-ray spectrometer appear simply as counts per unit of counting time; for example, the total number of counts acquired in 100 sec. This may be converted to peak count rate, such as counts per second and background counts subtracted in the same units. The wavelength dispersive x-ray spectrometer simply employs a single-channel energy discriminator so that low and high energy pulses may be rejected and only those in the energy range desired may be counted. The situation is entirely different with energy dispersive x-ray spectrometers. First, although the silicon detector has a resolution capable of separating the K_α lines of element Z from the K_α lines of element $Z + 1$, there is overlap between K_α lines of element Z and K_β lines of element $Z - 1$. When the sample is complicated by the presence of many elements of comparable atomic number, this overlap can become a serious problem. Thus the spectra must be deconvoluted prior to any acquisition of accurate intensity data. In other words, we must attempt to separate the spectrum into its various component spectra of individual elements. This deconvolution is generally accomplished by two principal approaches. Various mathematical techniques are used in order to verify the deconvolution. One technique employed is to enter into a computer program parameters such as detector resolution and energy central positions of the spectra of all elements to be considered. Once this information is entered into the program, the operator may simply command that simulated or calculated spectra for each element be generated from the program. This can be done by assuming certain peak shape functions in the simulation. Whether these peak shape functions represent the peak shapes that are in fact observed experimentally is a serious question which must be monitored by the operator. As long as there are no serious changes in the energy calibration of the spectrometer and the resolution of the detector, and if the fundamental assumptions that the peak shapes calculated are comparable to those observed experimentally, multielement spectra may be simulated by the linear combination of the individual elemental spectra so computed. There are mathematical algorithms that may be used to ascertain, by some statistical definition when the best fit has been achieved. Essentially, by multiplying the relative area of each individual spectrum required for the synthesis of the experimental spectrum by a sensitivity parameter for each element, which must be experimentally ascertained at some point, the concentration or amount of each element present in the sample may be estimated.

A second method is similar in principle; however, rather than relying upon computed line shapes, it employs experimental ones. In this method single element spectra are taken from samples prepared from foils, metal oxides, or other sources generating single element x-ray emission and these spectra are stored in a reference file. This reference file should contain all the elements that are anticipated to be in the experimental spectra, whether or not they are to be determined. When an experimental spectrum is acquired it may be synthesized by a linear combination of the spectra present in the reference file. The reason for having elements that are expected to be in the spectrum present in the reference file is that the spectral fitting process will be affected by the presence of these elements. Again, some established statistical techniques such as minimizing the sum of the squares of the differences between the experimental and computed spectra may be employed for determining the so-called best fit. The advantage of this method is that experimental line shapes are employed rather than those calculated from some line shape function. However, if energy calibration, detector resolution, or other significant instrumental parameter changes after the reference file has been created, problems similar to those mentioned for the synthetic line shapes will be encountered. If the reference file is generated from standard materials, the percentage of the peak area from the reference file required to fit the contribution of a particular element to the experimental spectrum may be taken as a quantitative measure of the amount of that element present in the sample. For example, if only 0.5 of the reference file spectrum for cobalt were required to fit the cobalt intensity in a particular experimental spectrum, then one might say that the amount of cobalt present in the sample was 0.5 of the amount of cobalt present in the standard used to create the reference file. However, this author prefers to use more than one standard in making a quantitative measurement. The reference file can therefore be fitted to a series of several standards, and the ratio or fraction of the reference file intensity required to fit each of the standards in the spectra may be plotted versus the concentration or amount of element present in each standard. In this way a more familiar calibration curve may be established from which the analytical result may be drawn. In most modern energy dispersive x-ray spectrometers linear regression programs and other valuable programs are available so that this process can be carried out directly within the computer. Remember, energy dispersive x-ray spectrometry is an instrumental method capable of simultaneous multielement determinations. This means that many calibration curves may need to be prepared for a multielement determination. It is extremely convenient that these calibration curves may be stored as linear regression parameters within the computer memory. Therefore, we see that one of the disadvantages of energy dispersive x-ray spectrometry, that is, the need to deconvolute and fit the experimental spectra with single element spectra, may be turned into an advantage if a computer data system is available.

E. Comparison of Wavelength and Energy Dispersive X-Ray Spectrometry

Jenkins has presented a comparison between wavelength dispersive and energy dispersive x-ray spectrometry [22]. He points out that in the early development of the wavelength dispersive methods, the designs were based upon experience in x-ray diffraction. Inspection of the design of even current wavelength dispersive spectrometers still shows this history. Jenkins says that the instruments currently in use may be classified into four major categories: (1) the single-channel or scanning wavelength spectrometers, in which elements are measured one at a time at a moderately slow rate; (2) the simultaneous multichannel instruments, in which 20 to 25 elements may be measured at a high rate; (3) the bremsstrahlung-tube-excited, energy dispersive spectrometers, which can in principle analyze all elements at a very high rate; and (4) the secondary fluorescer energy dispersive x-ray spectrometer. As Jenkins points out, all of these instruments have gained substantial flexibility with the addition of the minicomputer and more recently, microprocessor based data systems. In the case of the single-channel or scanning wavelength dispersive x-ray spectrometers, minicomputers have been interfaced to control the spectrometer even to the degree of selecting the analyzing crystal, collimator, and other parameters. In addition, of course, these computers may be used to process the x-ray data. One of the questions most frequently asked of x-ray spectrometrists is whether to purchase wavelength dispersive or energy dispersive. Many claims have been made about the advantages and disadvantages of both instrumental techniques. For example, energy dispersive spectrometers are often claimed to be faster than the single-channel scanning wavelength dispersive spectrometers. The cost of the overall system is generally claimed to be less than for a wavelength dispersive system. The energy dispersive system does allow for the detection of unexpected elements with ease and the display of the data using the computer based system. Some of the disadvantages of the energy dispersive system over the wavelength dispersive are: (a) in most cases the resolution is not as good: (b) a computer is needed for quantitative analysis, and (c) different artifacts may appear in the spectrum. The requirement of a source of liquid nitrogen has already been mentioned. These advantages and disadvantages can be argued with considerable expertise, based upon scientific points of merit. However, to argue these points in a general sense is perhaps not satisfactory. The particular analytical problem to be solved must be given prime consideration in the discussion of which of the instrumental types will best serve the purpose. The fact that the simultaneous multielement capability of energy dispersive x-ray spectrometry lies under software programming as opposed to hardware reorientation as in the simultaneous wavelength dispersive system is a distinct advantage. That is, the job to be done may be changed by simply loading a new computer program. The simple disadvantage of a computer being required for energy dispersive x-ray spectrometry is somewhat weakened by the fact that most modern wavelength dispersive x-ray spectrometers also employ computers for control of the functions and acquiring the data. Criticisms of misleading artifacts appearing in energy dispersive spectra are weakened by the fact that often these artifacts can provide insight into the characteristics of the sample. An advantage not mentioned for energy dispersive x-ray spectrometry is the fact that all of the data available in the spectrum are made available to the analyst in the computer core and can be used to advantage in the computer memory.

IV. Matrix Effects and Correction Methods

In spite of the fact that x-ray spectrometry can be one of the most precise techniques of instrumental methods of analysis, and the precision can be effectively controlled by the counting time, it does have disadvantages that must be considered. The effect of the matrix composition upon the intensity of the x rays emitted by the elements within that matrix is an important problem. Many of these effects are well understood and either theoretical or empirical methods have been developed for their compensation. There are two general classes of effects that will be discussed here. Certain general effects result from the absorption of x rays by the mass of the matrix. These are to be considered different from specific absorption effects due to the presence of certain elements in the sample and often referred to as interelement effects. Furthermore, certain effects that result from sample preparation must also be considered.

A. Matrix and Sample Condition Effects on Intensity

1. Methods of Sample Preparation

The sample types that may be conveniently analyzed by x-ray spectrometry consist of liquids and solids. Gases are not routinely analyzed by x-ray spectrometry. Although there are certain inconveniences in analyzing liquids, it is being done with increasing frequency. In early spectrometers an optical design was used in which the sample was placed beneath the x-ray tube and the associated optics. In these spectrometers liquid samples were analyzed with difficulty because of the uncertainty of the position of the surface of the liquid. The surface of the sample must be accurately located in the designed optical plane of the spectrometer. In order to better accommodate liquids, manufacturers began to design spectrometers that have inverted optics. In these spectrometers the sample is located above the x-ray tube and other optical components, thereby positioning the bottom of the sample precisely and reliably at the proper plane. A disadvantage is that it is necessary to employ an x-ray-transparent window in the bottom of the sample holder. Usually such windows consist of a thin polymer film such as Mylar or a similar material. This permits liquid sample preparation to be performed in a simple manner. One must be concerned with the fact that such thin films may allow leaks or in fact rupture, spilling the solution or liquid into the spectrometer. Care must be taken to ventilate sealed sample cups and not to employ vacuum when liquid samples are to be analyzed. Corrosive liquids have been exploded into a spectrometer as a result of applying a vacuum on a sealed sample cup with a thin film support at the bottom of the cup. Liquid samples often offer the disadvantage of contributing to a large background simply because liquids are composed of elements of low atomic number, leading to a high degree of scattering. Calibration of the spectrometer is simple, however, inasmuch as standard solutions are readily prepared.

Historically, it is much more common to analyze solid samples by x-ray spectrometry. This is perhaps a result of the obvious advantage of x-ray spectrometry for solid samples. In cases such as metals, minerals, refractory materials that cannot be dissolved, and other solid samples it is a much simpler process to analyze the solid directly when possible rather than put the material into solution. Obviously, x-ray spectrometry is often the technique of first choice when analyses must be performed on solids. A variety of sample preparation techniques may be employed for solid samples. First consider a material that is in the form of a fine powder or that may be ground into such a fine powder. In this case, the simplest sample preparation technique is simply to pour the powder into a conventional x-ray sample cup containing a Mylar window bottom, and to tap the cup to pack the powder into the sample area. If this process can be done sufficiently reproducibly, and if the particle size is sufficiently constant as discussed in the next section, quantitative analyses may be performed with such a simple preparation technique to within approximately 2 to 3% precision. It may be necessary to grind the sample reproducibly to obtain sufficiently small particle size to avoid particle size effects, as discussed later. If the powder cannot be reproducibly packed into the sample cup, or if preservation of the exact sample that was analyzed is required, the powders may be pressed into pellets. Usually the pellets are on the order of 25 mm in diameter and contain ~0.5 g of material. A variety of commercial pressing devices is available for such sample preparation. Some powdered samples will not cohere to form a stable pellet. In these instances a binding material that is inert to the x-ray spectrometer may be added to form a more stable pellet. Such materials as potassium chloride, sodium borate, and organic materials such as cellulose have been employed for this purpose. The analyst should ascertain that these pellets are reasonably stable in their geometry toward such factors as relative humidity or other aging processes. Even when pellets are employed, it may be necessary to grind the sample in order to obtain sufficiently small and reproducible particle size. A variety of aids for such preparation of samples is available such as aluminum cups that may be placed in the pressing die, filled with the sample, and then pressed in such a way that the edges of the cup roll into or compress the sample, forming a rigid support material. The analyst may need to consider impurities and major elements present in these sample cups. A similar technique is employed with sample cups made of lithium borate.

It is common in geological and mineralogical samples to encounter particulate material or bulk samples that can neither be easily ground nor dissolved. These refractory materials could present a difficult problem in sample preparation for x-ray spectrometric determinations. Fortunately, methods of sample preparation that eliminate or minimize these problems have been developed. These methods are based primarily on the use of fusion techniques [23]. In these techniques, ground or powdered forms of the sample or the smallest particle-size form that can be obtained is mixed in a crucible with a flux material. An example of a flux material that might be used is lithium tetraborate. When the flux–sample mixture is heated to an elevated temperature with a gas burner, a melt is obtained, which may be thoroughly mixed by shaking. In fact, a commercial device is available that will perform this process automatically [24]. Once this melt is homogenized, it may be poured onto a warm ceramic plate and allowed to cool. The result is a pellet or button that is flat on one side and may be further polished if necessary to obtain a smooth surface. The preparation of these buttons requires substantial experience on the part of the analyst to avoid cracking or other distortions that lead to unstable samples. Once such samples are prepared, they last indefinitely.

The above techniques represent only a few of the wide variety of sample preparation techniques that have been developed for x-ray spectrometry. It is beyond the scope of this chapter to discuss in great detail all the methods that have been employed. However, it is important to mention that techniques such as the use of ion-exchange-resin-impregnated filter papers and other methods for the extraction of ions from solution [25], and precipitation and co-precipitation techniques have been employed [26, 27], and a variety of other techniques has been reported. It must suffice here to say that the limitations in the methods and techniques of sample preparation for x-ray spectrometry lie only in the limitations of the analyst.

2. *Particle Size and Surface Effects*

As discussed above, there are many sample forms that may be suitable for x-ray spectrometry. One of these sample forms is powder or other small granulated particles. Generally, x rays useful for x-ray spectrometry penetrate samples of a typical size of only a few tens of micrometers. If the particle size in a powdered sample is on the order of this penetration depth, then both the emitted and excitation x rays may be seriously attenuated by the particle. For very small diameter particles the x rays penetrate completely through the particle, yielding not only fluorescence from within the entire particle, but from particles laying behind it. As shown in Fig. 20, the intensity may be severely attenuated in the size range in which the particles are suffi-

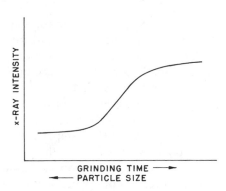

Fig. 20 X-ray intensity as a function of grinding time.

ciently large that the x rays cannot escape their dimensions. Thus we see, for extremely small diameter particles, that the x-ray intensity diminishes with increasing particle diameter; and with larger particle size, each particle becomes essentially an infinitely thick particle and no further effect is observed. Changes of as much as 35–50% in intensity may be observed as a result of particle size variation. Thus size is a significant problem in the analysis of powders or granulated particles by x-ray spectrometry. In order to avoid this problem, empirical techniques have been employed, such as grinding the powder sample until no further increase in x-ray intensity is observed. In most cases, once the minimum grinding time required so that no further change in intensity is detected has been established, this grinding time may be used on all furture samples with reasonable assurance.

The second method is to use a particle size correction factor such as the one given by

$$P_a = \frac{1 - \exp(-\mu_a d)}{\mu_a d} \tag{7}$$

where μ_a is the sum of the linear absorption coefficients for the composition of the particles for the absorption of both fluorescent and excitation x rays and d is the diameter of these particles. In fact, many particle size correction algorithms have been developed. The problem is that in many kinds of real samples, such as aerosols collected on filters, the particle size or the particle size distribution is not always accurately known.

The effects of surface roughness can also be important in the accurate determination of composition from x-ray intensity measurements. As an example of these kinds of effects, one can consider the use of polished metal samples. As one polishes a metal surface with an abrasive material, several things may happen, which may not be obvious to the analyst, that will affect the x-ray intensities. One of these is that, if the metal consists of a combination of a soft and a hard component, the softer component may be smeared across the surface. This essentially coats the surface of the sample with the softer component, yielding a higher result for that element than for the harder material. If a single hard material is polished in one direction, as it should be, parallel grooves and ridges will be created on the surface. The x-ray intensity observed from the surface of the sample will be dependent upon the orientation of these grooves with respect to the x-ray beam. Unless the sample is rotated in the sample holder, these grooves should always be located or positioned parallel to the direction of the incident beam. There are many detailed discussions of these types of phenomena that are beyond the scope of this chapter. However, they are discussed in detail in books such as Bertin's [3].

3.   *Matrix Absorption Effects*

As discussed earlier, x rays are attenuated as they pass through a sample matrix. The extent to which the x rays are attenuated depends upon the composition of the sample and the energy of the x radiation. There are two types of interaction between the x-ray energy and matter that lead to this attenuation of x-ray intensity. One is of a general nature; that is, interaction with the loosely held electrons in the orbitals of the atoms in the material. The second type is that which leads to photoejection of core electrons, thereby resulting in excited-state atoms that may emit, in turn, x radiation. The former type of interaction, which leads to variations in the intensity of x rays emitted by a sample, is often called the matrix absorption effect. It is this type that will be dealt with in this section; the latter type will be discussed immediately following. There are several factors to consider when x radiation enters the sample and leads to emission. As a result of the interaction with the loosely bound electrons, x-ray energy is lost, leading to Compton scattering or total annihilation of the x-ray energy. By and large, holding the x-radiation energy constant and increasing the atomic number of the matrix results in a decrease in the x-ray intensity as it travels through the sample. On the other hand, holding the average atomic number of the matrix or sample constant and decrasing the energy of the x-ray radiation leads to the same result. At any rate, the mass absorption coefficient for a low-atomic-number matrix is generally smaller than that for a high-atomic-number matrix. Thus elements present in a matrix of high atomic number have less chance of being excited to fluoresce than those present at the same concentration in a low-atomic-number matrix. This is a result of the fact that the intensity of the x radiation is more highly attenuated before arriving at an atomic site to bring about excitation. By the same argument, once an atom has been excited and, as a result, emitted x radiation, the probability of that radiation escaping the matrix decreases as the absorption coefficient, the average atomic number of the matrix, increases. Because of these effects, there is a matrix dependence of the intensity of x radiation emitted from an atom within the matrix. An element present in the identical concentration in a carbohydrate matrix will therefore be expected to emit more intense radiation than one that is present in a steel matrix. These general interactions therefore lead to a difficulty in the use of standards prepared in one matrix as calibration standards for a different matrix. The most direct way to avoid errors as a result of these absorption effects is to employ standards prepared in a matrix identical with that of the samples to be analyzed. There is, however, a clever approach to minimizing this problem. The intensity of the Compton scattering increases with the decrease in the average atomic number. Therefore, as the matrix atomic number increases, and thereby the intensity of emitted radiation decreases, the intensity of the Compton scattering also decreases. By taking the ratio of the analyte-element x-radiation intensity to the intensity of the Compton peak of the excitation radiation, one finds a relatively matrix-independent intensity relationship with concentration. The Compton scattering is typically very easy to measure, simply by taking the intensity of the Compton peak from a characteristic line of the x-ray tube emission. The Compton peak is usually easily identified as a peak occurring at an energy just less than the energy of the main tube characteristic line that is scattered from the sample. In summary, these matrix absorption effects alter the intensity of the emitted radiation from a sample as the composition of the matrix is altered. These general effects may be at least in part compensated for by recognizing that the effects also alter the intensity of Compton scattering. A simple measure of the Compton scattering intensity may then be utilized to partially, and in some cases extensively, correct for this problem.

4.   *Interelement Absorption-Enhancement Effects*

In addition to the general interaction of x radiation with loosely bound electrons, leading to attenuation of the x-ray intensity, specific interactions may also occur. These interactions are those resulting from the absorption of radiation, the energy of which is just greater than the absorption edge of atoms present within the sample. Consider, for example, the case of a mixture or alloy of the elements iron, cobalt, and nickel. The K absorption edge energies and the $K_{\bar{\alpha}}$ energies are shown in Table IV. As may be seen from this table, the K_{α} emission from iron and cobalt excite the K spectra of nickel, since the energies of the K_{α} lines are in both cases below

Donald E. Leyden

TABLE IV
Absorption Edge and $K_{\bar{\alpha}}$ Energies
for Fe, Co, and Ni (keV)

Element	K_{abs}	$K_{\bar{\alpha}}{}^a$
Fe	7.11	6.40
Co	7.71	6.73
Ni	8.33	7.47

^{a} $K_{\bar{\alpha}} = (2K_{\alpha_1} + K_{\alpha_2})/3$.

the absorption edge for nickel. However, it can also be observed from this table that the $K_{\bar{\alpha}}$ emission of the nickel atoms may excite the iron K_α emission. As a result, there will be a high specific absorption of the nickel K_α radiation by iron atoms present in the sample. In other words, the nickel may act as a secondary fluorescer for the iron. As a result of this phenomenon, it is easy to understand that, as the iron concentration increases in the sample, the nickel x-ray intensity may be seen to decrease substantially because of the specific absorption effects. In addition, the absorption of the nickel radiation by iron may result in enhancement of the iron x-ray intensity because of this additional excitation mechanism. Thus, all else being held constant, if a plot of the iron x-ray intensity versus concentration of nickel is made, an increase in the x-ray intensity of iron with respect to an increase in concentration of nickel is observed. On the contrary, if a plot of the x-ray intensity of nickel versus the concentration of iron in the sample is made, a decrease with respect to iron concentration is observed. These absorption-enhancement effects are rarely seen except in cases in which more than 1% by weight concentration of elements in the sample matrix is present. As a rule of thumb, only those elements that are separated in atomic number by two will have substantial interelement interaction. As in the above example, iron with atomic number 26 acted as an absorber for nickel with atomic number 28, whereas conversely nickel served as an enhancer of iron x-ray intensity. It should be rather obvious from the example presented that these interelement effects are commonly encountered in samples of metals, alloys, and minerals that contain relatively high concentrations of transition elements in which the absorption edges are commensurate with excitation by x radiation of concommitant elements within the sample. These effects, if left unaccounted for, may lead to errors on the order of 30–50%. Fortunately, the theory underlying these absorption-enhancement effects is fairly well understood and correction methods, although frequently empirically derived, have been developed to account for these effects.

B. Corrections for Interelement Effects

The interelement effects discussed above may lead to serious errors unless properly taken into account. Several attempts have been made to develop empirical methods for the correction of these interelement effects on x-ray intensity. Table V shows several algebraic relationships that have been proposed for these corrections. If we begin with the assumption that there are no interelement effects, the relationship between x-ray intensity and concentration for a particular element i is given by

$$I_i = k_i c_i \tag{8}$$

In this equation, k_i is a proportionality factor that is in fact the sensitivity of a particular instrument and instrument operating parameters for element i. This is a simple linear model for intensity–concentration relationships. However, in cases in which the specific interelement effects discussed above are present, it is necessary to develop relationships that include parameters that are often called influence coefficients. These parameters allow for the compensation of absorption-enhancement effects due to elements present in the matrix. The relationships given in Table V are some of those that have been presented and have been critically reviewed in several articles such as that by Rasberry and Heinrich [43]. If one is interested in the successful determination of i elements in a sample also containing j elements that may influence the result, then at least j standards must be analyzed in order to evaluate the influence coefficients a_{ij}. By measuring the intensity I_i of each element and using intensity or concentration depending upon the particular algorithm to be used, one may by the solution of simultaneous equations obtain these influence coefficients. Once the influence coefficients are so obtained, future samples may be run and corrections made from these coefficients. In this way, substantially more accurate data may be obtained than if no corrections at all were applied. It is through these influence-coefficient correction methods that successful application of x-ray spectrometry has been made to a variety of sample types such as alloys, minerals, ores, and other samples that contain relatively high concentrations of transition elements.

Some recent advances in computer programs have been made, not only for interelement corrections, but for estimation of elemental concentrations in a sample without the use of standards and including in the calculation effects of matrix and absorption-enhancement processes. Such programs must contain data relative to the particular instrument geometry, source of excitation, fluorescence yield, and perhaps an input related to the general composition of the sample. For example, is it a CH material (polymer), an H_2O

TABLE V

Comparison of Previous Procedures for Empirically Correcting Interelement Effects[a]

Reference	Equation in notation of the reference	Equation in notation of Rasberry and Heinrich [43]	Constant replaced by A_{ik}	Equation after solving for C_i/R_i	Form
Beattie and Brissey [28]	$(1 - R_a)W_a + \sum_{b \neq a} A_{ab}W_b = 0$	$(1 - 1/R_i)C_i + \sum_{k \neq i} \alpha_{ik}C_k = 0$	$\alpha_{ik} - 1$	$C_i/R_i = 1 + \sum_{k \neq i} A_{ik}C_k$	Hyp
Burnham et al. [29]	$(a_{11} - t_1)C_1 + a_{21}C_2 + a_{31}C_3 = 0$	$\left(\dfrac{1}{I_i'} - \dfrac{1}{I_i}\right)C_i + \sum_{k \neq i} \alpha_{ik}C_k = 0$	$(\alpha_{ik}I_i) - 1$	$C_i/R_i = 1 + \sum_{k \neq i} A_{ik}C_k$	Hyp
Guinier [30]	$E/E_0 = p_1/(p_1 + kp_2)$	$R_1 = C_1/(C_1 + \alpha C_2)$	$\alpha - 1$	$\left[\begin{array}{c} C_1/R_1 = 1 + A_{12}C_2 \\ \text{(Given only for binaries)} \end{array}\right]$	Hyp
Lucas-Tooth and Price [31]	$p_{nm} = a_n + I_{nm}(k_0 + \sum k_{nx}I_{xm})$	$C_i = a_i + I_i(b + \sum \alpha_{ik}I_k)$	$\alpha_{ik}I_k I_i'$	$C_i/R_i = \dfrac{a_i}{R_i} + b + \sum A_{ik}R_k$	Not hyp
Lucas-Tooth and Pyne [32][b]	$\left[\begin{array}{l} p_{nm}^{\text{X-ray}} = a_n + p_{nm}^{\text{ap}}(k_0 + \sum k_{nx}p_{xm}^{\text{ap}}), \\ p_{nm}^{\text{ap}} = \mu I_{nm} \end{array}\right]$	$C_i = a_i + \beta I_i(b + \sum \alpha_{ik}\beta I_k)$	$\alpha_{ik}\beta^2 I_k I_i'$	$C_i/R_i = \dfrac{a_i}{R_i} + b + \sum A_{ik}R_k$	Not hyp

		α_i		
Marti [33]	$aC_i + U_i = i_i(\alpha_i C_i + \sum_{j\neq i}\alpha_j C_j)$	$\left[(\alpha_{ik}I_i/a) - 1 \atop \text{(Background precorrection assumed)}\right.$	$\left. C_i/R_i = 1 + \sum_{k\neq i} A_{ik}C_k \right]$	Hyp
Gillieson et al. [34]	$P_{ni} = \alpha_n + I_{ni}(k_0 + \sum k_{nl}I_{li})$	—	Same as Lucas-Tooth and Price [31] —	Not hyp
Lachance and Traill [35, 38]	$R_A = C_A/(1 + \sum_{B\neq A} a_{AB}C_B)$	α_{ik}	$C_i/R_i = 1 + \sum_{k\neq i} A_{ik}C_k$	Hyp
Claisse and Quintin [39]	$C_A/R_A = 1 + \sum_{B\neq A}\alpha_B C_B + \sum_{j\neq A}\sum_{i\neq A}\alpha_{ij}C_iC_j$	α_{ik}	$C_i/R_i = 1 + \sum_{k\neq i} A_{ik}C_k + \sum_{k\neq i}\sum_{l\neq i} B_{ikl}C_kC_l$	Not hyp
Criss and Birks [19]	$R_i = C_i/(\sum \alpha_{ij}C_j)$	$\alpha_{ik} - 1$	$C_i/R_i = 1 + \sum A_{ik}C_k$	Hyp
Tertian [40, 41]	$Y = x/(1 + \phi x)$ $R_1 = C_1/(1 + \alpha x)$	α	$\left[C_1/R_1 = 1 + A_{12}(1 - C_2) \atop \text{(Given only for binaries)} \right]$	Not hyp
Thiele [42]	$C_i = C_{E,i}(k_0 + \sum k_{ij}C_{E,j})$	$\alpha_{ik}\beta^2 I_k I_i$	$\left[C_i = \beta I_i(b + \sum \alpha_{ik}\beta I_k) \atop \text{Similar to Lucas-Tooth and Pyne [32]} \right]$ $C_i/R_i = b + \sum A_{ik}R_k$	Not hyp
Rasberry and Heinrich [43]	—	—	$C_i/R_i = 1 + \sum_{k\neq i} A_{ik}C_k + \sum_{k\neq i}\dfrac{\beta_{ik}C_k}{1 + C_i}$	Not hyp

[a] Reprinted with permission from Rasberry and Heinrich, *Anal. Chem.* **46**, 81 (1974). Copyright by the American Chemical Society.

[b] In the original papers the concentrations are expressed in per cent. Further, the intensity term in these two equations is not background corrected; a_i is the correction.

material (aqueous solution), or an SiO_2 material (rock or ore). From these data and the fundamental parameter calculations, estimates of the elemental composition are made. Often results with 5–10% relative error are obtained. In many cases this is adequate for a quick answer, especially when no standards are available. By using advanced programming techniques in which the entire program is never in the computer at the same time (overlay), these programs have been executed by small minicomputers such as those sold as control-acquisition computers for the spectrometers. Substantial progress has been made toward an absolute analytical method that requires no standards.

V.   Applications of X-Ray Spectrometry

The applications of x-ray spectrometric analyses are many and varied. It is not within the scope of this introductory chapter to go into detailed discussion of these applications. However, an overview will be presented. First, one may consider the qualitative elemental identification for which x-ray spectrometry is useful. With wavelength dispersive x-ray spectrometry using a sequential spectrometer, scanning techniques may be employed to acquire spectra similar to that shown in Fig. 16; using the 2θ tables for the particular crystal employed, one may determine the 2θ angles for each element. Conversely, when a peak is observed at a given angle, the tables may be used to identify the element present. When elements are present in concentrations greater than ~ 1 ppm, it is usually a simple task with the scanning instrumentation to make the identifications. With the advent of energy dispersive x-ray spectrometers, qualitative applications became very much faster, simpler, and more accurate. With an energy dispersive x-ray spectrometer a spectrum of a reasonably wide energy range may be acquired in a simultaneous mode. Most of the instruments include qualitative identification capability. This is achieved by means of a marker representing a histogram type of presentation of the theoretical x-ray spectrum for each element, which may be superimposed upon the spectrum directly on the CRT display. Not only may the line positions and relative intensities be displayed (and on some spectrometers a calculated line shape), but in addition, the element for which the line was calculated and the identification of the line is also displayed. With these capabilities rapid identification of elements with atomic number greater than ~ 12 may be identified with ease and reliability. It is important that one make a match of as many lines as possible for the suspected element. Several of the commercial spectrometers have identification routines available in the computer software. These routines are capable

of scanning the spectrum and looking for matching sets of lines corresponding to the K_α/K_β pair for elements as well as L lines. These identification
routines have a high reliability factor and in most cases print warnings or
alert the user when the identification is not certain. Using these identification routines, programs have been prepared for the purpose of sorting alloys.
These sorting routines are very useful in cases in which a rapid identification
of materials on a routine basis is required. These programs offer the possibility in the future of preparing classification routines based upon the principles
of computerized learning machines. For example, it would be extremely
useful to be able to classify ore samples based upon their composition,
sources of origin, and other factors. By identification of elements present in
these ores and using classification theories, it is conceivable that pattern
recognition concepts can be applied to x-ray spectrometric classification. It is
important to realize then that energy dispersive x-ray spectrometry has
made a major contribution to the rapid, reliable, and nondestructive qualitative identification of elements present in a variety of sample forms. Using
scanning electron microscope techniques in conjunction with energy dispersive x-ray detection, the presence of elements on surfaces of samples may
be detected at a microscopic level and the surface element distribution
mapped. Modern instruments are capable of displaying elemental maps
as multicolor images on a color CRT display. Thus copper might be displayed in blue, iron in red, and zinc in yellow. In this way the analyst can
very rapidly get a visual image of the distribution of these elements across
the surface, using color as a dimension.

  Much more important is the quantitative capability of x-ray spectrometry.
The linear relationship between the x-ray intensity of elements present in the
sample and the concentration of those elements is the starting basis for the
quantitative application of x-ray spectrometry. In many samples such a
linear relationships holds true. Particularly in samples of low-atomic-
numbered matrix composition in which the analyte elements are present in
relatively low amounts (less than about 1%) the linear relationships generally
hold true. Such simple applications to quantitative determinations are
trivial and obvious. However, the real power of x-ray spectrometry was
found to lie in the quantitative determinations of elements present in samples
such as metals, ores, a variety of refractory materials, and glasses. In these
cases it is more often the situation that some type of correction of the x-ray
intensity data is required. As discussed earlier in the chapter, matrix absorption effects due to variations in the overall composition of the sample become
important and, in addition, interelement effects come into play. Early
application of x-ray spectrometry to these types of samples gave at best
semiquantitative results. As is the case with many forms of instrumentation
and technology, the full value of x-ray spectrometry in these areas was only

realized after the availability of inexpensive laboratory computers became a reality. In order to perform the interelement corrections using relationships such as those outlined in Table V, a computer was necessary. In 1960 a computer system in the laboratory would cost more than the instrumentation. In 1980 a computer system capable of performing these calculations was a small fraction of the cost of the instrument. In addition, with energy dispersive x-ray spectrometers, the computer system had to be an integral part of the instrument system. As a result of the availability of computers, better programs for intensity corrections, and a general better understanding of the power and potential of x-ray spectrometry, new and more frequent applications of x-ray spectrometry are being made. The computers have become so efficient that even small, desktop computers designed for home and small business use have been employed to execute the intensity-correction programs. Obviously, one can use large computer systems by remote job entry, on-line terminal access, or batch entry techniques. As a result of all of these facilities now being available, x-ray spectrometry has taken an important position as an analytical instrumental method capable of accurate, precise, and, very importantly, nondestructive quantitative elemental determinations. It is important that the reader be re-alerted to the fact that in general, x-ray spectrometry can provide little or no information concerning the chemical form or oxidation state of the elements that are detected. It is also important to realize that, although x-ray spectrometry is considered a bulk analytical method that gives information relative to the bulk composition rather than the surface composition, for example, in cases of high-atomic-number matrix composition, the observation depth may be only a few microns.

X-ray spectrometry is also beginning to find a role as an on-line or process-control instrument. For example, the determination of sulfur in petroleum can be and has been performed by the use of x-ray spectrometry. A simple system can be established for the detection of sulfur. In a flow process system, a sample bypass is constructed so that petroleum products flow at low pressure past a thin window through which the x rays may pass. In this way, the sulfur content may be monitored. In other cases such as those in which surfaces are coated with materials containing heavy elements, such as molybdenum or tungsten lubricants, these heavy elements may be detected by x-ray spectrometry directly on the process line. For purposes of quality control, high accuracy is very often not required and therefore extremely short x-ray counting times may be employed. In these cases high throughput may be attained, permitting essentially 100% quality-control sampling. The material being analyzed need never be touched or removed from the process operation. There is a growing number of examples of this kind of application of x-ray spectrometry.

X-ray spectrometry may also be used in a nondirect sense. For example, a coating on the surface of a sample will affect the intensity of x rays emitted from this substrate material. Quantitative results may be obtained by exploiting this phenomenon. In the case of tin plating of steel, the thickness of the tin layer deposited on the surface will affect the intensity of the iron x rays emitted from steel. By preparing a calibration set of standards in which the iron intensity is plotted versus the thickness of the tin-plate layer, a working curve may be obtained. Although this working curve is obtained only with considerable effort, from that time on, the thickness of the tin plating may be determined in a matter of minutes. The iron x-ray intensity is attenuated as the thickness of the tin film becomes larger. Thus by simply counting the iron x-ray intensity and using the working curve, one may obtain the thickness of the tin plate. A second example that illustrates the use of x-ray spectrometry in a nonstraightforward way is the determination of hydrogen in petroleum products. It should be obvious to the reader that the detection of hydrogen x rays is not possible. In a sample with a composition of only carbon and hydrogen, the application of x-ray spectrometry to a measurement of the hydrogen-to-carbon ratio is obviously not a straightforward one. However, recalling the fact that the Compton intensity increases with a decrease in the average atomic number of the matrix material, one may begin to develop the concept of this measurement. By taking samples of known hydrogen-to-carbon ratio, and plotting the intensity of the Compton scattering tube peak versus this ratio, again a working curve may be established. Some independent technique such as an elemental analyzer or CHN analysis is required to initially establish the calibration curve. The hydrogen-to-carbon ratios may be determined in a matter of minutes from that time on. The two examples presented here simply show that, with some imagination on the part of the analyst, x-ray spectrometry may act as a probe for interesting and difficult problems, even though they may not be approached in a direct way by measuring fluorescent intensity.

In summary, x-ray spectrometry is an instrumental analytical method that has found wide application in such diverse fields as geology and geochemistry, and the metals, ores, cement, petroleum, coal, and chemical industries. In addition, there have been applications in the fields of medicine and biology [44]. Recent years have found increasing application in the fields of environmental analysis and environmental studies. The technique is well established as one capable of providing accurate and precise analytical results with moderate costs and a more than acceptable time frame. In considering potential methods of solving an analytical problem, the analytical chemist should give x-ray spectrometry special consideration in those cases in which solid samples are to be analyzed and in particular in those cases in which elements with atomic numbers greater than approximately twelve are to be determined. Powders, solid homogeneous materials, and refractory materials are usually more difficult to analyze by methods other than x-ray spectrometry. As a general rule, x-ray spectrometry should not be considered as a technique of prime choice in cases of solutions in which the elemental composition is less than several per cent unless the analyst is willing to perform some type of chemical pretreatment. In such cases, atomic absorption spectrometry or some other analytical technique may be better suited to the problem. There is no doubt, however, that x-ray spectrometry will continue to solve more analytical problems.

References

1. American National Standards Institute, American National Standard: Radiation Safety for X-Ray Diffraction and Fluorescence Analysis Equipment. *NBS Hand. (U.S.)* No. III (1972).
2. R. Jenkins and D. J. Haas, *X-Ray Spectrom.* **2**, 135 (1973).
3. E. P. Bertin, "Introduction to X-Ray Spectrometric Analysis," p. 38. Plenum, New York, 1978.
4. R. Jenkins, "An Introduction to X-Ray Spectrometry," p. 20. Heyden, New York, 1974.
5. K. Siegbahn, C. Nordling, A. Fahlman, R. Nordberg, K. Hamrin, J. Hedman, G. Johnson, T. Bergmark, S. E. Karlson, I. Lindgren, and B. Lindberg, "ESCA: Atomic, Molecular and Solid State Structure by Means of Electron Spectroscopy." Almqvist & Wiksell, Stockholm, 1967.
6. L. A. Harris, *Anal. Chem.* **40**, 24A (1968).
7. L. S. Birks, "Electron Probe Microanalysis," 2nd ed. Wiley (Interscience), New York, 1971.
8. E. P. Bertin, "Principles and Practice of X-Ray Spectrometric Analysis," 2nd ed. Plenum, New York, 1975.
9. L. S. Birks, "X-Ray Spectrochemical Analysis," 2nd ed. Wiley (Interscience), New York, 1969.
10. Columbia Scientific Instruments, Austin, Texas.
11. Princeton Gamma-Trch, Princeton, New Jersey.
12. J. L. Duggan, W. L. Beck, L. Albrecht, L. Munz, and J. D. Spaulding, *Adv. X-Ray Anal.* **15**, 407 (1972).
13. W. D. Coolidge, *Phys. Rev.* **2**, 409 (1913).
14. W. Soller, *Phys. Rev.* **24**, 158 (1924).
15. E. W. White, B. V. Gibbs, G. G. Johnson, Jr., and G. R. Zechman, Jr., X-Ray Emission Line Wavelength and Two-Theta Tables. *ASTM Data Ser.* **DS 37** (1965).
16. R. Jenkins, "An Introduction to X-Ray Spectrometry," Chap. 6. Heyden, New York, 1974.
17. L. A. Currie, *Anal. Chem.* **40**, 586 (1968).
18. B. C. Clark, X-ray spectrochemical analysis on Mars, the Moon, and Earth. Presented at the 28th Annual Conference on Applications of X-Ray Analysis, Denver, Colorado, 1979.
19. J. W. Criss and L. S. Birks, *Anal. Chem.* **40**, 1080 (1968).
20. M. Singh, A. J. Dabrowski, G. C. Huth, J. S. Iwanczyk, B. C. Clark, and A. K. Baird, *Adv. X-Ray Anal.* **23**, 249 (1980).

21. G. C. Huth, A. J. Dabrowski, M. Singh, T. E. Economou, and A. L. Turkevich, *Adv. X-Ray Anal.* **22,** 461 (1979).
22. R. Jenkins, A comparison of wavelength dispersive and energy dispersive x-ray fluorescence analysis. *Int. Spectrosc. Colloq., 20th, Prague* (1977).
23. C. H. Drummond, *Appl. Spectrosc.* **20,** 252 (1966).
24. Claisse Fluxer, Corporation Scientifique Claisse, 7-1301 Place de Merici, Quebec, Canada.
25. W. J. Campbell, E. F. Spano, and T. E. Green, *Anal. Chem.* **38,** 987 (1966).
26. C. L. Luke, *Anal. Chim. Acta* **36,** 318 (1964).
27. J. W. Mitchell, C. L. Luke, and W. R. Northover, *Anal. Chem.* **45,** 1503 (1973).
28. H. J. Beattie and R. M. Brissey, *Anal. Chem.* **26,** 980 (1954).
29. H. D. Burnham, J. Hower, and L. C. Jones, *Anal. Chem.* **29,** 1827 (1957).
30. A. Guinier, *Rev. Univers. Mines, Metall., Mec.* **17,** 143 (1961).
31. H. J. Lucas-Tooth and B. J. Price, *Metallurgia* **64,** 149 (1961).
32. H. J. Lucas-Tooth and C. Pyne, *Adv. X-Ray Anal.* **7,** 523 (1964).
33. W. Marti, *Spectrochim. Acta* **18,** 1499 (1962).
34. A. M. Gillieson, D. J. Reed, K. S. Milliken, and M. J. Young, *ASTM Spec. Tech. Publ.* **STP 376,** 3 (1964).
35. G. R. Lachance, (1964). *Geol. Surv. Pap. (Geol. Surv. Can.)* No. 64-50.
36. R. J. Traill and G. R. Lachance, (1964). *Geol. Surv. Pap. (Geol. Surv. Can.)* No. 64-57.
37. G. R. Lachance and R. J. Traill, *Can. Spectrosc.* **11,** 43 (1966).
38. R. J. Traill and G. R. Lachance, *Can. Spectrosc.* **11,** 63 (1966).
39. F. Claisse and M. Quintin, *Can. Spectrosc.* **12,** 129 (1967).
40. R. Tertian, *Adv. X-Ray Anal.* **12,** 546 (1969).
41. R. Tertian, *Spectrochim. Acta, Part B* **24B,** 447 (1969).
42. B. Thiele, *Siemens-Z* **44,** 707 (1970).
43. S. D. Rasberry and K. F. J. Heinrich, *Anal. Chem.* **46,** 81 (1974).
44. D. E. Leyden and W. K. Nonidez, *CRC Crit. Rev. Clin. Lab. Sci.* **7,** 393 (1977).

Transform Techniques in Chemistry*

Alan G. Marshall†

Departments of Chemistry and Biochemistry
The Ohio State University
Columbus, Ohio

* Work supported by grants from Natural Sciences and Engineering Research Council of Canada (A-6178), University of British Columbia (21-9879), the Petroleum Research Fund (11458-AC6), and the Alfred P. Sloan Foundation.

† Alfred P. Sloan Research Fellow, 1976–1980.

57

I. General Advantages and Uses for Transforms

The purpose of this chapter is to provide the reader with a rapid understanding of the advantages of transform methods in chemistry. Although historical, mathematical, and instrumental detail is deliberately suppressed, a small number of introductory pictorial examples suffice to indicate the sorts of problems for which transforms provide a solution. With this approach, a wide variety of superficially different applications can be analyzed in the same way, simply by changing the names of the variables in going from one experiment to the next.

It is essential to have a grasp of Fourier transform fundamentals in order to make optimal practical use of the high sensitivity, resolution, and versatility of today's commercial FT-IR, FT-NMR, and FT-MS spectrometers. This basic knowledge will then apply to the many future applications of Fourier and other transforms in areas spanning almost the whole range of modern analytical chemical techniques.

The *multichannel* advantage (Section I.A) of Fourier or Hadamard transform spectroscopy has already been demonstrated for ion cyclotron resonance mass spectroscopy (FT-ICR, Section II.A), rotational spectroscopy (FT-microwave, Section II.B), vibrational spectroscopy (FT-IR, Sections II.C and II.D), Faradaic admittance (FT-electrochemistry, Section II.F), dielectric relaxation (FT-dielectric, Section II.G), and electron–nuclear double resonance (FT-ENDOR, Section II.I). Even though the multichannel advantage no longer holds for electronic spectroscopy (FT-visible, Section II.E), multielement analysis can still be achieved accurately and rapidly using Fourier methods in the visible–UV. Two-dimensional Hadamard IR (Section II.D) and two-dimensional FT-NMR (Section II.H) furnish new means for obtaining rapid cross-sectional images of inanimate or living objects. Two-dimensional FT-NMR and FT-ENDOR produce spectra in which different parameters are separated in different plotted directions: for example, proton chemical shift on one axis versus carbon-13 chemical shift on the other axis, for the same compound.

Apodization (Section I.B) consists of ways to emphasize particular aspects of a data set (e.g., signal-to-noise ratio, resolution) at the expense of others. *Convolution* (Section I.C) can be used to separate the nonuniform spectral intensity of a spectral source from the genuine intensity variations due to the sample. In fact, if the experimental line shape in any experiment is accurately known and is the same for all the "peaks," it is possible to narrow the displayed peaks to zero width (in principle). Finally, Hilbert transforms can be used (a) to generate spectra that are not observed directly (e.g., dispersion spectrum from an absorption spectrum), and (b) to provide a simple graphical analysis for spectral line shape (Section II.J).

A.   Multichannel Advantage: Improved Speed and/or Signal-to-Noise Ratio in Spectra

1.   Single-Channel Spectrometers

The multichannel advantage [1] is readily understood from detailed scrutiny of the spectrometers shown schematically in Fig. 1. Consider first the single-channel spectrometer (Fig. 1a,b). The "slit" of Fig. 1b may represent an actual aperture (as in optical spectrometers) or a mixer–filter detector [1] whose filter bandwidth corresponds to the slit width (as in continuous-wave NMR spectrometers). A spectrum of N amplitudes or intensities corresponding to N slit positions may then be acquired by scanning sequentially from one slit position (channel) to the next. The obvious disadvantage of this one-at-a-time scanning procedure is that we suppress the information in all but one channel during any given measurement (see below).

2.   Multidetector Spectrometers

An obvious improvement would be to open the slit width to the full spectral "window" of the desired spectral range, and replace the single broad-band detector of the single-channel spectrometer of Fig. 1b with an array of many narrow-band detectors, each positioned (or tuned) to dedect radiation from just one channel (Fig. 1c). Because of the conceptual simplicity of the multidetector spectrometer of Fig. 1c, it is logical to investigate its feasibility. It is desirable to be able to resolve spectral detail as narrow as the width of a typical spectral absorption line; therefore, the minimum number of channels required in any multichannel spectrometer is just the width of the entire spectral range of interest divided by the width of a single spectral line. The resultant necessary number of channels for various forms of spectroscopy is shown in Table I.

From Table I, electronic (visible–UV) spectroscopy would appear to be the least likely candidate for success with a multidetector spectrometer. Nevertheless, the resolution of a fine-grain photographic plate is sufficient to provide the huge number of required channels, since the desired spectrum may be dispersed over the necessary distance (a few meters) without undue effort. If the required number of channels is greatly reduced (say, to a few thousand), as in applications involving lower resolution and/or more limited spectral range, any of several types of electronic multidetector array [2, 3] (e.g., vidicon or self-scanning photodiode array) detectors can replace the photoplate. For example, in the vidicon (television camera) type, dispersed photons of different energy are simultaneously recorded at different regions of the focal plane of the camera. Although the electronic multidetector has relatively few channels compared to the photoplate multidetector, the

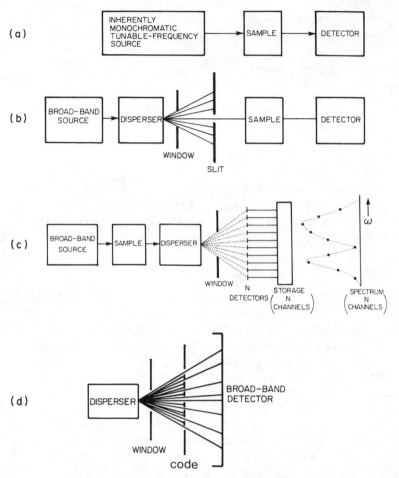

Fig. 1 Schematic diagrams of (a), (b) single-channel, (c) multichannel, and (d) multiplex spectrometers. (a) Source provides inherently monochromatic radiation. (b) Source produces inherently broad-band radiation, which is made monochromatic by passage through a disperser–window–slit combination as shown. Although the single detector may be either *broad-band* (detects radiation anywhere within the overall spectral range) or *narrow-band* (detects radiation only from a single channel), it is usual to associate a broad-band detector with a broad-band source (b) and a narrow-band detector with a narrow-band source (a). (c) This multidetector spectrometer is similar to that of (b) with the slit removed, so that the full *N*-channel spectrum impinges upon *N* individual detectors, each positioned (or tuned) to detect the radiation from just one spectral channel. The spectral intensity accumulates in the storage unit, and is ultimately displayed in the point-by-point intensity spectrum shown at the far right of this diagram. (d) Schematic encoding of spectral amplitudes of an *N*-channel spectrum, by insertion of a coding mask between the spectral window and the (single, broad-band) detector. In this particular code, the detected amplitude from any one channel is coded by a zero (slit closed for that channel) or unity (slit open for that channel). Particular codes are discussed in Section I.A.

TABLE I

Minimum Number of Channels Required for Various Types of Multidector Spectrometers

Type of spectroscopy	Largest usual frequency (Hz)	Typical spectral frequency range (Hz)	Width of typical line (Hz)	Approximate minimum number of channels[a]
Mossbauer	6×10^{18}[b]	10^8	10^7	10
ESCA	3.5×10^{17}	10^{17}	10^{14}	1,000
Photoelectron	5×10^{15}	3×10^{15}	10^{12}	3,000
Electronic	1.5×10^{15}	1.2×10^{15}	10^9	1,250,000
Vibrational	2×10^{14}	1.5×10^{14}	3×10^9	50,000
Rotational	4×10^{10}	3×10^{10}	10^5	300,000
ENDOR	5×10^{7}[c]	5×10^7	5×10^4	1,000
^{13}C NMR	8×10^{7}[d]	2×10^4	0.2	100,000
ICR[e]	2×10^6	2×10^6	10^2	20,000

[a] Number of channels is obtained by dividing the typical spectral range by the width of one line.

[b] ^{119m}Sn.

[c] ^{1}H.

[d] Magnetic field strength is approximately 75 kG (7.5 Tesla).

[e] Magnetic field strength is approximately 2 Tesla; minimum ionic mass-to-charge ratio is 15.

electronic device furnishes an immediate spectrum without the delays of photographic processing and densitometric reading of the developed photoplate.

In ESCA (electron spectroscopy for chemical analysis) [4, 5] and photoelectron spectroscopy [5, 6], electrons are dislodged from atoms or molecules by x-ray or UV radiation, respectively, and the released electrons possess a translational energy that depends on the energy of the bound state occupied by that electron in the original atom or molecule. By scanning the energy of the observed dislodged electrons, the energies of the original molecular electronic states may be determined. By passing the electrons through a perpendicular electric field, the dislodged electrons may be dispersed in space according to their velocity to achieve the arrangement shown schematically in Fig. 1c. In this case an electronic multidetector can again be based on the vidicon. Since electrons of different velocity can be dispersed to strike different regions of the screen, their arrivals will be recorded independently by different elements of the television camera grid. Because of the relatively small number of required detector channels (Table I), the multidetector spectrometer of Fig. 1c is thus feasible [7] for ESCA and photoelectron spectroscopy.

For the other forms of spectroscopy listed in Table I, direct multidetector methods are less attractive. For rotational (microwave) spectroscopy, for example, there is no broad-band radiation source available: a blackbody radiation source such as that used for other radiation energies (e.g., xenon or hydrogen discharge for UV, hot tungsten wire for visible, globar for near infrared and infrared, mercury vapor for far infrared) would have to be operated at an unreasonably high temperature in order to obtain sufficient radiation flux for use as a radiation source. Alternatively, the cost of an array of individual narrow-band microwave transmitters (at about $1000 each) as the "broad-band" radiation source would be excessive. For infrared spectroscopy, on the other hand, the necessary broad-band source is available, but it would be necessary to disperse the spectrum over many meters in order to resolve the desired spectral detail with existing individual (thermopile) detectors of about 1-mm width apiece, and the cost ($\sim$\$200 per detector) is again too high. (Photographic detection does not extend beyond about 12,000 Å, and is thus unavailable.) Finally, for NMR and ICR spectroscopy broad-band sources are again available, but the cost of an array of tens of thousands of individual narrow-band mixer–filter detectors (as for ENDOR) is again unreasonably high. In conclusion, the multidetector approach of Fig. 1c is simply not feasible, either geometrically or economically, for infrared, microwave, or radio-frequency spectrometers.

3. Multiplex Spectrometers

Quantitative analysis of the single-channel spectrometer performance leads quickly to some immediate improvements. If the N desired spectral elements (i.e., amplitudes or intensities at N different slit positions) are designated as $x_1, x_2, \ldots, x_N$, and the N observed amplitudes or intensities are designated as $y_1, y_2, \ldots, y_N$, then the single-channel spectrometer measurements (Fig. 1b) may be categorized as shown below, for the illustrative example of a 4-channel spectrometer,

$$y_1 = 1 \cdot x_1 + 0 \cdot x_2 + 0 \cdot x_3 + 0 \cdot x_4 \quad \text{(1a)}$$

$$y_2 = 0 \cdot x_1 + 1 \cdot x_2 + 0 \cdot x_3 + 0 \cdot x_4 \quad \text{(1b)}$$

$$y_3 = 0 \cdot x_1 + 0 \cdot x_2 + 1 \cdot x_3 + 0 \cdot x_4 \quad \text{(1c)}$$

$$y_4 = 0 \cdot x_1 + 0 \cdot x_2 + 0 \cdot x_3 + 1 \cdot x_4 \quad \text{(1d)}$$

Number of times each unknown
element x_i is measured $= \quad 1 \qquad 1 \qquad 1 \qquad 1$

It is useful to think of the experiment as a *code* that connects the observed values y_i to the unknown desired spectral elements x_i. The traditional advantage of one-at-a-time scanning spectrometers is then seen to be the simplicity of the code:

$$
\begin{array}{ccc}
\text{observations} & \text{code} & \text{unknowns} \\
\begin{bmatrix} y_1 \\ y_2 \\ y_3 \\ y_4 \end{bmatrix} & = \begin{bmatrix} 1 & 0 & 0 & 0 \\ 0 & 1 & 0 & 0 \\ 0 & 0 & 1 & 0 \\ 0 & 0 & 0 & 1 \end{bmatrix} \begin{bmatrix} x_1 \\ x_2 \\ x_3 \\ x_4 \end{bmatrix} & ; \quad \mathbf{y} = \mathbf{A}\mathbf{x}
\end{array} \qquad (2)
$$

Because the code matrix is just the unit matrix, the desired inverse calculation is trivial:

$$ \mathbf{A}^{-1} = \mathbf{A}, \qquad (3) $$

$$
\begin{array}{ccc}
\text{unknowns} & \text{inverse code} & \text{observations} \\
\begin{bmatrix} x_1 \\ x_2 \\ x_3 \\ x_4 \end{bmatrix} & = \begin{bmatrix} 1 & 0 & 0 & 0 \\ 0 & 1 & 0 & 0 \\ 0 & 0 & 1 & 0 \\ 0 & 0 & 0 & 1 \end{bmatrix} \begin{bmatrix} y_1 \\ y_2 \\ y_3 \\ y_4 \end{bmatrix} & ; \quad \begin{array}{l} x_1 = 1 \cdot y_1 \\ x_2 = 1 \cdot y_2 \\ x_3 = 1 \cdot y_3 \\ x_4 = 1 \cdot y_4 \end{array}
\end{array} \qquad (4)
$$

Extraction of the N unknown spectral elements from the N observations (i.e., the *inverse* code) is thus effortless for the one-at-a-time scanning spectrometer. However, the great disadvantage [see Eqs. (1)] is that each unknown element is detected only once during the N observations. Furthermore, in any experimental measurement characterized by a certain level of random imprecision or noise, it is desirable to repeat the measurement many times in order to obtain a more accurate result. The signal (in this case, the spectral amplitude or intensity in any one channel) will accumulate as the number of repeated measurements, N. The random noise amplitude may be treated as a random walk about zero (the average noise level), and the root-mean-square distance away from zero after N steps in a random walk is proportional to $N^{1/2}$. Therefore, the problem with our single-channel spectrometer code is that there are too many zeros in the code matrix. If we could detect each channel in every measurement, then after N measurements we would have measured each unknown element N times, and our signal-to-noise ratio would be improved by a factor of $N/N^{1/2}$, or $N^{1/2}$.

(*a*) *Hadamard Code* This *multichannel* or *Fellgett* advantage [8] can be achieved or approached with a wide variety of different codes. The very special feature of the Hadamard and Fourier codes is that the desired inverse code is again obtained trivially from the original code. Consider again the four-channel spectrometer, but this time with the Hadamard code:

observations

$$y_1 = 1 \cdot x_1 + 1 \cdot x_2 + 1 \cdot x_3 + 1 \cdot x_4$$
$$y_2 = 1 \cdot x_1 - 1 \cdot x_2 - 1 \cdot x_3 + 1 \cdot x_4$$
$$y_3 = 1 \cdot x_1 - 1 \cdot x_2 + 1 \cdot x_3 - 1 \cdot x_4$$
$$y_4 = 1 \cdot x_1 + 1 \cdot x_2 - 1 \cdot x_3 - 1 \cdot x_4$$

Number of times each unknown element x_i is measured $= 4 \quad\quad 4 \quad\quad 4 \quad\quad 4$

observations   code   unknowns

$$\begin{bmatrix} y_1 \\ y_2 \\ y_3 \\ y_4 \end{bmatrix} = \begin{bmatrix} 1 & 1 & 1 & 1 \\ 1 & -1 & -1 & 1 \\ 1 & -1 & 1 & -1 \\ 1 & 1 & -1 & -1 \end{bmatrix} \begin{bmatrix} x_1 \\ x_2 \\ x_3 \\ x_4 \end{bmatrix} \tag{5}$$

Equations (4) show that with the Hadamard code, each unknown element x_i is observed N times with the same absolute weight factor; that is, *the absolute value of each of the numbers in the code is unity.*

$$\mathbf{y} = \mathbf{Hx} \tag{6a}$$

$$|H_{mn}| = 1 \tag{6b}$$

Equation 6b is thus the key to achieving the full multichannel advantage of an improvement of $N^{1/2}$ in signal-to-noise ratio for the multiplex spectrometer (i.e., one using a code, such as in Fig. 1d) compared to a single-channel spectrometer operated for the same length of time (i.e., same number of total observations).†

If we delete the first row and column of the Hadamard code of Eq. (5), is is apparent that each row of the remaining array differs from the preceding row by *cyclic permutation* (see Section II.D). This intended property makes

---

† As discussed in more detail in [1], the noise must be independent of the signal strength for this argument to be valid. The Fellgett advantage thus applies to measurements in the radio frequency, microwave, and infrared (unless cooled detectors are used), but fails for visible–UV or particle detection in which individual photons or particles are counted, so that noise is proportional to the square root of the signal strength. See Section II.E for additional discussion.

the calculation of the inverse matrix trivial:

$$\mathbf{H}^{-1} = \frac{1}{N}\mathbf{H} \tag{7a}$$

in this case,

$$
\begin{array}{c}
\text{desired} \\
\text{unknowns} =
\end{array}
\qquad
\begin{array}{c}
\text{inverse code}
\end{array}
\qquad
\begin{array}{c}
\text{observations}
\end{array}
$$

or

$$
\begin{bmatrix} x_1 \\ x_2 \\ x_3 \\ x_4 \end{bmatrix}
= \frac{1}{4}
\begin{bmatrix}
1 & 1 & 1 & 1 \\
1 & -1 & -1 & 1 \\
1 & -1 & 1 & -1 \\
1 & 1 & -1 & -1
\end{bmatrix}
\begin{bmatrix} y_1 \\ y_2 \\ y_3 \\ y_4 \end{bmatrix}
\tag{7b}
$$

For example,

$$
\begin{aligned}
x_2 &= (y_1 - y_2 - y_3 + y_4)/4 \\
&= \tfrac{1}{4}(x_1 + x_2 + x_3 + x_4 - x_1 + x_2 + x_3 - x_4 \\
&\quad - x_1 + x_2 - x_3 + x_4 + x_1 + x_2 - x_3 - x_4) \\
&= x_2 \qquad \text{Q.E.D.}
\end{aligned}
$$

A Hadamard code may be constructed whenever [9]

$$N = 2^m, \qquad m = 2, 3, 4, \ldots \tag{8}$$

Hadamard spectroscopy is discussed more fully in Section II.D. Actual Hadamard infrared spectrometers employ a code that is derived from (but not identical to) the above example, because in practice it is easier to block half of the slits (i.e., let $H_{mn} = 0$ for half of the elements of any one row of the code) than to collect reflected light from the back of the code mask (i.e., $H_{mn} = -1$ for half of the elements of any row of the code). See Section II.D for details.

(b) *Fourier Code* The Fourier code is based on the same principles developed above for the Hadamard case:

$$\mathbf{y} = \mathbf{Fx} \tag{9a}$$

$$|F_{mn}| = 1 \tag{9b}$$

The general formula for the code is given by

$$F_{mn} = \exp(2\pi imn/N) = \cos(2\pi mn/N) + i\,\sin(2\pi mn/N) \tag{10}$$

The Fourier code is illustrated by an $N = 4$ (i.e., 4-channel) example:

observations Fourier code desired unknowns

$$\begin{bmatrix} y_0 \\ y_1 \\ y_2 \\ y_3 \end{bmatrix} = \begin{bmatrix} 1 & 1 & 1 & 1 \\ 1 & \exp(2\pi i/4) & \exp(2\cdot 2\pi i/4) & \exp(2\cdot 3\pi i/4) \\ 1 & \exp(2\cdot 2\pi i/4) & \exp(2\cdot 2\cdot 2\pi i/4) & \exp(2\cdot 2\cdot 3\pi i/4) \\ 1 & \exp(2\cdot 3\pi i/4) & \exp(2\cdot 3\cdot 2\pi i/4) & \exp(2\cdot 3\cdot 3\pi i/4) \end{bmatrix} \begin{bmatrix} x_0 \\ x_1 \\ x_2 \\ x_3 \end{bmatrix} \tag{11}$$

$$= \begin{bmatrix} 1 & 1 & 1 & 1 \\ 1 & \exp(i\pi/2) & \exp(i\pi) & \exp(i3\pi/2) \\ 1 & \exp(i\pi) & \exp(i2\pi) & \exp(i3\pi) \\ 1 & \exp(i3\pi/2) & \exp(i3\pi) & \exp(i9\pi/2) \end{bmatrix} \begin{bmatrix} x_0 \\ x_1 \\ x_2 \\ x_3 \end{bmatrix} \tag{12}$$

or

$$\begin{bmatrix} y_0 \\ y_1 \\ y_2 \\ y_3 \end{bmatrix} = \begin{bmatrix} 1 & 1 & 1 & 1 \\ 1 & i & -1 & -i \\ 1 & -1 & 1 & -1 \\ 1 & -i & -1 & i \end{bmatrix} \begin{bmatrix} x_0 \\ x_1 \\ x_2 \\ x_3 \end{bmatrix} \tag{13}$$

The great calculational simplicity of the Fourier code is that the desired inverse code may again be computed trivially from the original code:

$$F_{mn}^{-1} = (1/N)F_{mn}^* = (1/N)\exp(-2\pi imn/N) \tag{14}$$

in which the asterisk denotes complex conjugate. For example, for $N = 4$,

desired unknowns inverse code observations

$$\begin{bmatrix} x_0 \\ x_1 \\ x_2 \\ x_3 \end{bmatrix} = \frac{1}{4} \begin{bmatrix} 1 & 1 & 1 & 1 \\ 1 & -i & -1 & i \\ 1 & -1 & 1 & -1 \\ 1 & i & -1 & -i \end{bmatrix} \begin{bmatrix} y_0 \\ y_1 \\ y_2 \\ y_3 \end{bmatrix} \tag{15}$$

so that (for instance),

$$x_3 = (\tfrac{1}{4})(y_0 + iy_1 - y_2 - iy_3)$$
$$= (\tfrac{1}{4})(x_0 + x_1 + x_2 + x_3 + ix_0 - x_1 - ix_2 + x_3$$
$$- x_0 + x_1 - x_2 + x_3 - ix_0 - x_1 + ix_2 + x_3)$$
$$= x_3 \quad \text{Q.E.D.}$$

For the physical applications described in subsequent sections, it is common to sample a spectrometer time-domain signal $y(t)$ at N different times. Since each time-domain data point represents a weighted sum of oscillations at all the frequencies in the detected range (e.g., we hear all the tones at once when an orchestra plays), we have the now-familiar situation of N observables (N time-domain data points), related to N desired frequency-domain amplitudes by some sort of code:

$$
\begin{aligned}
y(t_0) &= F_{00}x(\omega_0) + F_{01}x(\omega_1) + \cdots + F_{0,N-1}x(\omega_{N-1}) \\
y(t_1) &= F_{10}x(\omega_0) + F_{11}x(\omega_1) + \cdots + F_{1,N-1}x(\omega_{N-1}) \\
&\vdots \\
y(t_{N-1}) &= F_{N-1,0}x(\omega_0) + F_{N-1,1}x(\omega_1) + \cdots + F_{N-1,N-1}x(\omega_{N-1})
\end{aligned}
\tag{16}
$$

The unique feature of the Fourier code is that if we choose the time-domain data points to be equally spaced,

$$
t_n = nT/N, \qquad n = 0, 1, 2, \ldots, N - 1
$$

then the (Fourier) code for finding the N frequency-domain amplitudes at N equally spaced frequencies,

$$
\omega_m = 2\pi m/T, \qquad m = 0, 1, 2, \ldots, N - 1 \tag{17}
$$

is just

$$
F_{nm} = \exp(i\omega_m t_n) = \exp(2\pi i mn/N) \tag{18}
$$

whose inverse is

$$
F_{nm}^{-1} = \exp(-i\omega_m t_n) = \exp(-2\pi i mn/N) \tag{19}
$$

The mathematical representation of the code elements as complex numbers simply corresponds to specifying the phase of the wave at each component frequency. For example, a purely real F_{nm} value corresponds to a pure cosine wave at frequency $2\pi m/T$, while a purely imaginary F_{nm} value corresponds to a pure sine wave at the same frequency

In conclusion, we have demonstrated that in order to achieve better signal-to-noise ratio from most spectrometers, it is useful to observe the whole spectrum at once according to a special code that relates each observed amplitude (or intensity) to the N total desired spectral amplitude (or intensity) elements. Although many conceivable codes are possible, the optimal code will be one for which each element of the code satisfies $|H_{mn}| = 1$ (or $|F_{mn}| = 1$), so that it is as if each channel of the spectrum is observed with unit weight factor in each measurement (rather than just one at a time as with the scanning spectrometer). Finally, the code should be constructed so that it is easy to find the inverse code that recovers the desired spectrum from the observed amplitudes or intensities. The special features of the Hadamard code (i.e., cyclically permuted rows) and the Fourier code (i.e., equally spaced time-domain samples) accomplish both of these objectives [10].† We shall now look further at how these transforms can be applied to spectroscopic detection and data manipulation.

B.   Apodization: Enhancement of Signal-to-Noise Ratio
or Resolution

Apodization (literally, removal of feet) denotes any modification of the shape of an existing waveform before Fourier transformation. The term was originally coined for the smoothing of the auxiliary wiggles ("feet") of the "sinc" line shapes in infrared interferometry (see Section II.C). In order to see why one might want to change the shape of a waveform, it is first necessary to assemble a small (pictorial) library of common waveforms and their corresponding Fourier transforms. In physical applications the waveform represents a time-domain signal or interferogram, and the corresponding Fourier transform represents the spectrum of that signal.

The two basic concepts of apodization are illustrated by the (frequency-domain) spectrum corresponding to a rectangular time-domain pulse in Fig. 2. First, the shape of the spectrum is determined by the shape of the original waveform. Second, the narrower the signal (of a given shape, in this case a rectangular pulse) in one domain, the broader is its representation in the transform domain. This second property suggests that a sufficiently short electric field pulse can serve as a spectral radiation broad-band source: for example, excitation of NMR signals over a range of 10 kHz can be achieved by using a short magnetic field pulse of about 10 μsec duration [11]. As another example of a short pulse used to produce broad-band radiation, electron impact spectroscopy [12] is based on the rapid passage of an electron past a molecule—this passing electron produces a very short sharp pulse of

----

† There might appear to be an infinite number of possible codes (e.g., any matrix with linearly independent rows, so that its inverse matrix exists) that could offer the same advantages as the Hadamard and Fourier codes. An optimal code, for which no systematic generating system yet exists, should give equal weight to the signal and noise in each channel during the coding and decoding processes. However, once a given code has been proposed, its performance by this criterion (i.e., whether or not the matrix code is "well-conditioned") can be determined, and on this basis the Hadamard and Fourier codes are as good as any others known. In practice, the Hadamard and Fourier codes are preferred because of their simple physical identification and implementation, and for the rapid and simple algorithms available for calculating the inverse code (which therefore need not be stored in advance) for any allowed size data set (usually 2^n, where n is an integer ranging from about 8 upward).

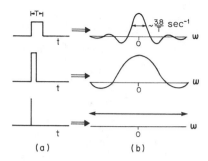

Fig. 2 (a) Time-domain and (b) frequency-domain representations of dc pulses of three different durations. Frequency spectra are the cosine Fourier transforms of the (real) signals shown in (a).

electric field at the molecule, and thus acts as a very broad-band, nearly flat source of radiation. The electron-impact frequency excitation range is sufficiently wide to excite the same sorts of energy-level transitions that are studied in conventional photoelectron and ESCA spectroscopy.

Although we are normally prejudiced by our eyes and ears to analyze our surroundings in the frequency domain (we judge light by its color and sound by its pitch), the preceding and ensuing examples show that it is equally desirable to be able to visualize the same response in the time domain. A small pictorial library of physically relevant Fourier transform pairs (i.e., a waveform and its corresponding spectrum) is shown in Fig. 3. This time we show both the cosine and sine transforms that correspond† to the steady-state conventional absorption and dispersion spectra that would be obtained using a slow-scanning single-channel spectrometer [13].

The most direct connection is that between a sinusoidal waveform and its frequency spectrum consisting of a single spike (Fig. 3a,b). In other words, a signal with just one frequency component must correspond to a sinusoidal wave whose phase is somewhere between a pure sine and pure cosine wave.

Figure 3c shows the familiar Fraunhofer spectral line shape that corresponds to a rectangular waveform. This line shape was originally encountered in analysis of diffraction from a slit [14], and is commonly encountered in FT-ICR, FT-IR, and FT-visible spectroscopy.

Figure 3d shows that the Lorentzian spectral line shape corresponds to an exponentially decaying waveform. This line shape is typically seen in FT-NMR, FT-ENDOR, FT-ICR, and many other types of spectroscopy.

Figure 3e shows that the Fourier transform of a Gaussian waveform gives again a Gaussian spectral line shape. Gaussian line shapes often arise from

---

† They correspond provided that the system is linear (e.g., that we are not saturating the spectroscopic line of interest).

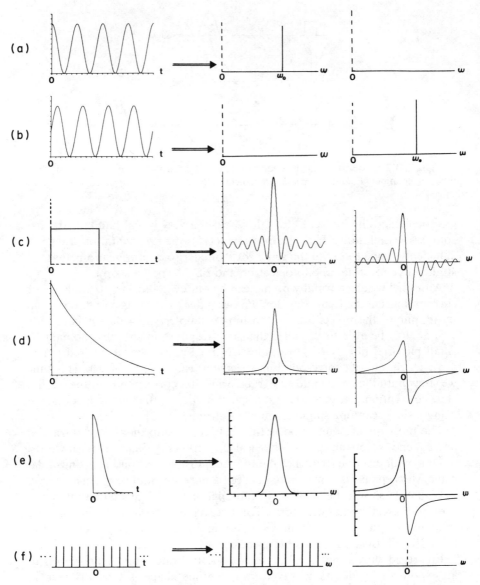

Fig. 3 Small pictorial library of time-domain waveforms (left) and their corresponding frequency-domain spectra (right) obtained by Fourier transformation. The left spectrum in each row represents the cosine Fourier transform and the right spectrum the sine transform. See text for interpretation and applications. [Computed curves kindly provided by Dr. D. C. Roe and Dr. R. E. Bruce.]

a distribution in line position, such as inhomogeneity either in the surrounding fields (NMR, ESR) or in the motion of the molecules themselves (rotational spectroscopy). It is also worth noting that for Gaussian and Lorentzian lines of equal width (at half maximum peak height), the Gaussian "wings" fall off much more rapidly than the Lorentzian.

Finally, Fig. 3f shows that the spectrum of a waveform composed of a series of equally spaced impulses is also a series of equally spaced impulses in the frequency domain. As we shall soon see (Section I.C), this hypothetical waveform (called shah) is useful in generating the effects of discrete sampling of the various continuous waveforms just discussed.

1. *Sensitivity Enhancement*

Many actual signals in Fourier transform spectrometers have the waveform of Fig. 3d, namely, an exponentially decreasing time-domain signal. By now we recognize that the longer the time-domain signal is acquired, the narrower is the corresponding frequency-domain spectral line, and the better is the spectral resolution. However, if the signal decreases with time while the noise level remains constant with time, it is clear that the signal-to-noise ratio decreases with longer acquisition period. The trade-off between signal-to-noise ratio and resolution is therefore simple and direct in Fourier transform spectroscopy: signal-to-noise ratio is optimized at short acquisition periods, and resolution is optimized at long acquisition periods.

Once a given data set of time-domain points has been acquired, it is still possible to enhance either signal-to-noise ratio or resolution. To enhance signal-to-noise ratio, we need simply weight the initial time-domain data points more than those near the end of the acquisition period. Convenient weight functions are either the boxcar truncation of Fig. 3c or the exponentially decreasing weight function of Fig. 3d. Both are in common use, and both will broaden the observed spectral line (we cannot increase signal-to-noise ratio without increasing the spectral line width and decreasing resolution) by the amounts shown in the frequency-domain plots of Figs. 3c and 3d.

2. *Resolution Enhancement*

Conversely, resolution can be increased by weighting the later time-domain points more than the initial points. A typical weight factor is an exponential with positive argument, $\exp(+t/\tau)$. Optimal resolution enhancement of a signal whose original form is that of Fig. 3d (decreasing time-domain exponential) will occur when the τ of the positive exponential is chosen to be

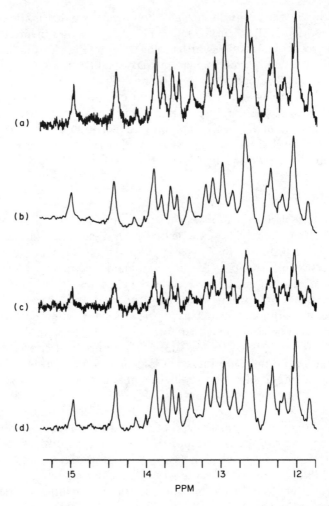

Fig. 4 Examples of apodization of an FT-NMR signal. (a) Fourier transform of the original unweighted free induction decay (FID) following a 90° pulse excitation. (b) Signal-to-noise enhancement: FID weighted by the factor $\exp(-\pi LBt)$, with LB = 3.0 Hz. Note improved signal-to-noise ratio and poorer resolution of (b) compared to (a). (c) Resolution enhancement: FID weighted by same factor as in (b), but with LB = −0.5. (Because of the large H_2O peak in this typical dilute macromolecular sample, even a small negative LB value degrades the signal-to-noise ratio significantly, so that exponential apodization is not a practical method for resolution enhancement.) (d) Exponential resolution enhancement (LB = −3.5) followed by Gaussian weighting by $\exp(-bt^2)$, in which $b = -a/(2GB \cdot AQ)$, with GB = 0.05 and AQ = 0.8192 sec acquisition time. Note that (d) gives resolution as good as (a), but with an improved signal-to-noise ratio approaching (b). 400-MHz ${}^1H$ FT-NMR spectra (297 K, 0.8192-sec duty cycle, one-pulse suppression of H_2O, 1000 transients, ~1-mM macromolecular concentration. [From A. G. Marshall, from a sample of *E. coli* transfer-RNA^{Val} kindly provided by Brian Reid.]

the same as the natural relaxation time τ of the time-domain signal itself: the weighted time-domain signal will then have the form of Fig. 3c.

A final apodization procedure becoming increasingly popular in FT-NMR [15] is based on a combination of the plots of Figs. 3c–e. Resolution is enhanced as just described to give a rectangular time-domain weighted signal of the form of Fig. 3c. Then, in order to improve the signal-to-noise ratio of this signal, a Gaussian weight function is applied to the rectangular signal to give a final time-domain signal whose shape is simply half of a Gaussian distribution curve. The Fourier transform of this final time-domain signal then gives a spectrum of lines with Gaussian shape, whose signal-to-noise ratio is about the same as if no apodization were used, but whose apparent resolution is better, because (see above) the Gaussian frequency-domain line shape has much narrower wings than the original Lorentzian line. Examples of these apodizations are shown in Fig. 4.

C. Convolution: Conversion of Integration into Multiplication

Convolution arises in the following common spectroscopic situation. Consider two functions, $h(t)$ and $e(t)$, as shown in Fig. 5. In physical applications, $h(t)$ might represent the response of a system to an ideal (infinitely brief) pulse

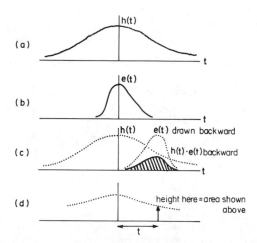

Fig. 5 Pictorial description of the convolution process (see text). To obtain the convolution of the two separate functions shown as traces (a) and (b) in the figure, one of the functions [in this case, $e(t)$] is first drawn backward (left-to-right) and then displaced with respect to the first function [in this case, $h(t)$]. The area under the product of these two functions is then shown as trace (c), and the numerical value of that area defines the amplitude of the convolution trace (d) for that displacement of $e(t)$ with respect to $h(t)$. [Adapted from Bracewell [16, Fig. 3.1].]

excitation; the Fourier transform $H(\omega)$ of that signal represents the frequency-domain response of the system to an excitation whose frequency spectrum is perfectly flat. Since the actual time-domain excitation waveform $e(t)$ is in practice never an infinitely sharp spike, the excitation frequency spectrum $E(\omega)$ is not perfectly flat. The actual system response $f(t)$ to such a nonideal excitation $e(t)$ will thus have a spectrum $F(\omega)$ that is distorted compared to the true spectrum $H(\omega)$ of the system.

Under very general conditions [16], the actual system response $f(t)$ can be expressed as the convolution of the system response $h(t)$ to an ideal (time-domain spike) excitation and the actual excitation $e(t)$:

$$f(t) = h(t) * e(t) \tag{20}$$

in which the symbol $*$ denotes the convolution process.

The convolution procedure can be understood entirely in pictorial terms, as shown in Figs. 5 and 6. To obtain the convolution of $h(t)$ with $e(t)$, we first redraw $e(t)$ backward (left to right) as shown in Fig. 5c, then superimpose $e(t)_{backward}$ onto the original $h(t)$ function, and multiply the two functions together. The area under that product then defines the value (i.e., the height) of the convolution $f(t)$ when the $h(t)$ and $e(t)_{backward}$ plots are displaced by a distance t relative to each other. The remaining values of $f(t)$ are obtained by sliding the $e(t)_{backward}$ curve along the abscissa, and repeating the above process for each relative displacement of $h(t)$ and $e(t)_{backward}$.

A simple graphical construction for this convolution process is shown in Fig. 6. One of the original functions is plotted backward on a separate piece of paper, and one then mentally constructs the area under the product of the movable and fixed curves as the lower curve is moved from left to right along the abscissa.

The general effect of convolution, seen in Figs. 5 and 6, is to broaden and smooth the original curve. The effect is analogous to that of using a broad

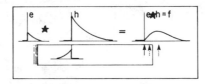

Fig. 6 Graphical construction for convolution. The movable piece of paper has one of the functions plotted backward. By passing the (bottom) movable function from left to right with respect to the other (fixed, upper) function, and mentally noting the approximate common area between the two functions, the qualitative form of the convolution function (upper right) can be found quickly without any calculations (see text). [Adapted from Bracewell [16, Fig. 3.6].]

slit in a scanning optical spectrometer: the sharp detail of an ideal spectrum (infinitely narrow slit) is smoothed out by the broad slit.

An extraordinary property of convolution leads immediately to a way to recover an ideal spectrum from one that has been distorted by a nonideal excitation. If $f(t)$ is the convolution of $h(t)$ and $e(t)$, then it can be shown [16] that

$$F(\omega) = H(\omega) \cdot E(\omega) \tag{21}$$

In other words, the convolution process (basically, an *integration*) in the *time* domain becomes a much simpler *multiplication* in the *frequency* domain The great value of convolution can be understood by analogy to logarithms: just as logarithms convert *multiplication* into *addition* (a much simpler operation), convolution effectively converts *integration* into *multiplication*:

$$a \cdot b \rightarrow A + B \quad \text{(log domain)} \tag{22}$$

$$a * b \rightarrow A \cdot B \quad \text{(Fourier transform domain)} \tag{23}$$

With the increasing use of small dedicated computers for analytical chemical tasks, convolution seems certain to become much more widely used in data reduction (see Section II.J for an example).

Equation (21) indicates that if the excitation profile in either the time domain $e(t)$ or frequency domain $E(\omega)$ is known, we can recover the ideal spectral response $H(\omega)$ of the system simply by dividing the actual frequency-domain response $F(\omega)$ by the spectrum $E(\omega)$ of the actual excitation:

$$H(\omega) = F(\omega)/E(\omega) \tag{24}$$

Examples of this "deconvolution" process will be encountered in Section II.

Finally, since multiplication in one domain corresponds to convolution in the Fourier transform domain [Eqs. (20) and (21)], we can immediately expand our library of Fourier transform pairs (Fig. 3) to include any others that are formed by multiplying any two of the original waveforms together. For example, suppose we multiply the time-domain waveforms of Figs. 3a and 3c together, to give a cosine wave that is truncated to zero after a specified time. Convolution of the frequency-domain plots of Figs. 3a and 3c (again using the method of Fig. 6) then gives a spectrum with the same shape as Fig. 3c, but now centered at the ω_0 of Fig. 3a.

If we multiply the time-domain waveforms of Figs. 3d and 3f together, we produce a time-domain signal consisting of a set of equally spaced discrete data points whose envelope is a decreasing exponential. (This is exactly the sort of data set acquired in FT-NMR spectrometers.) The spectrum of such a time-domain signal consists of equally spaced frequency points whose envelope is that of Fig. 3d. In other words, the effect of equally spaced sampling in the time domain is to produce a spectrum that is sampled (without any distortion) at equally spaced points in the frequency domain.

We are now ready to address a variety of applications of the basic transform methods introduced in Section I.

II.   Specific Applications of Transforms

A.   Fourier Transform Ion Cyclotron Resonance (FT-ICR) Mass Spectroscopy

The two most chemically useful analytical spectroscopic techniques are probably NMR and mass spectrometry. Although extended discussion of conventional mass spectrometers is not appropriate here, the basic procedure involves ionizing gaseous neutral molecules directly (with a beam of accelerated electrons or a high electric field) or indirectly (by reaction with previously ionized species), followed by deflection of ions of a particular mass-to charge ratio using electrically charged plates or magnetic fields (much as with the focus or convergence controls in a color television picture tube), so that those ions pass through a small slit to an ion detector. Then, by varying the electric or magnetic (or both) deflection fields, ions of different mass-to-charge ratio can be observed in turn, essentially in the spirit of the single-slit photon spectrometer of Fig. 1b.

There is one type of mass spectrometer in which the various mass-to-charge ions are simultaneously dispersed in space and recorded on a photographic plate, as in the multichannel detector of Fig. 1c, but then it is necessary to process the photograph before seeing the spectrum, removing much of the time-saving advantage of the multichannel detection. One is therefore led to consider possible multiplex (coded detection, with decoding to obtain the spectrum) mass spectrometric detection schemes. The Hadamard method will not help, because noise is in this case source-limited [1] and increases as the square root of the signal, thereby canceling the Fellgett advantage. However, there is another kind of mass spectrometer, the *ion cyclotron resonance* (ICR) spectrometer that is susceptible to the use of Fourier methods in much the same way that they are used in NMR spectroscopy.

Figure 7 is a schematic diagram of an ICR mass spectrometer. An ion moving in a magnetic field is constrained to a circular path; the frequency ν

of this circular motion is proportional to the charge-to-mass ratio q/m for that ion:

$$v = qB/2\pi m \qquad (25)$$

in which B is the applied magnetic field strength, and all units are mks. When such ions are irradiated by a circularly polarized (i.e., rotating) electric field whose frequency is close to the ion "cyclotron" frequency of Eq. (25), the resulting ion motion becomes spatially coherent (i.e., all the ions move in a group) as the ions absorb energy from the irradiation by increasing the radii of their cyclotron orbits (see Fig. 7). Once the ions are all moving essentially together, their composite cyclotron motion will induce a *macroscopic* voltage in the surrounding plates. For example, at an applied magnetic field strength of 1 Tesla (10 kG), ICR frequencies for singly charged ions of mass 16 to mass 400 fall in the radio-frequency range between about 35 kHz and 1 MHz (see bottom of Fig. 7).

The ICR spectrometer thus produces a signal whenever the irradiation frequency matches the ion cyclotron frequency of ions of a given mass-to-charge ratio present in the sample. In other words, the device can function as a mass spectrometer to detect ions spanning a range of mass-to-charge ratios, by irradiation of the sample with an oscillating electric field whose frequency is scanned slowly over the range required by Eq. (25), with simultaneous monitoring of power absorption, in the same spirit as a conventional single-slit scanning spectrometer (Fig. 1b). Finally, since the detector no longer counts individual ions (as in a conventional non-ICR mass spectrometer), noise is independent of signal strength, and the Fellgett advantage of Fourier transform methods is now available.

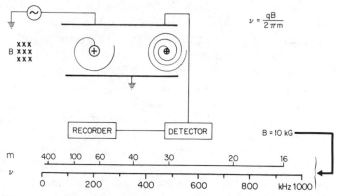

Fig. 7 Schematic diagram of an ion cyclotron resonance mass spectrometer. See text for operation.

1. Broad-Band Excitation

The two major experimental problems in implementing Fourier methods in ICR spectroscopy are broad-band excitation and broad-band detection. Table I and Fig. 7 show that it is necessary to excite signals over a range of about 1 MHz. Since the ICR signal, once excited, will decay in a few milliseconds (at typical operating pressure of 10^{-7}–10^{-9} Torr), the excitation must be achieved in a millisecond or less.

The excitation used to produce the very first FT-ICR spectrum [17] was a simple rectangular radio-frequency pulse near the ICR frequency of interest. However, Fig. 2 shows that in order to excite a wide frequency range the pulse duration must be very short. A quick calculation [18] shows that the required pulse length is about 0.1 μsec to excite ICR signals over a range of about 1 MHz. However, in order to excite the ions to the desired radius (~ 1 cm), the amplitude of such a pulse would have to be about 10 kV [18].

In order to avoid the high-voltage problems associated with pulsed excitation, Comisarow suggested the use of a fast frequency-sweep excitation, in which the desired frequency range is irradiated with a linearly time-varying frequency sweep whose duration is of the order of 1 msec [19]. All subsequent FT-ICR experiments published to date have been based on the frequency-sweep excitation, and the ICR response to this excitation has recently been analyzed [18].

2. Broad-Band Detection

Since the high frequency in a typical ICR spectrum may be as large as 1–2 MHz, it is necessary to sample the time-domain ICR signal at a rate of at least 2–4×10^6 data points (i.e., at least two points per sinusoidal cycle) in order to detect oscillation at that frequency. This rapid sampling requirement is one reason that Fourier methods were applied much later to ICR (1974) [17, 19] than to NMR (1966) [20], because the largest sampled frequencies in NMR time-domain signals are 10–100 times lower than in ICR, and it took many years to develop fast digitizers.

Representative theoretical FT-ICR time-domain signals and their corresponding frequency-domain spectra are shown in Fig. 8. At low pressure the line width depends on the acquisition time; at high pressure the line width depends on the ion–molecule (reactive and nonreactive) collision rate constant. Theoretical FT-ICR line shape at arbitrary sample pressure has recently been analyzed, and the results shown to agree precisely with experimental FT-ICR line shapes [21]. The principles of operation of the detector electronic circuitry have recently been discussed [22].

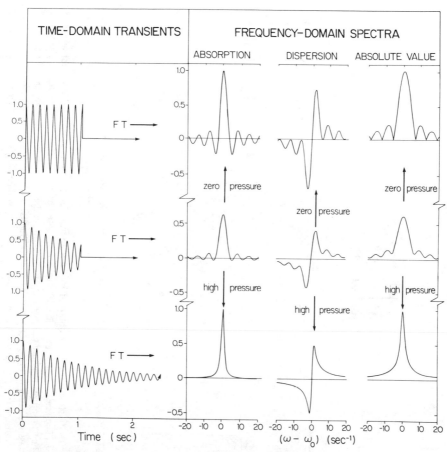

Fig. 8 Cosine transform (absorption), sine transform (dispersion), and magnitude (absolute-value) frequency-domain ICR spectra obtained by analytical Fourier transformation of the continuous time-domain signals shown at the left of each set of spectra. The magnitude-mode spectrum is the square root of the sum of the squares of the absorption and dispersion spectra. All frequency-domain spectra have the same horizontal and vertical scale. The top row of graphs corresponds to the zero-pressure limit $(TA \ll \tau)$, the bottom row to the high-pressure limit $(TA \gg \tau)$, and the middle row to an intermediate pressure for which $TA = \tau$, in which TA is the data acquisition period and τ is the relaxation time for (exponential) decay of the FT-ICR time-domain signal. [Adapted from Marshall [13], p. 679.]

3.   Performance

Signal-to-noise ratio: Improvement (see Table I) by a factor of up to 100 compared to conventional scanning ICR spectrometer, because up to 10,000 times as many "scans" can be accumulated in the same time it would take to acquire a single conventional ICR spectrum.

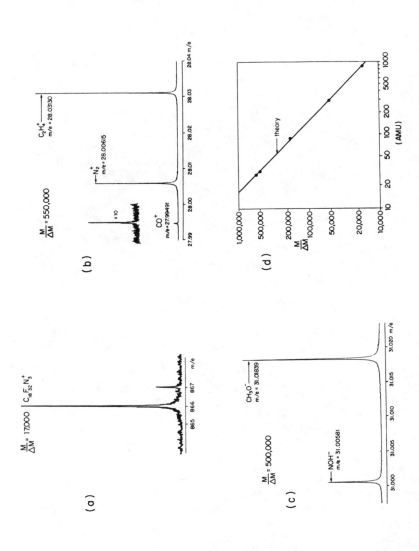

Fig. 9 Ultrahigh-resolution FT-ICR mass spectra. Each spectrum was obtained by Fourier transformation of a digitized transient acquired for 1 sec. (a) Region near $m/e = 866$ peak of Fig. 10. (The peak at $m/e = 867$ is the 18% carbon-13 satellite, $^{12}C_{17}\ ^{13}CF_{32}N_3^+$, of the peak at $m/e = 866$.) (b) Ternary mixture of CO, N_2, and C_2H_4. The CO sample pressure was 50 fatm, demonstrating the simultaneous achievement of high sensitivity and high mass resolution. (c) Negative ion mass spectrum from CH_3O^- and NOH^-. (d) Finally, theory predicts [21] that for transient ICR signals with the same decay time constant and same acquisition time, mass resolution should vary inversely with ion mass, as confirmed experimentally by this data, taken from (a), (b), and (c). [From Comisarow [26].]

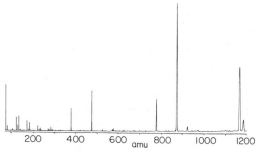

Fig. 10 Wide mass range FT-ICR mass spectrum of tris-perfluoroheptylazine, $C_3N_3(C_7F_{15})_3$, mol wt = 1185, from $m/e = 60$ to $m/e = 1200$. This spectrum was obtained by Fourier transformation of a three time zero-filled [27] transient signal, which was in turn produced from the sum of 100 individual transients, each acquired for 16 msec. Total data acquisition time is thus only 1.6 sec. Note the increase in mass resolution as mass decreases (see Fig. 9). [From Comisarow [26].]

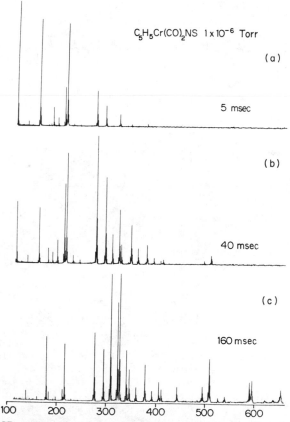

Fig. 11 FT-ICR mass spectra of $CpCr(CO)_2NS$ as a function of reaction time. (a) FT-ICR mass spectrum at short reaction time. It is essentially the same as the mass spectrum produced by a conventional mass spectrometer. Parts (b) and (c) show mass spectra at longer reaction times. [From Comisarow [26].]

Speed: Improvement (see Table I) by a factor of up to 10,000 compared
to conventional ICR, making it possible to combine a gas chromatograph with an FT-ICR spectrometer.

Mass resolution: Improvement by a factor of up to 1000 or more compared
to conventional ICR (see Fig. 9).

Mass range: Improvement by a factor of about ten compared to conventional ICR (see Fig. 10).

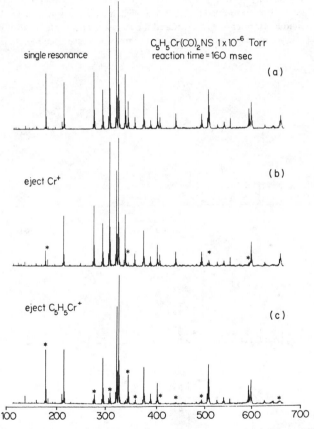

Fig. 12   FT-ICR double resonance mass spectra. (a) is the same as Fig. 11c. Plot (b) is the
FT-ICR spectrum at 160-msec reaction time with Cr^+ ion ejected from the ionic sample as soon
as it is formed. The differences between the two spectra are indicated by asterisks. Plot (c) is
the FT-ICR mass spectrum at 160-msec reaction time with $CpCr^+$ ion ejected as soon as it is
formed. Each of the missing peaks (*) is a chemical offspring of the ejected ion. Thus a series of
ion–molecule reactions can be studied by selectively removing any particular ionic species.
[From Parisod and Comisarow [28].]

Ion–molecule chemistry: Retains all capability of conventional ICR for studying connectivity of ion–molecule reactions [23], with the advantage that fewer experiments are required [24]. This feature is in addition to the multichannel advantage, and is illustrated in Figs. 11 and 12.

Mass calibration: Automatic, since (to a good approximation) the measurment of the ICR frequency for one peak of known m/e ratio determines the magnetic field strength and thereby provides [Eq. (25)] for the determination of all other detected m/e values from measurement of their ICR frequencies.

Operating pressure: 10^{-7}–10^{-10} Torr, so that ICR (and especially FT-ICR) is possible with samples of extremely low volatility, such as nucleotides [25].

Negative ions: Identical operation as for positive ion detection, except for sign of charge on cell plates.

The performance features listed above show that FT-ICR offers a number of powerful advantages over conventional ICR mass spectrometry and perhaps over other types of mass spectrometers in certain applications. With the introduction of, the first commercial FT-ICR spectrometer (Nicolet FT/MS-1000) in 1981, this performance potential should begin to be realized.

B. Fourier Transform Pure Rotational (FT-Microwave) Spectroscopy

FT-ICR, FT-NMR, and FT-microwave experiments are all based on coherent excitation that produces coherent rotational or precessional circular motion. All the excited molecules or ions thus move together (i.e., with the same angular "phase"), so that their individual (small) electric or magnetic dipole moments add coherently to give a macroscopic (large) moment whose rotation induces oscillation in a fixed detector. Fourier transformation of these time-domain detected oscillations then yields a spectrum of the amplitudes of the motions excited in that frequency range. Of the three experiments, FT-ICR was discussed first because it can be described wholly in classical mechanical terms. FT-NMR and FT-microwave experiments, on the other hand, require a quantum-mechanical treatment.

A pictorial representation of the FT-NMR and FT-microwave experiments is given in Fig. 13. In the more familiar FT-NMR case, a sample is placed in a strong static magnetic field (amplitude H_0), and an equilibrium magnetization builds up in the $H_0(\mathbf{z})$ direction as the nuclear spins "polarize" (align their directions) along $\mathbf{z}$. A second magnetic field (amplitude H_1) rotating with angular frequency

$$\omega_0 = \gamma H_0, \qquad \gamma = \text{nuclear magnetogyric ratio} \qquad (26)$$

Alan G. Marshall

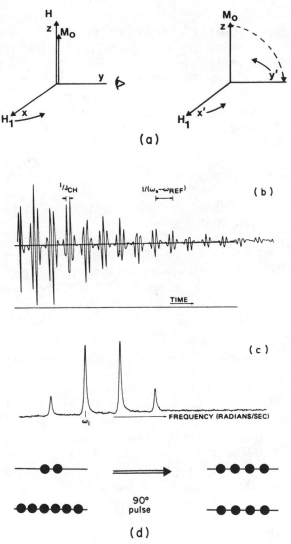

Fig. 13 Schematic features of pulsed NMR (a)–(d) and pulsed microwave (d) experiments:
(a) Equilibrium nuclear magnetization M_0, aligned along applied magnetic static field H_0,
just as a rotating magnetic field H_1 is applied as shown in a laboratory frame (left). Subsequent
motion of M_0 in a coordinate frame rotating at H_1 rad/sec (so that H_1 is stationary) appears
at right. (b) Following a 90° H_1 pulse (see text), the signal seen by the observer stationed as in
(a, left) is shown here. Different frequencies (due to spin–spin couplings J_{CH} or sample and
reference peaks ν_0 and ν_{ref}) produce a beat pattern as shown. (c) Fourier transform of (b), to
give FT-NMR carbon-13 spectrum of CH_3OH. (d) Populations for a single spin-one-half
system (vertical scale = energy) before (left) and after (right) the 90° pulse discussed above
(see text). [(b) and (c) adapted from Farrar [29, Fig. 8.1].]

in the $x-y$ plane then appears stationary, and H_0 disappears if we go to a coordinate frame that also rotates at ω_0 synchronously with H_1 [29]. In the rotating frame, the magnetization initially along z will now precess (rotate) about the H_1 direction at an angular frequency $\omega_1 = \gamma H_1$. After the H_1 field has been on for a time $(\pi/2)/(\gamma H_1)$, the original magnetization will have rotated by 90° and will lie in the x-y plane; H_1 is then turned off, and an oscillating signal appears (Fig. 13b) in a fixed coil positioned as shown in Fig. 13a as the magnetization rotates in the x-y (laboratory) frame. Finally, Fourier transformation of the sampled time-domain oscillating signal then gives a spectrum (Fig. 13c) of the Larmor frequencies excited in the sample. (The spectral lines have finite width because the time-domain signal eventually decays due to interactions of the spins with each other or with their surroundings.)

A more general picture of this experiment is based on the populations of the energy levels in a two-level (e.g., spin one-half) system (Fig. 13d). The effect of the 90° pulse is to equalize the populations of the upper and lower levels, but in such a way that the spins have a common phase (i.e., all the spins point in the same direction in the x-y plane). FT-NMR applications have been reviewed so often and so recently that none are given here. More recent two-dimensional FT-NMR experiments are treated in Section II.H.

In proceeding to the microwave case, we need simply recognize that the magnetic moment of the sample and the applied rotating magnetic field (H_1) that produces a detectable oscillating magnetic polarization are replaced by an applied rotating electric field that polarizes the electric dipole moments of the sample to produce a detectable oscillating electric polarization [30]. The remaining conceptual differences between the two experiments involve only the different natural frequencies and the different time-domain signal decay rates.

Broad-band excitation (see preceding section) can be achieved using a sufficiently short rectangular pulse of (in this case) microwave radiation. In order to excite a bandwidth of approximately 50 MHz, a pulse of 10 nsec has been used [31]. Detection requires sampling at least twice per cycle of the largest frequency to be observed, to give a sample rate in this case of 10 nsec/point (100×10^6 points/sec). In 1976 digitizers operating at this speed could provide only one bit of information (e.g., whether the detected signal voltage is positive or negative), so that it was necessary to accumulate many digitized transient microwave signals in order to build up the signal-to-noise ratio. However, because the microwave relaxation time is so short, it is possible to repeat the experiment 30,000 times/sec [31]. Thus an initial signal-to-noise ratio of less than 0.01 per transient can be improved to about 5:1 by signal averaging for about 10 sec. A full discussion of signal averaging may be found in [32].

A time-domain, digitized, accumulated microwave transient and its Fourier transform are shown in Fig. 14 [31]. Exponential apodization (Section IB) was used to enhance the signal-to-noise ratio. The spectrum is for the $1_{11} \rightarrow 1_{10}$ rotational transition in CD_2O at 6083 MHz for very low pressures and $T = -77°C$. The signal was sampled at 10^6 points/sec to give

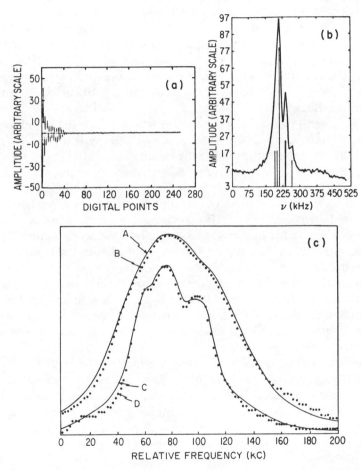

Fig. 14 Experimental pure rotational spectra. (a) Time-domain transient emission microwave signal from deuterated formaldehyde. (b) Fourier transform of digitized transient of part (a). The markers indicate the frequencies measured with a molecular-beam spectrometer. (c) Experimental (dotted lines) and theoretical (solid lines) steady-state microwave absorption spectra of the same transition in CD_2O: upper trace at $T = -78°C$ and lower trace at $T = -110°C$. Note the greatly increased resolution of the Fourier transform spectrum of (b) compared to the steady-state spectrum of (c) at the same temperature. [From Ekkars and Flygare [31] and Flygare [33].]

512 time-domain data points per transient. A comparison with older measurements using the steady-state (i.e., single-channel, scanning) method [33] (Fig. 6c) shows a marked increase in resolution.

Performance

Signal-to-noise ratio: Factor of twenty improvement over conventional microwave spectrometer (assuming 50-MHz spectral bandwidth, typical line width of 500 kHz, and average dipole transition moment of 1 D).

Small dipole moments: Can detect weak lines for which a prohibitively high Stark voltage would be required for the steady-state experiment.

Resolution: No power broadening is introduced by overmodulation—optimum signal-to-noise ratio in the steady-state experiment requires modulation amplitude that produces line broadening of a factor of $\sqrt{3}$.

Speed: Factor of at least 100 over conventional steady-state spectrometer. Even though only one peak may be present in the 50-MHz bandwidth, the multichannel advantage still applies because the position of the peak is not known in advance, so that it would still be necessary to scan the whole bandwidth with a conventional spectrometer.

Although a variety of pulsed microwave phenomena have been demonstrated beginning in 1955 [34], theoretical [35] and experimental [31, 36] Fourier transform microwave spectroscopy is a quite recent development. Commercial FT-microwave spectrometers are not yet available, but the very promising results summarized above augur well for wider future interest in this area.

C. *Fourier Transform Vibrational (FT-IR) Spectroscopy*

Fourier data reduction in infrared spectroscopy is essentially identical to that in the FT-ICR, FT-NMR, and FT-microwave examples of the previous sections, except that the desired spectral frequency information is dispersed in space rather than in time. The two advantages of this different approach are: (a) it is no longer necessary to use a coherent source, so that conventional IR sources are suitable, and (b) the necessary sampling rates are much smaller than the 10^{14} points/sec that would be required for time-domain sampling.

The interferometer (Fig. 15) used in FT-IR is simply a means for adding the electric fields of two beams of light that have traveled a different path length since leaving the sample. Whenever the beams travel the same path length (i.e., equal distances between the beam splitter and either the fixed or movable mirrors of Fig. 15), the two beams add to give a maximum

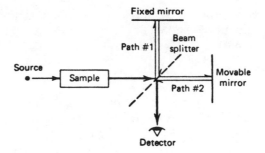

Fig. 15 Schematic diagram of a Michelson interferometer. The beam splitter divides the incident beam from the source into two beams of equal amplitude that travel out and back from the two mirrors; the two reflected beams then recombine and are sent to the detector by the beam splitter. (Half the intensity is lost at the beam splitter.) The device thus provides a means for adding together two sample beams that have traveled different path lengths on their way to the detector.

amplitude, because electromagnetic waves of any frequency recombine with the same phase. However, when the distance between the beam splitter and the fixed mirror is different from the distance between the beam splitter and the movable mirror, then the amplitude of the recombined light wave depends upon the frequency of the wave (see Fig. 16). As the two light beam paths are made increasingly different, more and more frequency components become out of phase in the recombined beam, resulting in an "interferogram" pattern of recombined intensity as a function of movable-mirror position (Fig. 17). As for the preceding NMR, ICR, and microwave examples, the interferogram represents a sum of waves of all frequencies from the source, and the various frequency components become out of phase with increasing path difference in the same way that the NMR (or ICR or microwave) time-domain signal components become out of phase with increasing time. The Fellgett advantage (Section I.A) therefore applies, and the spectrum (of the source) can be recovered by Fourier transformation of a digitized interferogram (Fig. 17). The main different feature in the FT-IR case is that the interferogram is two-sided (that is, symmetric about the mirror position that gives zero path difference), while the NMR (or ICR or microwave) transients are defined only after time zero and are thus one sided. The FT-IR absorption spectrum of a sample is obtained by generating an interferogram with a sample located between the interferometer and the detector.

Performance

Multichannel advantage: For measurements taken at equal resolution in equal measurement time with the same detector on an instrument with

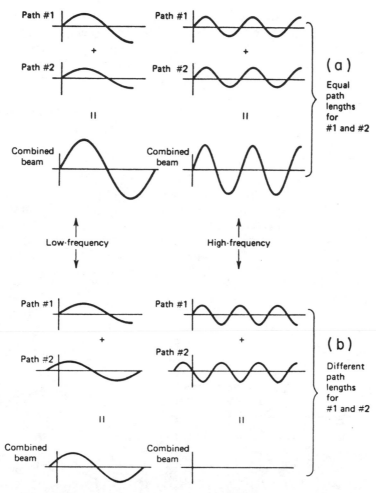

Fig. 16 Schematic pictures of the combined beam electric field at the detector (Fig. 15) for low-frequency (left) and higher frequency (right) beams that have traveled (a) equal or (b) different path lengths on their way to the detector. When the path lengths are equal, beams of both frequencies add coherently to produce a resultant maximal amplitude at the detector (a). When the path lengths are slightly different, the low-frequency component beams still add with nearly the same phase to produce a combined wave of large amplitude (b, left), but the higher frequency component beams add with appreciably different phase [here shown as 180° phase difference at (b, right)] to produce a resultant combined beam of much reduced amplitude. As the path difference is increased (by moving the movable mirror of Fig. 15), even the relatively lower frequency components add with quite different phase, and therefore the amplitude of the interferogram (Fig. 17) decreases from a maximum value (for equal paths) to smaller values as the path difference between the two beams increases, just as the amplitude of an NMR (or ICR or wave) time-domain signal decreases with time when the spectrum consists of signals at many different frequencies. [From Marshall [13, p. 688].]

the same optical throughput and efficiency, the signal-to-noise ratio for a spectrum from the Fourier spectrometer will be $\sqrt{N}$ times greater than for the corresponding spectrum from a grating instrument, where N is the number of resolution elements (see Table I). Alternatively, for equal signal-to-noise ratio, the Fourier spectrometer will produce a spectrum in a factor of $(1/N)$ less time.

Throughput advantage: The Fellgett (multichannel) advantage corresponds to opening the exit slit of an optical spectrometer. There is in addition a Jacquinot (thoughput) advantage corresponding to opening the entrance slit: this leads to an additional throughput of a factor ranging between about 10 at 500 cm^{-1} to about 200 at 3500 cm^{-1} [37].

Difference spectra: Because the final FT-IR spectrum is stored in digital form, it is a simple matter to compute the difference spectrum between two data sets. In the conventional optical absorption experiment, absorption from the solvent is eliminated by taking the difference

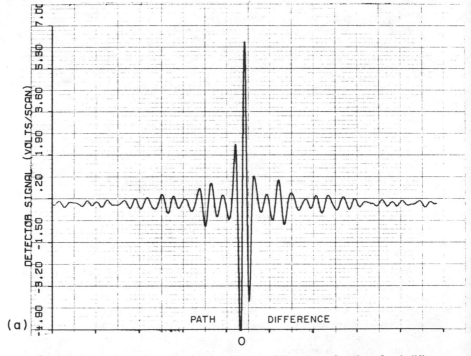

Fig. 17a Infrared interferogram signal at detector of Fig. 15 as a function of path difference between the two split beams. The Fourier transform is shown in Fig. 17b.

between the absorption from the dissolved sample and the absorption from the solvent. However, if the solvent absorption happens to be very strong (as for H_2O solvent), it then becomes necessary to take the difference between two very small numbers (the transmitted intensity through the sample minus that through the solvent), and the effect of a slight amount of stray light in a conventional single-slit instrument renders the comparison essentially useless. The interferometer is much less sensitive to stray light problems, and can thus provide useful difference spectra in such cases, as shown in Fig. 18. Although virtually no useful detail can be seen in the aqueous FT-IR spectrum of neomycin (Fig. 18a), subtraction of the spectrum of the solvent (Fig. 18) leaves a highly resolved difference spectrum (Fig. 18c).

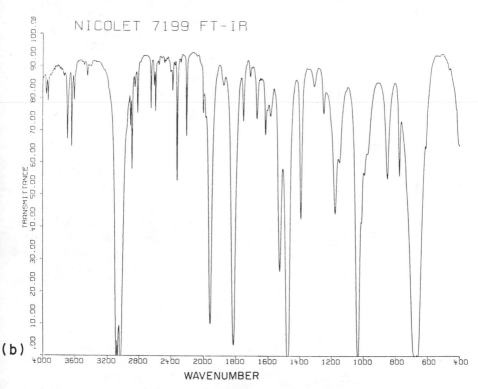

Fig. 17b Infrared spectrum of benzene at 0.025-mm path length. Spectrum was obtained by discrete Fourier transformation of the digitized interferogram. [Courtesy of Dr. Robert L. Julian, Nicolet Instrument Corporation, Madison, Wisconsin.]

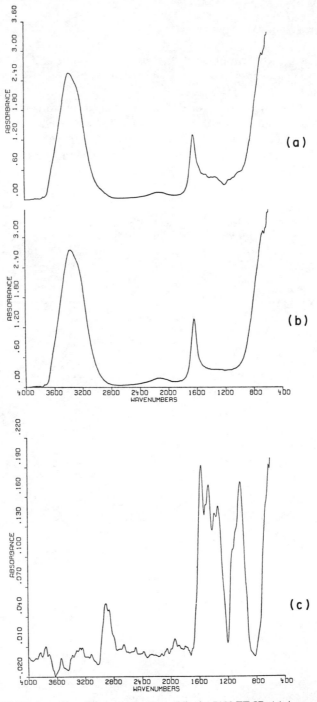

Fig. 18 FT-IR absorption difference spectrum: Nicolet 7199 FT-IR. (a) Aqueous neomycin. (b) Spectrum of water reference sample. (c) Subtraction of water spectrum from neomycin spectrum. Note the greatly enhanced detail in the difference spectrum compared to the top spectrum. Neomycin is at submicrogram concentration. [Spectra courtesy of Dr. Robert L. Julian, Nicolet Instrument Corporation, Madison, Wisconsin.]

Astronomical sources: Because planetary and stellar sources are so weak, the signal-to-noise advantage of FT-IR has revolutionized IR characterization of these sources. Ammonia is easily detected from the far-IR spectrum of Jupiter [38], and the $^{12}C/^{13}C$ ratios of various star types have been determined from the relative intensities of $^{12}C^{16}O$ and $^{13}C^{16}O$ vibrations [39].†

Resolution: Resolution is determined by the "retardation," or maximum difference in path length from the interferometer (typically ~ 10 cm in commercial FT-IR instruments). Because of the wide range of frequencies involved, the number of interferogram data points required varies with wavelength region. For example, low-resolution far-IR spectroscopy (~ 4 cm^{-1} resolution) might require 1000 data points or less, while mid-IR spectroscopy at medium resolution (say, 0.5 cm^{-1}) might require 32,000 data points, and ultrahigh resolution (10^{-3} cm^{-1}) could require more than 10^6 points and thus some external (as opposed to on-line minicomputer) data reduction [37].

Reaction kinetics: The speed advantage of FT-IR makes possible the rapid acquisition of IR spectra at selected time intervals following the mixing of various reactants, so that it becomes possible to monitor simultaneously the relative concentrations of several components as a function of reaction time. See Fig. 19 for use of FT-IR to monitor the curing of an epoxy resin.

GC-IR: The speed advantage of FT-IR also makes possible the useful coupling of an FT-IR spectrometer to the output of a gas chromatograph (Fig. 20). This technique is complementary to GC–mass spectroscopy, and offers the entire IR spectrum of a given eluent fraction (rather than just the absorbance at a single wavelength), with somewhat lower sensitivity (~ 100 ng versus < 1 ng) than GC–mass spectroscopy.

LC-IR: FT-IR may also be combined with liquid chromatography (see Fig. 21). In this example, the 2990–2850-cm^{-1} window is specific for petroleum oil, while the 1100–1040-cm^{-1} window is specific for silicone oil. The 680–660-cm^{-1} and 3100–3000-cm^{-1} windows characterize the silicone oil as aromatic and the petroleum oil as aliphatic. All this information can be recovered from the IR characteristics, even though the GC peaks themselves overlap.

Remote sensing of stack gases: The high resolution of FT-IR makes it possible to distinguish a large number of lines in a spectrum, and thus to identify many components in a mixture. An important class of applications of FT-IR is the remote sensing of plumes stack gases [37].

---

† Johnson and Mendez (39) present IR spectra for 32 stars.

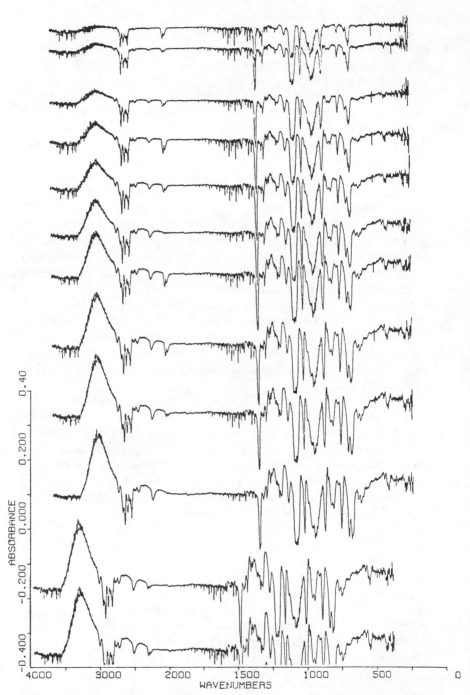

Fig. 19 Chemical reaction monitoring by FT-IR. Successive spectra are taken mostly at 30-sec intervals (zero time at top), with each spectrum ratioed to the time-zero run. Thus, each spectrum represents the IR absorbance change since time zero. The system is a curing epoxy resin. [Courtesy of Dr. Robert L. Julian, Nicolet Instrument Corporation, Madison, Wisconsin.]

Quantitative estimates of the amounts of several component gases can be provided at the sub-ppm level by FT-IR detection.

A fuller listing of recent FT-IR applications, and a summary of instrumental and calculational aspects may be found in [37]. The above examples should, however, suffice to show why Fourier methods have vastly increased the range of problems accessible to solution by IR analysis.

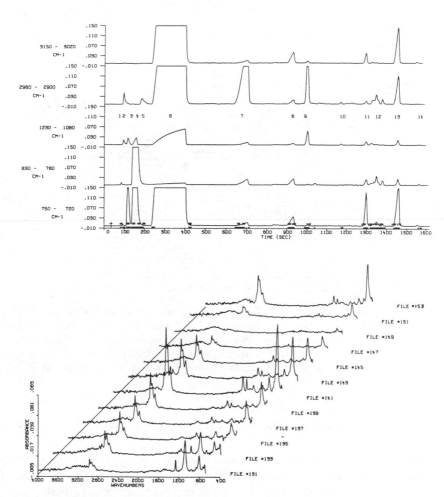

Fig. 20   Gas-chromatograph–FT-IR analysis of a reaction mixture. The reaction is among *t*-butyl α-ethylcaproate, toluene, and carbon tetrachloride. The top plots show IR absorbance as a function of time, while the lower "chemigram" gives the complete IR spectrum at 8-sec intervals looking "through" the bands labeled 12. [Courtesy of Dr. Robert L. Julian, Nicolet Instrument Corporation, Madison, Wisconsin.]

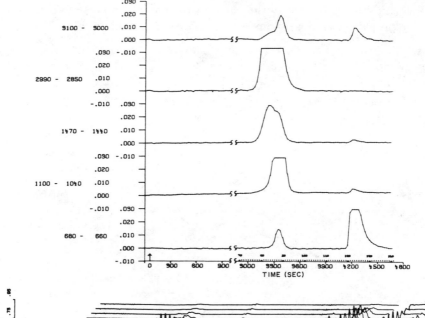

Fig. 21 Liquid-chromatograph–FT-IR analysis, showing a separation of petroleum oil and silicone oil of overlapping molecular weight ranges. The top plots show IR absorbance as a function of time for fixed wavelength, while the bottom plots show spectra collected at different times during elution of the two overlapped chromatographic peaks. See text for interpretation. [Courtesy of Dr. Robert L. Julian, Nicolet Instrument Corporation, Madison, Wisconsin.]

D. Hadamard Transform Vibrational (HT-IR) Spectroscopy

The basic Hadamard code was described in Section I.A.3. Such codes are directly useful in so-called error-correcting binary codes, and a code based on a rank-8 Hadamard matrix was used to code the pictures that were transmitted back from Mars in 1969 by a Mariner spacecraft.

Although it is conceivable to construct a spectrometer in which the radiation falling on each slit position of the mask is either transmitted directly to the detector (weight factor of +1) or subtracted from the detector output

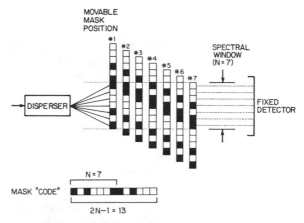

Fig. 22 Schematic diagram of the use of a single $(2N - 1)$-slit mask to generate the N different linear combinations of spectral intensities from an N-channel spectrum, for the case that $N = 7$. The 13-slit mask is first placed between the spectral window and the detector as shown at the far left (see also Fig. 1d), to give a pattern of open and shut slits described by the sequence 0 1 0 1 1 1 0, and the transmitted total intensity is recorded by the broad-band detector. Next, the mask is translated by one position across the window, to generate the new pattern of open and shut slits described by the sequence 1 0 1 1 1 0 0, and the total transmitted intensity is again recorded. Proceeding in this way, one obtains $N = 7$ different observed total transmitted intensities, corresponding to 7 linearly independent combinations of the (desired) unknown spectral intensities in each of the 7 spectral channels. By the method described in Section I.A.3, these $N = 7$ equations in 7 unknowns may then be solved to yield the desired N-channel intensity spectrum from the sample. [From Marshall and Comisarow [1, p. 51].]

(weight factor of -1), it is much simpler to employ a mask with open and closed slits such as that shown in Fig. 1d. The detector total signal y then consists of a weighted sum of all the desired spectral intensity elements x_i, each weighted by a factor of either zero or one, depending on whether that particular slit was shut or open, respectively. (The inverse code is again derived simply by replacing 0 by -1 in the original code—see [1] for an example with $N = 3$.)

Experimentally, the cyclically permuted rows of the Hadamard code provide for a great simplification in instrumental operation. For any given code, we would expect to use one mask (whose pattern of open and shut slits corresponds to the first row of the code matrix, then switch to a second mask (whose pattern is given by the second row of the code), etc., until we have produced N detected intensities corresponding to N individual masks. However, because each row of the Hadamard code is generated by cyclic permutation of the preceding row, we do not need N individual maks, but can get by with just a single mask of $(2N - 1)$ slits, as shown for the $N = 7$ case in Fig. 22. Successive slit arrangements (i.e., successive rows of the code matrix) are then produced simply by translating the mask across the spectral window by one slit per move.

On both computational and mechanical grounds, then, the Hadamard scheme of Figs. 1d and 22 conveniently accomplishes an improvement of $\sqrt{N}/2$ in signal-to-noise ratio compared to an experiment involving the same total observation time with a single-slit scanning spectrometer (Fig. 1b). The reason is that half the N possible slits are open during each measurement, rather than just one (so the signal is $N/2$ times larger), and the noise increases as $\sqrt{N}$ because the noise is effectively added N times in a random-walk fashion. Alternatively, the Hadamard spectrometer can provide the same signal-to-noise ratio in a factor of $4/N$ as much time as the conventional single-slit scanning spectrometer. Although the FT-IR scheme might thus seem better than HT-IR, because effectively all the slits are open in the FT-IR case, the interferometer splits the incident beam and thus loses half the possible signal also, so that theoretical signal-to-noise ratio for HT-IR and FT-IR should be the same (when all other instrumental parameters are equal).

The key to the Hadamard operation is the ability to construct a code with approximately half zeros and half ones in any one row, such that the rows are linearly independent and differ from each other by cyclic permutation. It turns out that these conditions are compatible only for certain choices of N: for example, $N = 4$ will not satisfy all the conditions at once. A simple rule of thumb is that the Hadamard scheme just described can be accomplished whenever $N = 3, 7, 15, \ldots, 2^n - 1$, where n is an integer [40].

Figure 23 gives a direct experimental comparison between HT-IR and conventional scanning IR spectra obtained under the same conditions for the same length of time. The enhanced signal-to-noise ratio of the HT-IR spectrum is readily apparent, and agrees with the theoretically predicted multichannel advantage discussed above.

1.   *Performance*

Multichannel advantage: HT-IR spectrometers should provide the same signal-to-noise ratio or time-saving advantage as FT-IR spectrometers operating under the same conditions with the same detector (see Fig. 23).

Resolution: HT-IR instruments disperse the IR spectrum with a grating, and resolution is ultimately limited by diffraction effects. Diffraction in turn is determined by the length of the grating (or in an interferometer, by path length differences between the two beams). Since maximal

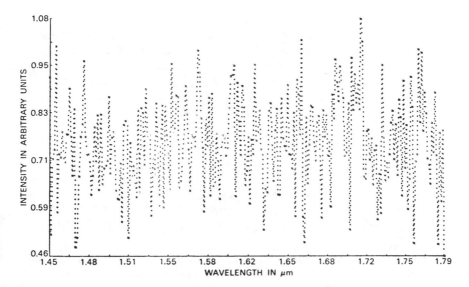

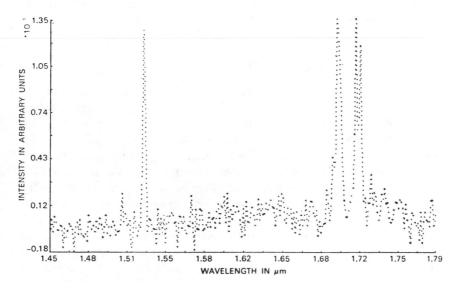

Fig. 23   Infrared spectra for 1.5-μm mercury emission lines under high-noise conditions. Top: conventional scanning spectrometer of the single-channel type shown in Fig. 1b. Bottom: Hadamard transform spectrometer with 255-slot code; * denotes computed spectral values, while dotted lines denote computer's linear interpolation between computed values. Conditions and total observation time are the same for both spectra. [From Decker [41, p. 133A].]

grating lengths are of the order of 20 cm, while maximal path length differences in an interferometer approach 2 m, the FT-IR spectrometer resolution is potentially about ten times greater than for HT-IR.

Strong and weak lines: When strong and weak emission lines are present in the same spectrum, both signals add in the interferogram, leading to severe dynamic range limits in the digitization process. However, in HT-IR it should be possible to blank out any given strong peak(s) with a fixed mask, thereby providing increased dynamic range for detection of the weak peaks.

Mechanical aspects: Because the mask motion in a Hadamard spectrometer need be controlled only to within the order of a slit width rather than the order of a wavelength as with interferometric spectrometers, mechanical constraints are less severe in HT-IR than FT-IR. (This is of course because the wavelength accuracy is built into the grating of the HT-IR instrument.)

HT-IR versus FT-IR: From the above discussion, it appears that for point sources or samples, FT-IR is the more generally useful approach, but HT-IR is particularly attractive when the system must be portable, or in relatively low-cost applications involving medium resolution (200–300 resolution elements), 1% accuracy, and low cost. A full discussion of history and applications may be found in [42] and [43].

2. *Two-Dimensional Imaging with HT-IR*

Perhaps the most promising application for HT-IR is its unique ability to provide images (and even a complete IR spectrum for each point in a two-dimensional image) of a source. The basic idea is shown in Fig. 24, which illustrates a portion of the code based on a 63-element (31 × 33 array) two-dimensional mask that is interposed between the source and the entrance aperture of the spectrometer, at the focal plane of a camera or telescope [44]. A 31 × 33 point image of the source can then be obtained by measuring the transmitted intensity through 63 independent masks, located on the rim of a disk as shown in Fig. 24a. If the successive masks differ from each other according to a Hadamard code in each dimension, the decoding calculation is rapid and simple as before. If, in addition, we now allow the mask-encoded intensity to serve as the source for an HT-IR spectrometer whose exit slit also coded with a Hadamard (one-dimensional) mask, we can then obtain a complete (say, 1023-point) digitized HT-IR spectrum for each of the 63 points of the two-dimensional image of the source (i.e., about 64,000 total data points).

There are two particularly useful ways to display selected portions of such

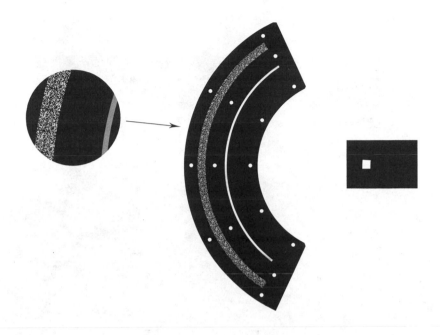

Fig. 24 Spatial encoding mask and field stop (blocking mask) used both for the Hadamard transform imager and the spectrometric imager (see text). [From Swift *et al.* [44].]

a huge data set. The first is to plot the 31 × 33 array image for just one selected wavelength channel, as in Fig. 25a. Figure 25a also demonstrates (a) the speed with which a single image can be obtained, and (b) the enhancement in image quality on integration over several individual image data sets. A second display is to choose a single point in the image, such as the indicated point in the image of a flame in Fig. 25b, and plot the IR intensity as a function of wavelength [44].

The combination of spatial mapping at high speed with detailed IR spectral profile provided by the HT-IR "triple" code (two-dimensional spatial coding at the entrance slit and one-dimensional wavelength coding at the exit slit of the spectrometer) seems particularly attractive for time-resolved studies (e.g., displaying spectra of different portions of a flame at different times to show combustion sequences) and for meteorological profiles: a downward-looking high-resolution ozone mapper is currently planned for the Space Shuttle [42]. HT-IR instruments are not currently available commercially, but seem likely to reemerge for imaging applications.

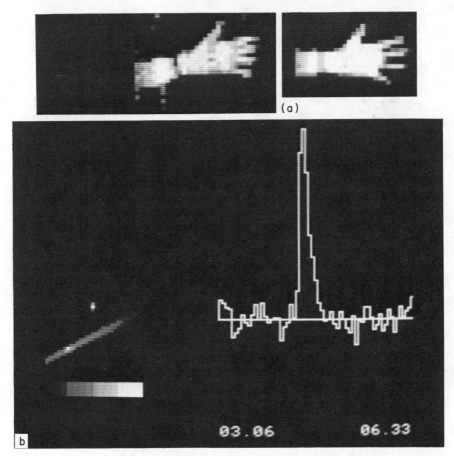

Fig. 25 Displays of IR two-dimensional spectral images of a source. (a) Two-dimensional images of the thermal emission of a human hand, reconstructed by the Hadamard transform imager (entrance slit), in the 8–14-μm region, using a cooled (Hg,Cd)Te detector. Left: picture obtained in a single frame time of 25 msec. Right: image integrated over 16 frame times, each 25 msec long. Note that along the arm one can resolve the shirt sleeve, bare arm, wristwatch, and hand' (The palm of the hand is warmer than the digits.) (b) Two-dimensional image (left) of a flame and a small (nearly point source) blackbody (bright spot just above flame), together with a gray scale of intensity and a spectrum (right) of a point in the flame. The selected point is indicated by the cursor (bright) spot on the flame image. The wavelength range is from 3.06 to 6.33 μm. [From Swift *et al.* [44].]

E. Fourier Transform Electronic (FT-Visible; UV) Spectroscopy

Generation of a visible–uv spectrum from the Fourier transform of a digitized interferogram is achieved in the same general way as for FT-IR (Section II.C). However, there are two critical differences between FT-IR and FT-visible experiments. The first is that the $\sqrt{N}$ Fellgett (multiplex) advantage in FT-visible spectroscopy is exactly cancelled by the property that noise in visible spectroscopy is proportional to the square root of the signal (whereas noise is independent of signal in FT-IR). There is thus no signal-to-noise or speed advantage for FT-visible versus scanning visible spectroscopy. Figure 26 shows that the appearance of the two spectra is different, since noise in the scanning spectrometer is concentrated at or near the peaks, while noise in the FT-spectrometer is spread out uniformly across the spectral range. Hirschfeld [45] has called this property the "distributive advantage" of Fourier spectroscopy [46].

The second difference between FT-IR and FT-visible spectra is that in the visible range it is no longer possible to sample the interferogram at a spacing of at least two points per cycle of the optical waves. For a standard He–Ne reference laser, the basic sampling interval for the interferometer is 0.6328 μm (one cycle of the laser line frequency) [47]. Thus the shortest light wavelength that can be sampled properly is 2(0.6328) = 1.266 μm (7901 cm^{-1}). Higher frequencies will appear "folded back" (aliased) into the range 0 to 7901 cm^{-1}, for essentially the same reason that a spinning wagon wheel sampled by the frames of a motion picture appears to speed up, then slow down as the wheel spins faster and faster [32].

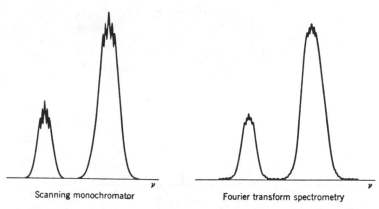

Scanning monochromator   Fourier transform spectrometry

Fig. 26 Distributions of photon noise for emission spectra (visible range). [From Mertz [46, p. 10].]

An experimental interferogram and FT-visible flame emission spectrum for several elements are shown in Fig. 27. The aliasing effect in Fig. 27 is evident from comparison of the various spectra in Fig. 28 for the same system. When the spectrum is digitized at a rate sufficiently high to locate correctly all the observed peak frequencies (Fig. 28a), most of the data points are wasted in covering the uninteresting low-frequency range. As a result, digital resolution is insufficient to resolve the two potassium peaks.

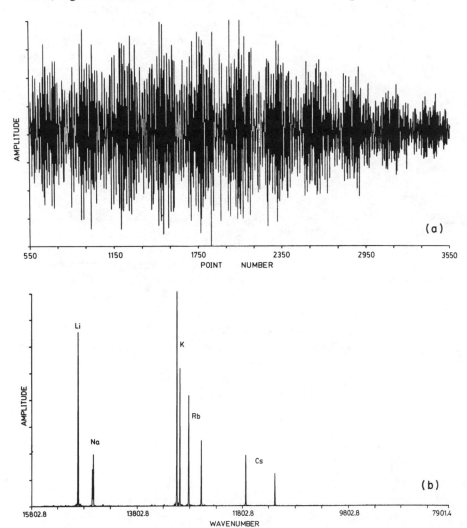

Fig. 27 (a) Digitized interferogram, and (b) corresponding Fourier transform spectrum for flame emission of the alkali metals. [From Horlick *et al.* [47].]

By using progressively slower digitization rates (Fig. 28b–d), various peaks become "folded over" into the (narrow) band of correctly located frequencies, but the digital resolution increases proportionately and the potassium peaks are now clearly resolved (Fig. 28d).

Performance

Multiplex advantage: None. See, however, the distributive "advantage" of Fig. 26.

Wave number axis: Precise and accurate, with no calibration required.

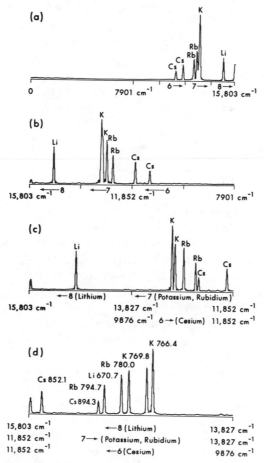

Fig. 28 Fourier transform flame emission spectra of Li, K, Rb, and Cs using sampling rates given by the following multiples of the direct rate corresponding to the 632.8-nm line of a He–Ne laser: (a) ×2, (b) ×1, (c) ÷2, (d) ÷4.

Resolution: High, achieved with relatively compact system, and with resolution easily controlled and line shape easily manipulated using apodization techniques (Section I.B).

Computerized spectrometer: Since Fourier data reduction can be carried out by an on-line computer, the facility for other computerized data and/or spectrometer operations is readily implemented (e.g., difference spectra, baseline smoothing, peak height or area calculations, signal averaging).

Wavelength range: The FT-visible spectrometer is inherently capable of covering a wide wavelength range, an advantage over array-based multidectors [48] whose finite length and detector element density limit the possible wavelength range.

The preliminary results from [47] have been extended in more recent work [49], to which the interested reader is referred. In spite of the disappearance of the multiplex advantage, FT-visible spectroscopy appears to offer several important and distinct advantages over conventional scanning or array-based spectrometers [47, 49].

F.   Fourier Transforms in Electrochemistry

Most modern electroanalytical techniques are based on observation of the time-varying response (often electrical current) to an applied perturbation (e.g., potential), as in dc and ac polarography [50]. However, conventional electrochemical assay procedures are based on cell response at just one frequency (fundamental and second harmonic ac polarography), or a single time (pulse polarography, dc polarographic response at end of drop life), or a time-integrated response (dc polarographic average current, chronocoulometry) [51]. Therefore, Fourier transformation of the time-domain response to a brief excitation can again (as for the spectroscopic examples of the preceding sections) provide a multichannel advantage compared to conventional steady-state scanning over the same frequency range. Moreover, because any particular frequency-domain spectrum can be obtained so quickly by FT methods, it becomes feasible to acquire the complete response versus potential profile, rather than (as with conventional methods) the response at just a single potential (e.g., peak potential) or in a potential region where the response is potential-independent (e.g., plateau current).

The most useful observed parameter is the Faradaic admittance $A_F(\omega)$, defined as the Faradaic current $I_F(\omega)$ divided by the potential $E(\omega)$ across the double layer:

$$A_F(\omega) = I_F(\omega)/E(\omega) \qquad (27)$$

There are three major conceptual differences between the generation and interpretation of FT electrochemical results compared to Fourier transform spectrometry. First, it is as if the system has a "natural" frequency of zero: the frequency-domain "peak" is always centered at the same (zero) frequency. It is the line width that gives the desired (kinetic) information. Second, the preferred excitation mode is no longer the sharp pulse of FT-NMR or FT-microwave spectroscopy, but rather a series of pulses whose phase and amplitude vary in a pseudorandom (i.e., known) fashion, to give a frequency excitation spectrum with approximately constant amplitude and pseudorandom phase across the frequency excitation range [52]. Third, it is usual to express the response in terms of its magnitude (compare to magnitude mode in spectral line shape) and its phase (or more usually, the cotangent of the phase of the response), where the phase is defined by [53]

$$\phi = \cot^{-1}\{\operatorname{Im}[A_F(\omega)]/\operatorname{Re}[A_F(\omega)]\} \qquad (28)$$

As explained in detail in [51–53], the advantages of $\cot(\phi)$ as a data format are: (a) since $\cot(\phi)$ is the ratio of the in-phase and 90°-out-of-phase (quadrature) components of the response, many complicated contributions to the admittance magnitude rate law cancel out (e.g., electrode area, electrode geometry, and mode of control of the applied dc potential), and (b) for a quasi-reversible process, a plot of $\cot(\phi)$ versus $\sqrt{\omega}$ gives a straight line whose slope is proportional to the electrode reaction's heterogeneous rate constant k_s (see Fig. 29).

Performance

Multichannel advantage: A Faradaic admittance spectrum (the most sought-after component of the cell admittance) requires about 2 sec for measurement and computation, or about at least three orders of magnitude less time than for a conventional point-by-point procedure [52].

Large rate constants: Since the slope of a plot of $\cot(\phi)$ versus $\sqrt{\omega}$ varies as $(1/k_s)$, data of high precision are required to determine the relatively small slope for a system with a large k_s. Figure 29b (left) shows such a plot (for a single 2 sec measurement and FFT calculation), and corresponds to the best precision available from a conventional (non-FFT) measurement. However, because of the multichannel advantage, it is possible to accumulate many passes in the time it would normally take to obtain a single conventional spectrum, and Fig. 29b (right) shows the greatly improved signal-to-noise ratio obtained by signal averaging 64 replicate measurements for the same system. Thus, because the data are now more precisely determined, it is possible to measure accurately

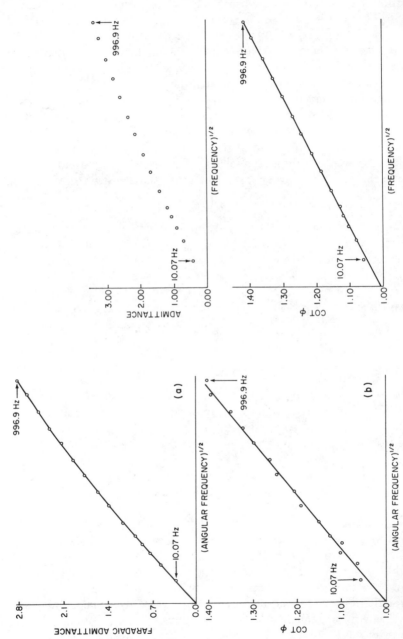

Fig. 29 (a) Frequency spectrum of peak Faradic admittance magnitude and (b) phase angle cotangent obtained by one-pass measurement of 2-sec duration (left) and by ensemble averaging of 64 replicate measurements (right). System: 2.00×10^{-3} M $Cr(CN)_6^{-3}$ in 1.00 M KCN at 25°C; excited by 15-frequency pseudorandom noise waveform with 2.0-mV amplitude for individual components; dc potential of -1.300 V versus Ag/AgCl; Faradic admittance in units of amperes per millivolt. Solid curves are the theoretical response for a heterogeneous charge transfer rate constant $k_s = 0.40$ cm/sec. Uncertainty in k_s for replicate measurement $= \pm 0.31\%$ (relative standard deviation). [From Smith [52, Figs. 4 and 5].]

smaller slopes and thus larger k_s values than have been possible by conventional methods ($k_s > 1$ cm sec^{-1}).

dc cyclic voltammogram: Since the FT ac polarogram contains profiles of admittance versus potential at all frequencies within the detected range, it contains in particular the dc cyclic voltammogram, which is thus generated without separate experiments (see Fig. 30 for examples).

Semiintegral and semiderivative voltammetry: The asymmetric shape of an ordinary voltammogram is inconvenient for simple interpretation (see Fig. 31a). Two improvements are readily achieved with the aid of the FT approach and knowledge of convolution properties (section I.C). Semiintegral voltammetry [54] can be shown to be identical to convolution of the observed current with a $t^{-1/2}$ function and yields the much simpler sigmoidal line shape of Fig. 31b. Alternatively, deconvolution

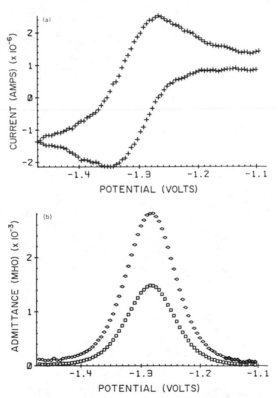

Fig. 30   (a) Ac cyclic voltammogram and (b) typical ac FT-polarograms of $Cr(CN)_6^{-3}$/ $Cr(CN)_6^{-4}$ couple at HMDE in aqueous cyanide media: all data for any one potential obtained simultaneously. The ac polarograms depict results from 5 replicate measurements at each E_{dc} value, 3.0 sec into drop life: □, 1840 rad/sec; ◇, 7977 rad/sec. [From Bond *et al.* [59, p. 237].]

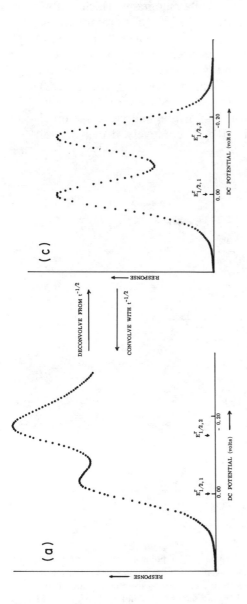

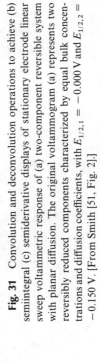

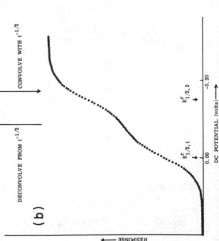

Fig. 31 Convolution and deconvolution operations to achieve (b) semiintegral (c) semiderivative displays of stationary electrode linear sweep voltammetric response of (a) two-component reversible system with planar diffusion. The original voltammogram (a) represents two reversibly reduced components characterized by equal bulk concentrations and diffusion coefficients, with $E_{1/2,1} = -0.000$ V and $E_{1/2,2} = -0.150$ V. [From Smith [51, Fig. 2].]

with $t^{-1/2}$ from a reversible linear sweep voltammogram (semidifferentiation) produces the much narrower symmetrical shape of Fig. 31c. (The diffusion-controlled voltammogram can be considered to be distorted by a broadening function arising from the $t^{-1/2}$ dependence of a planar diffusion process [51].)

Data manipulations: With the on-line minicomputer used for FT calculations, a variety of apodization and convolution procedures are possible. The displayed ac polarogram can be smoothed by Fourier transformation, followed by zeroing of all but the first several data points, then reverse transformed to give a "digitally filtered" spectrum [51]. Recently, a deconvolution procedure that corrects for the distortions produced by the instrument (using a dummy cell to discover the instrument response) has been demonstrated [55]. It has further been proposed that because of the speed with which the full spectral response can be obtained, and because of the greater information content of the full spectrum (compared with a conventional measurement at just a single frequency or single time point), the observed response might be used for automatic monitoring and/or correction of electroanalytical assays. For example, the computer might be programmed to test for nonlinearity in a plot of $\cot(\phi)$ versus $\sqrt{\omega}$ in the case of quasi-reversible charge transfer, and to sound a warning to the analyst that the nature of the electrode reaction is inappropriate [55]. Very recently, a simple interpolation procedure based on adding zeros to an existing time-domain data set before Fourier transformation has been used to produce admittance polarograms with the same shape, but greatly increased digital resolution [56], by a method previously established in FT-IR [57] and FT-NMR [58].

G.   Fourier Transforms in Dielectric Relaxation

In dielectric relaxation, the typically displayed quantities are the dielectric constant and dielectric loss, which represent the in-phase and 90°-out-of-phase components of the dielectric constant for a system contained between the plates of a capacitor and subjected to an oscillatory applied voltage excitation (see [13], pp. 438–444). The schematic form of these two quantities is illustrated in Fig. 32. It is clear that for a complete characterization of the line shapes, measurements over a wide frequency range are required. Such frequency-domain measurements in practice typically require different apparatus in different frequency ranges, and are time-consuming and difficult to control. In 1969, Fellner-Feldegg [60] described the first applications of a

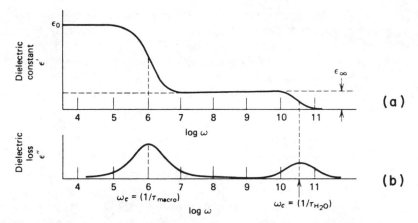

Fig. 32   (a) Frequency dependence of dielectric constant ϵ' and (b) dielectric loss ϵ'' with frequency plotted on a log scale. This response might represent a macromolecule dissolved in water: for each peak, the critical frequency ω_c corresponds to the half-way point in ϵ' versus ω (or the maximum value for ϵ'' versus ω) and is the reciprocal of the dielectric relaxation time for that process. The left-hand peak (ϵ'' versus ω) represents macromolecular relaxation, and the right-hand peak relaxation of the solvent water molecules themselves. Experiments are usually conducted at frequencies below about 10^{10} Hz. [From Marshall [13, p. 441].]

time-domain "reflectometry" method based on Fourier transformation of the time-domain response to a step excitation.

As for the previous examples we have examined, the dielectric response (say, the electric current following a voltage step excitation) is obtained from the convolution of the excitation and the response to an impulse excitation (compare Section I.C). Representative excitation and response are shown in Fig. 33. Experimentally, the dielectric constant and loss are extracted from measurements of the incident and reflected (or transmitted) voltage waves in a transmission line containing the sample (see [61] for details).

Successful extraction of the dielectric response depends on accurate knowledge of the excitation waveform, in order that the Fourier transform of the excitation can be computed for use in deconvolution of the observed response. Two general approaches have been suggested. First, when very

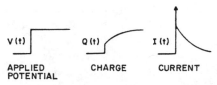

Fig. 33   Schematic charge and current of a dielectric sample following application of an ideal step voltage $V(t)$. [From Cole [61, p. 285].]

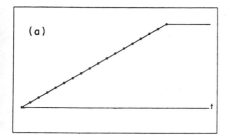

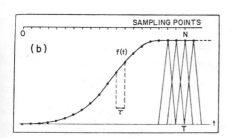

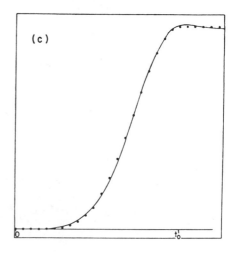

Fig. 34 Approximations to the actual experimental step voltage that serves as the excitation waveform in time-domain dielectric measurements. (a) Linear ramp; (b) polygon approximation (i.e., a sum of triangular pulses whose envelope is matched to the experimental solid curve); (c) $\sin^5(\pi t/2t_0)$ approximation (O) to experimental (——) waveform, with $t_0 = 80$ psec. [From Gestbloom and Noreland [62, pp. 1632–1633].]

fast step rise times are needed (for measurements of high-frequency response), the step shape can be approximated by a simple linear ramp (Fig. 34a), a polygon approximation (Fig. 34b), or (most recently) by a $\sin^5(\pi t/2t_0)$ waveform (Fig. 34c) [62]. The latter appears particularly accurate in representing the experimental excitation profile (Fig. 34c), and its discrete Fourier transform is readily evaluated for use in deconvolution of the response from the excitation. Very recently, analytical fits to the incident pulse have been replaced by numerical transforms of the actual pulse [63] resulting in fewer artifacts and irregularities and also more consistent time referencing for incident and multiple reflection pulses. These advantages are particularly useful for work with dilute solutions [63]. A representative example of the dielectric response to a step excitation is shown in Fig. 35 [64]; the data are plotted in the usual Cole–Cole display [65], as described further in Section II.J.

A second excitation mode, based on a pseudorandom noise excitation, has been proposed for low frequencies (6–7 decades below 1 MHz) [66]. (Another alternative in this frequency range is lumped circuit electronic methods to generate data for transformation [67].) This frequency range includes

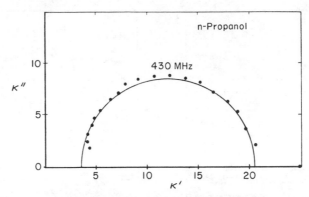

Fig. 35 Cole–Cole plot [65] (dielectric loss versus dielectric constant) for *n*-propanol at 25°C, based on time-domain reflectometry data using step pulse excitation. [From Fellner–Feldegg [64, p. 2122].]

relaxation processes for polymers (e.g., Fig. 36). The various time-domain dielectric methods thus cover essentially the full available range from about 1 to 10^{10} Hz.

Performance

Multichannel advantage: The very great time-saving advantage of Fourier data reduction of a time-domain response overcomes a principal problem of all dielectric measurements, namely, long-term drift in the electronics. The results for any given spectrum are thus more reliable than for the corresponding steady-state measurement, particularly since the tedious frequency resetting and bridge balancing at each frequency of the steady-state method are avoided [66]. However, long-term drift of the pulse occurrence times (20 psec or more over periods of seconds or minutes) degrades the signal-averaged response from a sum of many transients, and has so far limited the potential signal-to-noise enhancement that should accrue from the full Fellgett advantage.

Aqueous solutions: Accuracy of the derived results for relaxation of such solutes as amino acids, mono- and disaccharides, and peptides has been improved by time-domain reflection methods, by taking the difference between the time-domain responses for the solution and for pure water [68].

Deconvolution: Again, because an on-line minicomputer provides the necessary Fourier transform data reduction, it is possible to implement a variety of curve-fitting algorithms to simulate the excitation waveform, whose Fourier transform can then be used to deconvolute the observed response to give the true sample response [60, 62, 64].

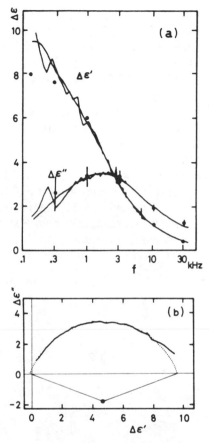

Fig. 36 Complex permittivity of poly(γ-benzyl-L-glutamate) in 1,2-dichloroethane at 50.0°C, using pseudorandom noise excitation. (a) Real and imaginary part of permittivity increment. The smooth curves are data of 100 accumulated responses. The rough curves are data of a single run, in which the measuring time per run is 28 msec. Small circles and dots are data from a conventional transformer bridge. (b) Cole–Cole plot of data from (a). [From Husimi and Wada [66, p. 218].]

H. Two-Dimensional Fourier Transform Nuclear Magnetic Resonance (2DFT-NMR) Spectroscopy

As previously noted, one-dimensional FT-NMR experiments and applications have been recently reviewed [29, 32]. A large number of the newer FT-NMR applications have been based on two-dimensional Fourier transforms of two-dimensional arrays of time-domain data, in which one of the time scales represents a varying delay time between two stages of the sample preparation period [69]. Another class of two-dimensional FT-NMR applications is based on converting a spatial distribution of nuclear spins to a frequency distribution, whose spectra can be obtained by Fourier means, and thereby furnish a two-dimensional image of the sample [70].

Most of the first class of 2DFT-NMR experiments are based on variations of the spin-echo idea first proposed by Hahn and Maxwell in 1952 [69]. This experiment uses two (or more) pulses that have the effect of refocusing the various nuclear magnetic moments to eliminate any initial differences in Larmor frequency (chemical shift). Alternatively, various decoupling irradiation modes have the effect of removing any spectral splittings due to scalar *J*-coupling between different nuclei. With suitable excitation modes, it is thus possible to suppress either chemical shift differences or *J*-coupling differences between different signals in an FT-NMR spectrum. Then, from a series of such measurements, a two-dimensional FT-NMR spectrum can be constructed with (for example) chemical shift in one dimension versus *J*-coupling constant in the second direction, as in Fig. 37 [69]. Another very promising display consists of a plot of proton chemical shift in one dimension

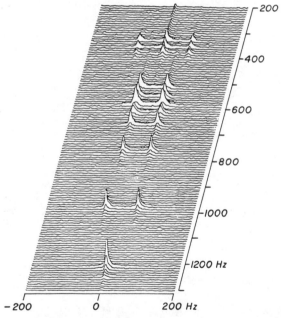

Fig. 37 Two-dimensional *J* spectrum of carbon-13 in sucrose, obtained by the gated decoupler method. Carbon-13 chemical shifts are displayed in the upward direction and proton–carbon multiplet structure in the horizontal direction. The spectrum shown here is in the absolute-value (magnitude) mode; no attempt was made to achieve high resolution in the *J* dimension. [From Freeman and Morris [69, p. 17].]

versus carbon chemical shift in the second dimension (Fig. 38) [69] This second experiment should be especially useful, because each carbon chemical shift is now correlated in position to the chemical shift of the proton(s) directly bonded to that carbon.

The 2DFT-NMR experiment offers a second multichannel advantage over the corresponding one-dimensional FT-NMR case. For example, for the chemical shift correlation experiment, the corresponding 1DFT experiment might be to measure an FT carbon-13 spectrum for each of (say) 128 different ^{1}H decoupling frequencies, while in the same time the 2DFT experiment samples 128 time-domain profiles. The 2DFT experiment thus gives

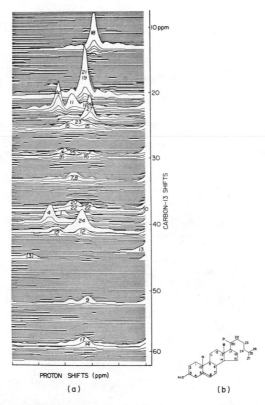

Fig. 38 (a) Chemical shift correlation map for the protons and carbon-13 in cholesteryl acetate (b). The carbon-13 shift scale has been referenced approximately to tetramethylsilane. The quaternary sites (10 and 13) show no correlation peaks, because those carbons have no directly bonded protons. The doubling of the peaks from sites 11, 12, 15, 16, and 22 is attributed to the chemical shift difference between axial and equatorial protons. [From Freeman and Morris [69, p. 25].]

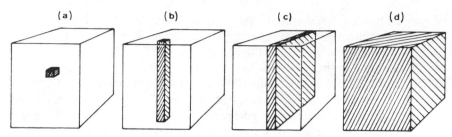

Fig. 39 Classification of various imaging schemes: (a) sequential point measurement, (b) sequential line measurement, (c) sequential plane measurement, and (d) simultaneous measurement. Different NMR frequencies at different points in the cubic array are produced by application of magnetic field gradients along each of the various independent directions. [From Brunner and Ernst [70, p. 88].]

a signal-to-noise advantage of [(spectral width)/(multiplet width)]$^{1/2}$ compared to the 1DFT series of experiments [71].

In the second class of 2DFT-NMR applications, a spatial distribution of spins (say, protons of water in a linear capillary) is converted to a distribution in frequency by applying a linear field gradient that shifts the resonant

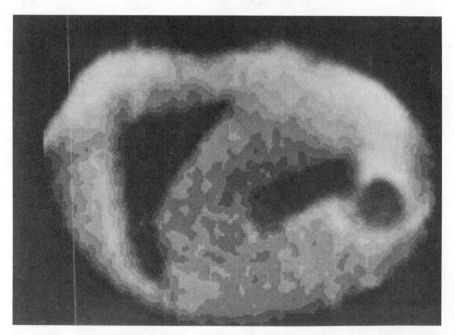

Fig. 40 NMR image of a slice of calf heart, 18 mm thick. The image was produced from 65 different field-gradient orientations, at a frequency of 4 MHz for protons. [From Lauterbur [73, p. 17].]

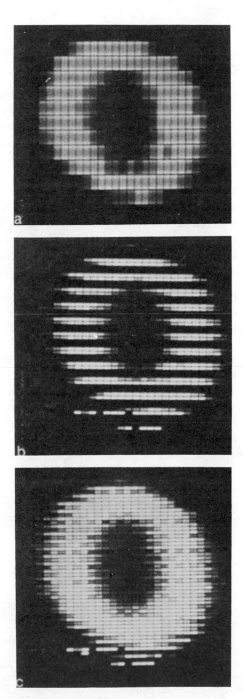

Fig. 41 Planar spin images from protons in a mineral oil annulus, showing buildup of the image with a fourfold interlace. (a) Coarse planar image: spot wobbling has been used to fill in the gaps between individual spots. (b) Second stage of the fourfold interlace pictures. Spot size is reduced to avoid overlap. (c) Complete fourfold interlace image. [From Mansfield and Maudsley [74, p. 112] (see [74] for more details).]

Larmor frequency of a given proton according to its spatial position along the gradient [72]. By applying different gradients in different directions, as shown in Fig. 39, it is possible to obtain images of single small regions (by collecting data at just one frequency at a time), linear profiles (from one-dimensional Fourier transformation of time-domain data corresponding to spins whose NMR frequencies are dispersed along a given field gradient direction), and two- or three-dimensional images (from various schemes for converting number of spins versus field into number of spins versus frequency) [70].

Figure 40 shows a remarkably detailed image of a slice of calf heart [73], based on the one-dimensional approach of Fig. 39b. More rapid data acquisition and reduction should follow from planar imaging schemes (Fig. 39c) of the type that produced the images of Fig. 41 [74]. Further examples are given in [13, Chapter 22]. Major applications are expected in the imaging of biological specimens, based on the distribution of the water in the sample.

I. Two-Dimensional Fourier Transform Electron–Nuclear Double Resonance (2DFT-ENDOR) Spectroscopy

Application of a series of three 90° microwave pulses (Fig. 42) to a paramagnetic sample produces electron spin echoes whose amplitudes vary with the time spacing between the pulses according to the frequencies of the hyperfine interactions between the electron and its neighboring nuclei. The modulation pattern was first explained and demonstrated by Mims and co-workers [75], and provides the same information that is present in ENDOR experiments in which a second (radio-frequency) irradiation is used to excite various electron–nuclear couplings [76]. Fourier transformation of the electron spin-echo decay envelope (i.e., the echo maximum amplitude as a function of delay time between (for example) the first and third 90° pulse) thus gives an ENDOR-type spectrum [77, 78].

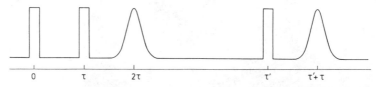

$$
\begin{array}{ccccc}
0 & \tau & 2\tau & \tau' & \tau'+\tau
\end{array}
$$

Fig. 42   The stimulated electron-spin-echo scheme. Transmitter (microwave) pulses are applied at $t = 0$, τ, and τ'. The Hahn echo occurs at $t = 2\tau$, and the stimulated echo at $t = \tau + \tau'$. Dephasing (decay rate) is faster for the τ echo than for the τ' echo, so that the stimulated echo is more suitable for measuring hyperfine frequencies via the modulation effect (see text). [From Merks and de Beer [77, Fig. 1].]

It should be recognized that there is no multichannel advantage in this type of Fourier transform experiment, because the data are not collected in real time; each data point corresponds to a separate delay time for a given pulse train, so that it is necessary to begin a new pulse train to create the next echo whose amplitude constitutes the next data point. The practical advantages of the Fourier approach are that no second radio-frequency irradiation is needed, eliminating microphonic complications, and extension of the spectral range to relatively low frequencies (less than 1 MHz) is readily accomplished.

Figure 43 shows a stacked series of spectra, each obtained by taking the Fourier transform of the echo envelope as a function of τ' for a given value of τ (see Fig. 42 for definition of τ' and τ). The most obvious feature of the spectra is that there is a sinusoidal variation in peak amplitude as τ is varied, so that in any one trace one or more peaks may actually be absent.

Some other features of the spectra and their generation deserve comment. First, since experimental data do not extend back to zero delay time, it is necessary to apodize the echo envelope by tapering its amplitude at longer

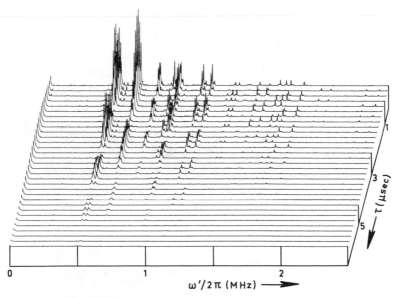

Fig. 43 Stacked one-dimensional Fourier transforms of the electron-spin-echo envelope modulation of Gd^{3+} in $Bi_2Mg_3(NO_3)_{12} \cdot 24H_2O$. The Fourier spectra are obtained by Fourier transforming the echo modulation with respect to τ' for different values of τ. The absolute-value mode is chosen for display to avoid problems in phasing the displayed spectrum. The peak amplitudes vary sinusoidally with τ (see text). [From Merks and de Beer [77, Fig. 4].]

delay time: this is one reason that the spectra are usually plotted in absolute-value mode. Second, because the decay in the τ dimension is much faster than the decay in the τ' dimension, the observed apparent line width in the ω dimension is about 50 times larger than in the ω' dimension. Third because of a large exponentially decaying dc component in the echo envelope (which would lead to a huge broad peak at the left of the spectra), it is usual to subtract the digitally determined dc decay component before Fourier transformation of the echo envelope. Finally, because the line width in the ω dimension is so much larger than in the ω' dimension, many fewer traces ($\sim$ 32 or 64 versus 2048 or more) are needed to disply the ω dependence of the spectra compared to the ω' dependence.

Performance

Multichannel advantage: None, because the experiment is not performed in real time (see text).

Frequency range: Extremely flat baseline and high sensitivity are much improved compared to conventional ENDOR experiments.

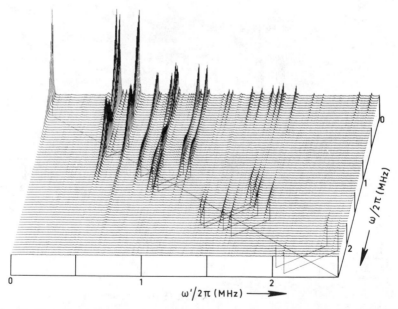

Fig. 44   Two-dimensional Fourier transform of the electron-spin-echo modulation envelope of the system of Fig. 43. The Fourier transform is done with respect to τ and τ' (see Fig. 42), yielding the frequency dimensions ω and ω'. Absolute-value mode is again chosen for display. Because of the symmetry in τ and τ' (apart from line width), the peaks have mirror symmetry about the line $\omega = \omega'$ (see text). [From Merks and de Beer [77, Fig. 5].]

Microphonic problems: Absent, because detection occurs in the absence of
  any microwave or radio-frequency irradiation.

Peak suppression: For any one-dimensional Fourier transform spectrum
  at a given τ (Fig. 43), one or more peaks may be absent, and peak
  amplitudes vary with choice of τ. It is thus necessary to perform a
  two-dimensional Fourier transform with respect to both τ and τ' in
  order to recover the desired spectrum with all the peaks (Fig. 44).

Peak assignment: Assignment of peaks is facilitated in the two-dimensional array by taking advantage of the reflection symmetry about the
  $\omega = \omega'$ line.

J. Hilbert Transforms in NMR and ESR Spectroscopy

   In virtually all types of spectroscopy the usual final data display is either
absorption versus frequency or magnitude versus frequency. However, for
any absorption spectrum, there is a corresponding dispersion spectrum [13],
as shown in Fig. 45, that contains the same information (although less
symmetrically displayed) as the absorption spectrum. Whenever the information content of two curves is the same, there is usually a recipe for obtaining one from the other, and the mathematical recipe for going from
absorption to dispersion is called a *Hilbert transform*. In fact, the Fourier

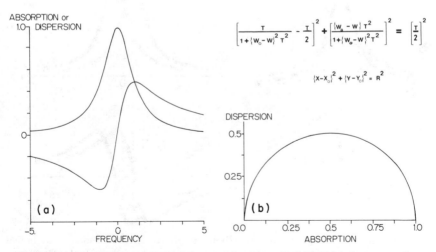

$$\left[\frac{\tau}{1+|w_o-w|^2\tau^2} - \frac{1}{2}\right]^2 + \left[\frac{|w_o-w|\tau^2}{1+|w_o-w|^2\tau^2}\right]^2 = \left[\frac{1}{2}\right]^2$$

$$|x-x_o|^2 + |y-y_o|^2 = R^2$$

Fig. 45   (a) Absorption and dispersion versus frequency, and (b) dispersion versus absorption (DISPA). For a given DISPA plot, a reference circle is constructed with diameter equal to
the maximum absorption peak height, centered at half the maximum absorption peak height
on the abscissa. The DISPA plot is a circle if and only if the original spectral line shape is
Lorentzian.

transform of a time-domain data set automatically generates both the absorption and dispersion modes, scaled to the same vertical axis. Ordinarily, the broader shape and asymmetrical form of the dispersion spectrum render it less interesting for characterization of spectra, and the dispersion data from the Fourier transform are usually discarded. However, an extraordinary and useful property results from combining the dispersion and absorption information, as will now be explained.

The two most fundamental types of inhomogeneous spectral line broadening are shown in Fig. 46: namely, two (or more) lines of different position (i.e., different resonant frequency) or two (or more) lines of different line width (i.e., different relaxation time). One of the most basic problems in spectroscopy is to distinguish between these (and other) mechanisms for line broadening, particularly since it is usually possible to "fit" a single observed broad line to any of several different mechanisms. Traditional approaches have required multiple experiments designed to preferentially shift (e.g., by change of solvent), broaden (e.g., by change in temperature), or disperse (e.g., by change in magnetic field in the case of NMR) the various individual components of the original inhomogeneously broad line. However, multiple experiments necessitate multiple controls (e.g., changing the

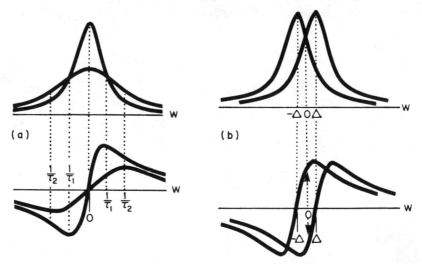

Fig. 46   Absorption (top) and dispersion (bottom) spectra. (a) Superposition of two Lorentzians of equal position (resonant frequency) but different line width (relaxation time). (b) Superposition of two Lorentzians of equal width but different position. The DISPA plot (see text) provides a simple means for distinguishing between these two general types of line broadening. [From Marshall [79, p. 522].]

applied magnetic field in NMR can change relaxation times as well as chemical shifts), and the whole problem can quickly get out of hand.

In recent theoretical [79, 80] and experimental [81–83] work, it has been shown that (a) a plot of dispersion versus absorption (DISPA) for a single Lorentzian line gives a circle (Fig. 45), and (b) a DISPA plot for a single spectral data set gives a curve whose displacement magnitude and direction from that reference circle are characteristic of (and can thus be used to diagnose) the line-broadening mechanism. Most generally, any distribution

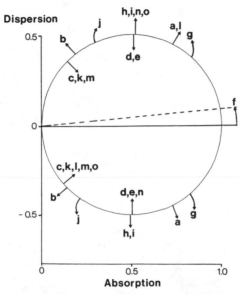

Fig. 47 Reference circle (diameter equal to maximum observed absorption peak height) and direction of displacement of a DISPA plot for various line-broadening mechanisms: (a) unresolved pair of Lorentzians of equal width and different position [80]; (b) Gaussian distribution (in position) of Lorentzians of equal width [80]; (c) pair of Lorentzians of equal position and different width [80]; (d) log–Gauss distribution in transverse relaxation time for Lorentzians of equal resonant frequency [80]; (e) log–Gauss distribution in correlation time for Lorentzians of equal resonant frequency [80]; (f) phase misadjustment (rotates circle about the origin [80]; (g) spectra obtained by Fourier transformation of a truncated time-domain transient [82]; (h) power broadening [84]; (i) overmodulation [83]; (j) chemical exchange between two peaks of different position [80]; (k) chemical exchange between two peaks of different width [80]; (l) distortion produced by one adjacent peak of equal intensity and width [84]; (m) distortion from two adjacent peaks of equal intensity and width, located at equal separation on either side of the peak from which the DISPA curve is drawn [84]; (n) effect of too-long time constant in recording of EPR derivative spectrum [83]; (o) effect of slight baseline drift in EPR derivative spectrum [83]. Finally, for a simultaneous distribution in both peak position and in peak width, the peak position distribution dominates DISPA behavior [84].

of two or more lines of equal line width but different position will give a DISPA curve displaced *outside* the reference circle, while any distribution of two or more lines of equal position but different width will give a DISPA curve displaced *inside* the reference circle [79]. Some 16 distinct line-broadening mechanisms have been analyzed, and their corresponding DISPA displacements mapped in Fig. 47 [84]. Experimental examples from NMR [79, 81, 82] and EPR [83] spectroscopy are shown in Figs. 48–50.

Figure 48a shows the DISPA plot for an unresolved ^{19}F NMR doublet, in which the two component peaks are separated by about 0.7 of one individual line width to give a single broad observed peak. The characteristic

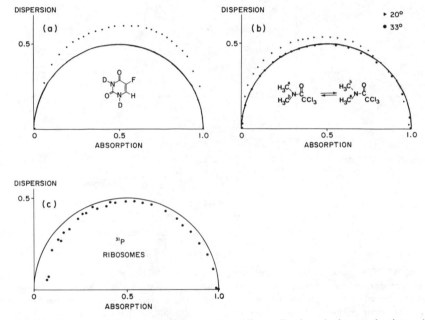

Fig. 48 Three experimental DISPA plots for different line-broadening mechanisms. (a) Unresolved doublet from ^{19}F NMR spectrum of 5-fluorouracil. The spectrum was acquired under conditions that the two peaks are separated by less than one line width, and thus appear as a single broad peak. The direction of displacement of the DISPA data points from the reference circle identifies the broadening mechanism (see Table II). (b) Chemical exchange between two sites of different chemical shift: DISPA plot for the N-methyl peak of N,N-dimethyltrichloroacetamide. A temperature of 20°C is just above coalescence; 33°C increases the exchange rate to the fast-exchange limit (Lorentzian line shape, with DISPA points on the circle). (c) Distribution in relaxation time from ^{31}P NMR of bacterial ribosomes: DISPA points inside the circle indicate a distribution in line width rather than a distribution in chemical shift as the dominant line broadening mechanism. [Plots (a) and (b) from Roe *et al.* [81, pp. 765, 766]; (c) from Marshall [79, p. 523].]

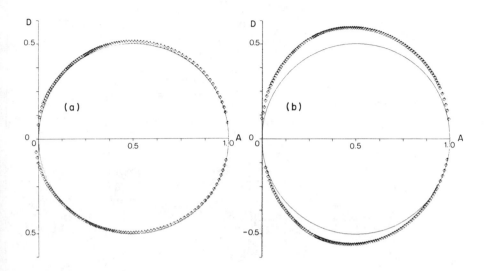

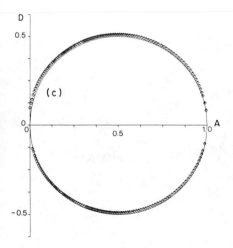

Fig. 49 EPR DISPA plots. (a) The $m = 0$ transition for peroxylamine disulfonate (PADS) at low modulation amplitude ($H_m = 13$ mG) and low concentration (10^{-4} M). (b) DISPA plot for the same system but at higher modulation amplitude ($H_m = 0.4$ G). (c) Same as (a), but at higher concentration (~ 0.01 M). Note that the DISPA plot easily distinguishes between the inhomogeneous line broadening of (b) and the homogeneous line broadening of (c). [From Herring *et al.* [83, Fig. 6].]

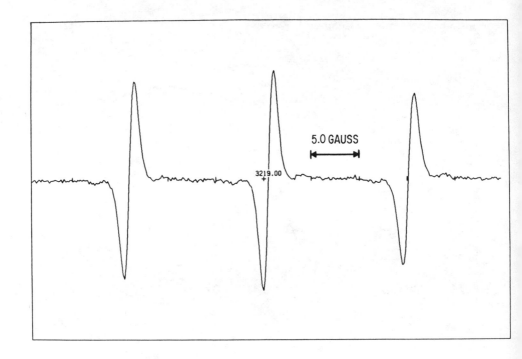

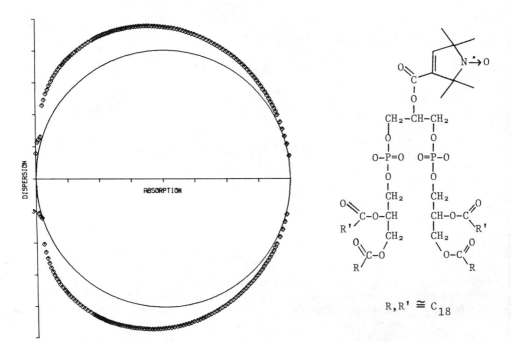

Fig. 50 EPR derivative spectrum (top) and corresponding DISPA plot (bottom left) for the $m = 0$ transition for a mitroxide spin-labeled cardiolipin (structure at bottom right). The pronounced DISPA deviation outside and to the left of the reference circle is due to unresolved hyperfine splitting of the EPR transition [85].

DISPA deflection outside and to the right of the reference circle identifies the line-broadening mechanism as an unresolved splitting [81]. Figure 48b shows that for a system involving chemical exchange between two peaks of different chemical shift (methyls a and b of the illustrated compound), the DISPA plot readily shows that the fast-exchange limit has been reached at 33°C (DISPA data points lie on the reference circle), but not at 20°C (DISPA points lie outside the circle) [81]. In other words, it is now possible to identify exchange limits using the data from a single temperature, without the usual temperature variation. A final NMR example is shown in Fig. 48c, for the single broad peak from ^{31}P NMR of bacterial ribosomes [79]. The displacement of the data points within the circle definitively identifies the line broadening mechanism as a distribution in relaxation time.

Examples from EPR are shown in Figs. 49 and 50. In these cases, the dispersion spectrum is not observed directly, but is readily computed from the Hilbert transform of the absorption spectrum [83]. Figure 49a shows the DISPA plot for a simple nitroxide free radical: the line shape is clearly Lorentzian because the data points fall on the reference circle. Figure 49b shows that the effect of modulation broadening is to displace the DISPA data points outside the reference circle: this displacement has been used to

TABLE II

Summary of the Effects of Various Mechanisms of Spectral Line Broadening
upon a Plot of Normalized Dispersion versus Normalized Absorption[a]

Origin of line broadening	Direction of displacement of the corresponding DISPA curve from its reference circle[b]
Unresolved splitting of two adjacent peaks	Outward and to the right
Gaussian distribution in line position	Outward and to the left
Two line of different width	Inside and to the left
Log–Gauss distribution in line width	Inside directly vertically
Exchange between two peaks of different position	Outward
Exchange between two peaks of different width	Inward and to the left
Phase misadjustment	Position of circle rotated about the origin (easily identified and rectified in practice)
Modulation broadening	Outward
Truncation of time-domain data set	Outward

[a] In each case the magnitude of the displacement is directly related to the line broadening parameter (e.g., peak separation, distribution variance).

[b] The reference circle is drawn with its diameter along the abscissa, centered on the abscissa at half the maximum absorption mode peak height, with a diameter equal to the maximum absorption mode peak height. See [79–86].

TABLE III

Summary of Some Major Aspects of Transform Applications in Spectroscopy and Relaxation Experiments

Application	Usual broadband excitation waveform	Usual form of displayed spectrum	Multiplex advantage?	Principal experimental advantages	Recent references
FT-ICR	Fast frequency sweep	Magnitude (absolute value)	10,000 in speed, 100 in S/N[a]	Speed; 1000 × better mass resolution; extended mass range; negative ions; ion–molecule reactions	[17–19, 21, 22, 24]
FT-microwave	Rectangular pulse	Absorption mode	400 in speed, 20 in S/N	Speed; absence of modulation broadening; small dipole moments	[31, 33–36]
FT-IR	[Interferometer]	Absorption mode (absorption or emission)[b]	Varies with wavelength range; up to 10,000 in speed	Speed; throughput advantage; difference spectra; weak sources; GC-IR; LC-IR; reaction kinetics	[37, 81]
Hadamard-IR	Lamp + mask	Absorption[b]	Same as FT-IR	Portability; spatial imaging	[42, 43]
FT-visible; UV	[Interferometer]	Absorption[b]	None	Multielement simultaneous analysis, wave-number accuracy, wide frequency range, high resolution	[47, 49]

			Speed^a		References
FT-electrochemical	Pseudorandom noise	Magnitude and phase (or cotangent of phase)	1000 in speed	Speed; determination of large rate constants; dc and ac voltammograms simultaneously; semiderivative and semiintegral display	[50–53]
FT-dielectric relaxation	Step pulse; or pseudorandom noise	Cole–Cole plot [64]	100 in speed	Speed (avoids drift in electronics); difference plots for aqueous solutions; deconvolution of pulse imperfections	[61]
2DFT-NMR	Multiple rectangular pulses	Magnitude mode	None	Spatial imaging; correlation of ^1H and ^{13}C chemical shift; correlation of chemical shift and coupling constant; automation	[69, 70]
2DFT-ENDOR	Three rectangular pulses	Magnitude mode	None	Flat baseline; high sensitivity (especially at low frequencies); absence of microphonics; improved peak assignment over 1DFT-ENDOR	[75, 77, 78]
Hilbert (NMR; EPR)	Conventional or pulsed	Dispersion versus absorption	None	Determination of line broadening mechanism using data from a single spectrum	[79–87]

^{a} Signal-to-noise ratio. Speed advantage is with respect to conventional detection to give a spectrum of the same signal-to-noise ratio; S/N advantage is with respect to conventional detection to give a spectrum with the same total data acquisition time.

^{b} "Absorption mode" here means the cosine Fourier transform of the interferogram, whether the interferogram represents physical absorption or emission.

calibrate the modulation amplitude of an EPR spectrometer, based on data from a single spectrum (rather than the many modulation amplitudes usually needed). Figure 49c shows the DISPA plot for a broader spectrum from a more concentrated sample: here the placement of DISPA points on the reference circle confirms that the line-broadening mechanism is exchange broadening, which retains the Lorentz line shape [83]. Finally, Fig. 50 shows the DISPA plot from the $m = 0$ transition of a nitroxide spin-labeled cardiolipin molecule. In this case, the unresolved hyperfine splittings are clearly evident from the pronounced DISPA displacement from the reference circle [85].

DISPA displacements of the type discussed above are perhaps most directly obvious from a plot of the radius of the DISPA curve versus frequency. Recent investigations [86] suggest that the most sensitive way to combine dispersion and absorption data may be to plot the square of the DISPA radius versus square root of frequency. The DISPA technique has recently been reviewed [87].

The EPR results have a further implication. Since the dispersion need not be measured directly, it should be possible to extend the DISPA analysis to any form of spectroscopy for which the elementary line shape is a Lorentzian line, even when only the absorption spectrum is observed directly. The diagnostic value of the DISPA plot is shown in Table II, which lists the effect on a DISPA plot of a number of selected line broadening mechanisms shown graphically in Fig. 47. Finally, it is important to remember that the diagnostic information is derived from the data from a *single* spectrum, often avoiding (e.g., Fig. 48c) the multiple experiments otherwise necessary to identify the line broadening mechanism. Because of the simplicity of the DISPA data reduction (one need simply plot dispersion versus absorption, where both digitized spectra are normally already in memory after a discrete Fourier transform of time-domain data), it seems certain to be applied widely in many types of spectroscopy.

III. Summary

Table III summarizes the highlights of the transform applications treated in Section II. Although the mathematical operations are common to virtually all the applications, it is clear that the reasons for using transform methods are qualitatively different among the various applications (beyond the obvious difference that the frequency ranges differ in the various experiments). Since Fourier transform software is almost universally available in most computer languages, and since many of the specific spectrometers involving Fourier transforms are now commercially available, the advantages available from transform methods are bound to become much more widespread in the very near future.

Acknowledgments

The author wishes to thank the following persons for invaluable discussions and/or proofreading: R. H. Cole, M. B. Comisarow, R. de Beer, W. H. Flygare, M. Harwit, T. Hirschfeld, G. Horlick, R. L. Julian, H. Mantsch, G. A. Morris, and D. E. Smith.

References

1. A. G. Marshall and M. B. Comisarow, *in* "Transform Techniques in Chemistry" (P. R. Griffiths, ed.), Chap. 3. Plenum, New York, 1978.
2. J. D. Winefordner, J. J. Fitzgerald, and N. Omenetto, *Appl. Spectrosc.* **29,** 369 (1975).
3. G. Horlick, *Appl. Spectrosc.* **30,** 113 (1976).
4. K. Siegbahn, C. Nordling, A. Fahlman, R. Norderg, K. Hamrin, J. Hedman, G. Johansson, T. Bergmark, S. Karlson, I. Lindgren, and B. Lindberg, "ESCA: Atomic, Molecular, and Solid State Structure Studied by Means of Electron Spectroscopy." Almqvist & Wiksell, Stockholm, 1967.
5. K. Siegbahn, C. Nordling, G. Johansson, J. Hedman, P. F. Heden, K. Hamrin, U. Gelius, T. Bergmark, L. O. Werme, R. Manne, and Y. Baer, "ESCA: Applied to Free Molecules." North-Holland Publ., Amsterdam, 1969.
6. D. W. Turner, A. D. Baker, C. Baker, and C. R. Brundle, "High Resolution Molecular Photoelectron Spectroscopy." Wiley, New York, 1970 .
7. H. Fellner-Feldegg, U. Gelius, B. Wannberg, A. G. Nilsson, E. Basilier, and K. Siegbahn, *J. Electron. Spectrosc.* **5,** 643 (1974).
8. P. Fellgett, *J. Phys. Radium* **19,** 187 (1958).
9. M. Hall, Jr., "Combinatorial Theory," p. 204. Ginn (Blaisdell), Boston, Massachusetts, 1967.
10. G. Forsythe and C. B. Moler, "Computer Solution of Linear Algebraic Systems." Prentice-Hall, Englewood Cliffs, New Jersey, 1967.
11. R. R. Ernst, *Adv. Magn. Reson.* **2,** 1 (1968).
12. C. E. Brion, *in* "MTP International Review of Science, Mass Spectroscopy, Physical Chemistry," Series One (A.D. Buckingham and A. Maccoll, eds.), Vol. 5, Chap. 3. Butterworth, London, 1972 .
13. A. G. Marshall, "Biophysical Chemistry: Principles, Techniques and Applications," Chaps. 13 and 20. Wiley, New York, 1978.
14. R. P. Feynman, R. B. Leighton, and M. Sands, "The Feynman Lectures on Physics," Vol. 1, p. 30-5. Addison-Wesley, Reading, Massachusetts, 1963.
15. A. G. Ferrige and J. C. Lindon, *J. Magn. Reson.* **31,** 337 (1978).
16. R. Bracewell, "The Fourier Transform and Its Applications," 2nd ed., Chap. 3. McGraw-Hill, New York, 1978.
17. M. B. Comisarow and A. G. Marshall, *Chem. Phys. Lett.* **25,** 282 (1974).

18. A. G. Marshall and D. C. Roe, *J. Chem. Phys.* **73,** 1581 (1980).
19. M. B. Comisarow and A. G. Marshall, *Chem. Phys. Lett.* **26,** 489 (1974).
20. R. R. Ernst and W. A. Anderson, *Rev. Sci. Instrum.* **37,** 93 (1966).
21. A. G. Marshall, M. B. Comisarow, and G. Parisod, *J. Chem. Phys.* **71,** 4434 (1979).
22. M. B. Comisarow, *J. Chem. Phys.* **69,** 4097 (1978).
23. L. R. Anders, J. L. Beauchamp, R. C. Dunbar, and J. D. Baldeschwieler, *J. Chem. Phys.* **45,** 1062 (1966).
24. M. B. Comisarow, V. Grassi, and G. Parisod, *Chem. Phys. Lett.* **57,** 413 (1978).
25. R. T. McIver, Jr., E. B. Ledford, Jr., and J. S. Miller, *Anal. Chem.* **47,** 692 (1975).
26. M. B. Comisarow, unpublished observations.
27. M. B. Comisarow and J. D. Melke, *Anal. Chem.* **51,** 2198 (1979).
28. G. Parisod and M. B. Comisarow, unpublished observations.
29. T. C. Farrar, *in* "Transform Techniques in Chemistry" (P. R. Griffiths, ed.), Chap. 8. Plenum, New York, 1978 .
30. J. C. Davis, Jr., "Advanced Physical Chemistry: Molecules, Structure, and Spectra," Chap. 8. Ronald Press, New York, 1965.
31. J. Ekkars and W. H. Flygare, *Rev. Sci. Instrum.* **47,** 448 (1976).
32. J. W. Cooper, *in* "Transform Techniques in Chemistry" (P. R. Griffiths, ed.), Chaps. 4 and 9. Plenum, New York, 1978.
33. W. H. Flygare, *J. Chem. Phys.* **41,** 206 (1964).
34. R. H. Schwendeman, *Annu. Rev. Phys. Chem.* **29,** 537 (1978).
35. J. C. McGurk, T. G. Schmalz, and W. H. Flygare, *Adv. Chem. Phys.* **25,** 1 (1974).
36. W. H. Flygare and T. G. Schmalz, *Acc. Chem. Res.* **9,** 385 (1976).
37. P. R. Griffiths, *in* "Transform Techniques in Chemistry" (P. R. Griffiths, ed.), Chaps. 5 and 6. Plenum, New York, 1978.
38. S. T. Ridgeway and R. W. Capps, *Rev. Sci. Instrum.* **45,** 676 (1974).
39. H. L. Johnson and M. E. Mendez, *Astronom. J.* **75,** 785 (1970).
40. N. J. A. Sloane, T. Fine, P. G. Phillips, and M. Harwit, *Appl. Opt.* **8,** 2103 (1969).
41. J. A. Decker, Jr., *Anal. Chem.* **44,** 127A (1972).
42. M. Harwit, *in* "Transform Techniques in Chemistry" (P. R. Griffiths, ed.), Chap. 7. Plenum, New York, 1978.
43. M. Harwit and N. J. A. Sloane, "Hadamard Transform Optics." Academic Press, New York, 1979.
44. R. D. Swift, R. B. Wattson, J. A. Decker, Jr., R. Paganetti, and M. Harwit, *Appl. Opt.* **15,** 1595 (1976).
45. T. Hirschfeld, *Appl. Spectrosc.* **30,** 68 (1976).
46. L. Mertz, "Transformations in Optics," pp. 9–10. Wiley, New York, 1965.
47. G. Horlick, R. H. Hall, and W. K. Yuen, *in* "Fourier Transform Infrared Spectroscopy" (J. R. Ferraro and L. J. Basile, eds.), Vol. 3, pp. 37–81. Academic Press, New York, 1982.
48. Y. Talmi, ed., "Multichannel Image Detectors." Am. Chem. Soc., Washington, D.C., 1979.
49. A. Y. Irfan, A. P. Thorne, R. A. Boklander, and H. A. Gebbie, *J. Phys. E.* **12,** 472 (1979).
50. D. E. Smith, *CRC Crit. Rev. Anal. Chem.* **2,** 247 (1971).
51. D. E. Smith, *Anal. Chem.* **48,** 517A (1976).
52. D. E. Smith, *Anal. Chem.* **48,** 221A (1976).
53. R. J. Schwall, A. M. Bond, and D. E. Smith, *J. Electroanal. Chem.* **85,** 217 (1977).
54. K. B. Oldham, *Anal. Chem.* **45,** 39 (1973).
55. R. J. Schwall, A. M. Bond, and D. E. Smith, *Anal. Chem.* **49,** 1805 (1977).
56. R. J. O'Halloran and D. E. Smith, *Anal. Chem.* **50,** 1391 (1978).
57. G. Horlick and W. K. Yuen, *Anal. Chem.* **48,** 1643 (1976).

58. E. Bartholdi and R. R. Ernst, *J. Magn. Reson.* **11,** 9 (1975).
59. A. M. Bond, R. J. Schwall, and D. E. Smith, *J. Electroanal. Chem.* **85,** 231 (1977).
60. H. Fellner-Feldegg, *J. Phys. Chem.* **73,** 616 (1969).
61. R. H. Cole, *Annu. Rev. Phys. Chem.* **28,** 283 (1977).
62. B. Gestbloom and E. Noreland, *J. Phys. Chem.* **80,** 1631 (1976).
63. R. H. Cole, unpublished results.
64. H. Fellner-Feldegg, *J. Phys. Chem.* **76,** 2116 (1972).
65. K. S. Cole and R. H. Cole, *J. Chem. Phys.* **9,** 341 (1941).
66. Y. Husimi and A. Wada, *Rev. Sci. Instrum.* **47,** 213 (1976).
67. R. H. Cole, private communication.
68. E. G. Finer, F. Franks, M. C. Phillips, and A. Suggett, *Biopolymers* **14,** 1995 (1975).
69. R. Freeman and G. A. Morris, *Bull. Magn. Reson.* **1,** 5 (1979).
70. P. Brunner and R. R. Ernst, *J. Magn. Reson.* **33,** 83 (1979).
71. G. A. Morris, private communication.
72. P. C. Lauterbur, *Nature (London)* **242,** 190 (1973).
73. P. C. Lauterbur, *Phys. Today* May, p. 17 (1978).
74. P. Mansfield and A. A. Maudsley, *J. Magn. Reson.* **27,** 101 (1977).
75. L. G. Rowan, E. L. Hahn, and W. B. Mims, *Phys. Rev.* **137,** A61 (1965).
76. G. Feher, *Phys. Rev.* **114,** 1219 (1959).
77. R. P. J. Merks and R. de Beer, *J. Phys. Chem.* **83,** 3319 (1979).
78. R. P. J. Merks and R. de Beer, *J. Magn. Reson.* **38,** 305 (1980).
79. A. G. Marshall, *J. Phys. Chem.* **83,** 521 (1979).
80. A. G. Marshall and D. C. Roe, *Anal. Chem.* **50,** 756 (1978).
81. D. C. Roe, A. G. Marshall, and S. H. Smallcombe, *Anal. Chem.* **50,** 764 (1978).
82. A. G. Marshall and D. C. Roe, *J. Magn. Reson.* **33,** 551 (1979).
83. F. G. Herring, A. G. Marshall, P. S. Phillips, and D. C. Roe, *J. Magn. Reson.* **37,** 293 (1979).
84. A. G. Marshall and R. E. Bruce, *J. Magn. Reson.* **39,** 47 (1980).
85. F. G. Herring, private communication.
86. R. E. Bruce and A. G. Marshall, *J. Phys. Chem.* **84,** 1372 (1980).
87. A. G. Marshall, *in* "Fourier, Hadamad, and Hilbert Transforms in Chemistry" (A. G. Marshall, ed.), pp. 99–123. Plenum, New York, 1982.

Electrochemical Characterization of Chemical Systems

Larry R. Faulkner

Department of Chemistry
University of Illinois at Urbana-Champaign
Urbana, Illinois

I. Introduction

This chapter is designed to introduce the principles of electrochemical measurements to scientists who suspect that these approaches might be useful but know little about them. Chemists have been wary about doing

137

electrochemistry, because they have perceived the field as a specialist's specialty, with arts and jargon that are rather foreign to other branches of chemistry. On the other hand, the concepts are important to a huge segment of chemical science and technology, and the techniques have become steadily more widely employed by investigators who are not electrochemical specialists, simply because they are useful—for trace analysis, for characterizing the redox properties and electronic structures of molecules, for studies of catalysis, for controlled generation of reactive intermediates, and so forth. Using electrochemistry productively—and understanding the results—is probably no more complicated than using gas chromatography knowledgeably. Our goal here is to create a systematic background for understanding electrochemical measurements and to introduce the more important techniques and the basic experimental aspects of their implementation.

This chapter begins with a very brief discussion of some important ideas about electrochemical systems that have proved to be common stumbling points on the road to understanding. Then we proceed to a more careful consideration of the basic elements of electrochemical systems, and we synthesize those elements into a variety of methods for attacking an experimental problem. Finally, we close with a discussion of the practical aspects of building cells and doing experiments.

Obviously a chapter of this size cannot cover a field as large as electrochemical characterization in subtle detail. Instead, we hope to attain a visceral feeling for the elements that go into an electrochemical experiment, the things that go on during the experiment, and the information that can come out of it. For deeper understanding more thorough sources must be consulted. The initial references [1–18] in the list at the end of the chapter are general, and will prove useful to the interested reader. The first six of them deal primarily with the fundamental physical phenomena involved in electrochemistry, whereas [7–16] are oriented toward the theory of experimental methods, and [17] and [18] cover details of implementing the methods in the laboratory.

II. Some Distinctive Concepts

At the outset, we need to face an important set of basic features of electrochemical systems, so that we can keep them clearly in mind during the remaining discussions. Confusion over these particular points is common among students approaching the field for the first time; therefore we shall draw attention to them directly and try to make them as clear as possible from the start.

(a) *The system is not homogeneous.*   We always must be aware of the location of action. The electrode can only interact with its immediate environment, which may differ from most of the solution because electrochemical products build up there, or reactants are depleted. It is useful to define four zones, as shown in Fig. 1. The *electrode* itself may act as a source or sink for species involved in the surface chemistry. The reduction of Zn^{2+} to form zinc amalgam and the oxidation of a copper electrode to form Cu^{2+} are simple examples. The boundary between the electrode and the electrolyte solution is a discontinuity where charge imbalances usually are found. That is, there is an excess of charge on the metal surface and an oppositely charged excess ionic charge in solution. This *double layer* region has a structure that bears importantly on electrochemical measurements. Moreover, it can also contain layers of substances chemisorbed onto the electrode surface, and they can either undergo electrode reactions or affect the rates of other reactions. The double layer has a thickness typical of molecular coulombic interactions, usually less than 100 Å. The *diffusion layer* comprises the zone where the concentrations of electroreactants or products have been affected by a reaction taking place at the surface. Excesses and deficiencies in this zone cause diffusion of material into the *bulk solution* or from the bulk into the diffusion layer. The bulk is the region far enough from the electrode to offer homogeneous composition. The thickness of the diffusion layer depends on the speed with which material can be transported and the time scale of the experiment, but usually it is on the order of 10 μm (100,000 Å).

(b) *Many things can happen at once.*   An *electrode reaction* is the whole body of chemical changes accompanying charge transfer across an electrode/electrolyte interface. The reaction may be fairly simple, possibly involving

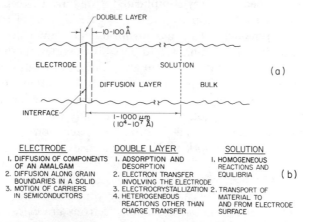

Fig. 1   (a) Zones of interest in an electrochemical system. (b) Some processes of interest categorized by location.

only the *heterogeneous* (at the interface) charge transfer step. The reduction of 9,10-diphenylanthracene (DPA) to its anion radical in an aprotic medium is such a case [19–21]. On the other hand, the reaction may involve several steps, some heterogeneous and others *homogeneous* (in solution). If the DPA molecule considered above is reduced in the presence of a proton donor, such as phenol, then the electrode process becomes fairly complicated [19–21]. After two heterogeneous and two homogeneous steps, one obtains 9,10-diphenyl-9,10-dihydroanthracene (DPAH$_2$):

$$DPA + e \rightleftarrows DPA^{\bar{\cdot}} \tag{1}$$

$$DPA^{\bar{\cdot}} + PhOH \rightarrow DPAH\cdot + PhO^- \tag{2}$$

$$DPAH\cdot + e \rightleftarrows DPAH^- \tag{3}$$

$$DPAH^- + PhOH \rightarrow DPAH_2 + PhO^- \tag{4}$$

Thus the electrode reaction, like an overall homogeneous reaction can be dissected into a *mechanism* of steps. Figure 1b lists some of the types of reactions and the locations where they take place. Each reaction has a rate constant and the overall rate of the reaction can be expressed in terms of the constants, as usual. Since the concentrations of various species are not uniform in the diffusion layer, the rates of some chemical reactions will be faster in some places and slower in others.

An important difference between an electrochemical reaction and the more familiar homogeneous process is the role of *mass transport*. Often the rate of the reaction is limited by the rate at which we can bring reactants to the electrode or remove the products, so it is useful to think of the transport process as a step in the whole mechanism.

Life obviously becomes very complicated if we have to keep track of the kinetics of the charge transfer reaction itself, the kinetics of homogeneous steps involving time-dependent and position-dependent concentrations of reagents, the rates of mass transfer, and possibly the rates of other steps such as adsorption or crystallization. Fortunately, one can usually simplify things because one or two of the steps are slower than the rest and are, therefore, *rate limiting*. Then we can focus on those steps and ignore the rest. Thus we may find a system whose behavior depends only on the charge-transfer kinetics, or on the rate of a single homogeneous reaction, or on the rate of mass transfer. We try to work with limiting cases, exactly as we do with homogeneous systems.

(c) *The current is an expression of the reaction rate.* An electrode reaction yields a net transfer of charge. For every mole of DPA reduced by the mechanism of Eqs. (1)–(4), two moles of electrons are extracted from the electrode and are replaced by charge drawn from the external circuit. To preserve electroneutrality, the external circuit obtains the electrons from the

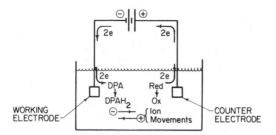

Fig. 2 Charge movements in a cell where DPA is reduced to $DPAH_2$ at one electrode. The ion movements will be such that the equivalent of 2 units of negative charge move from left to right per molecule of DPA reduced.

electrolyte itself via a second electrode (Fig. 2). Of course, an electrode reaction must take place at this second electrode in order to deliver that charge, and it must be an oxidation. Electron and ion movements transfer charge from place to place, so no appreciable charge builds up anywhere. The heterogeneous charge transfer reactions are *transducers* that allow communication across the boundaries between an *ionic conductor* (the solution) and the *electronic conductors* (the electrodes).

If N moles of reactant are consumed, and n electrons per molecule are involved, then the total charge passed Q is

$$Q = nFN \tag{5}$$

where F is the Faraday constant (96,485 coulomb per mole of electrons). Equation (5) is *Faraday's law*, a simple statement of stoichiometry. The *rate* of reaction in mol per second is dN/dt, thus the rate of change in the total quantity of charge is

$$dQ/dt = nF\,dN/dt = i \tag{6}$$

where we recognize that dQ/dt is the current i.

Since we are dealing with a heterogeneous reaction, whose rate is generally proportional to the available electrode area, the reaction rate is usually normalized by area to give dimensions of mol per second per square centimeter,

$$i/nFA = (1/A)dN/dt = \text{overall reaction rate per unit area} \tag{7}$$

Thus the current is an expression of the reaction rate, and the proportionality factor linking them is nFA.

If other electrode reactions take place simultaneously with the one of interest, they also contribute currents proportional to their rates and the sum of currents is observed. These currents are all *faradaic currents* because they arise from chemistry adhering to Faraday's law. Other currents derived from strictly physical processes are also present and called *nonfaradaic currents* (see Section III.G). One always tries to arrange conditions so that the single faradaic current of interest is much greater than the background nonfaradaic and faradaic contributions.

(d) *The potential is an expression of electron energy.* Most people are thoroughly uncomfortable with potential in an electrochemical context. The problem is always an unhappy experience with sign conventions that seemed arbitrary and signs that changed when reaction directions changed. Let us forget about that for the moment, and think of potential in the electrostatic sense as the work required to bring a positive test charge from the point at infinity to the interior of a phase. That work can be raised or lowered by controlling a very slight excess charge on the phase with some kind of power supply. It is obvious that this work is related to the energy required to bring an electron from a vacuum into the phase, hence the potential controls the energy of that electron within the phase. At more negative potentials, reached by increasing the negative excess charge, the electron is at relatively high energies. At more positive potentials, established with a less negative or a positive excess charge, one has low electron energies. The relationship is linear, and one can tune the energy at will over a range of 2–6 eV, which is ∼200–600 kJ/mol.

Electrochemistry happens because the energetics are favorable, just as ordinary chemistry does. The difference in electrochemistry is that one can *control* the energetics over quite a wide range. Consider the reduction of DPA as expressed in Eq. (1). Since the energy of the electron can be varied at will, we can anticipate that in some region of potential, the energy balance favors the oxidized form, DPA. In other words, the electron is more stable on the electrode than on DPA⁻ in solution. This region clearly lies at positive values with respect to some critical potential where equal energies apply

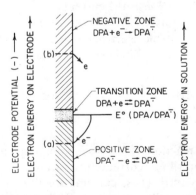

Fig. 3 Potential as a variable controlling relative electron energies. At potential (a) the electrode would oxidize any DPA⁻ at its surface and at potential (b) it would reduce any DPA.

(Fig. 3). At more negative potentials, the electron energy is very high with respect to DPA$^{\bar{}}$ and reduction occurs readily. The critical energy is the standard potential E^0 for the couple. At potentials far more positive than E^0, only DPA is stable at the surface, and at potentials far more negative than E^0, only the reduced form, DPA$^{\bar{}}$, can exist there. Within a range of $\sim 2kT$ on either side, mixtures of DPA and DPA$^{\bar{}}$ would be seen at the surface, because electrons are distributed among available orbitals on both sides of the interface approximately according to thermal–statistical equilibrium. Since kT at 25°C is 25.7 meV, we can expect the transition region to be perhaps $100/n$ mV wide. One can productively visualize E^0 as a solution "energy level." A "filled" state is the reduced form and an "empty" one is the oxidized species.

Taking the more familiar case of Ce^{3+} and Ce^{4+} in 1 M H$_2$SO$_4$, we refer our thoughts about the conditions near the electrode to the standard potential of the couple, which is effectively $+1.44$ V versus NHE. At much more negative values, the energy level is filled, and we have only the reduced form Ce^{3+} at the electrode, whereas at more positive potentials, the level is empty and one finds the oxidized form Ce^{4+}. Near $+1.44$ V, mixtures of the two forms would exist, and we could regard the energy level as partially filled.

(e) *One cannot simultaneously control current and potential.* Since the potential controls the energetics of reaction, it also controls the rate of reaction, and thus the current. Setting the potential sets the current. Controlling the current through the cell will force the system to the corresponding potential. There is a strictly functional, one-to-one relationship between these variables.

III. The Basic Elements

A. *Potential*

We have now begun to understand that the potential of an electrode has something to do with excesses of charge on the electrode and with the electrostatic potential energy of an electron residing there. Now we need to define this concept more carefully so that we can learn how to measure and control the potential.

1. *Cells†*

When experiments are done on electrochemical systems, currents usually flow. Charge is withdrawn from the electrolyte, and it must be replaced by

---

† See [1, 2, 6, 9].

an electrode reaction at a second electrode (Fig. 2). To carry out measurements, we need at least these two electrodes, which together with the electrolyte make up the *cell*. Usually we are interested only in one of the electrodes, called the *working electrode*; and the second *counter electrode*, serves to meet the requirements for electroneutrality. We cannot rigorously measure the electrostatic potential of either electrode on any absolute scale, so our next best option is to work with the difference between those two absolute potentials, which is simply the easily measured voltage across the electrodes.

Figure 4 shows that the voltage is actually composed of several important potential differences. At each interface, where double layers exist, there are sharp transitions. These interfacial electrostatic potential differences control the relative energies of charge carriers in an electrode and its adjacent solution, hence control the rates and directions of faradaic reactions. These differences, expressed on an experimental scale that adds a constant, are what we call the "potentials" of the electrodes. The overall cell voltage also includes a difference in potential between the parts of the solution just opposite the two electrodes. This contribution is the iR drop reflecting current flow through the solution resistance. When the iR drop is zero or negligibly small, the overall voltage is just the sum of the interfacial differences and is called the *cell potential*.

If we want to control the interfacial potential differences at the working electrode, we can proceed by controlling the overall cell voltage. Unfortunately, a change in that voltage is distributed among the working- and counter-electrode potentials and the iR drop in a manner that may not be predictable; hence control can be difficult, unless we can design the system so that the interfacial difference at the counter electrode is invariant.

2. *Reference Electrodes*†

The key to this problem is to construct an electrode with a composition that is not very sensitive to current flow. A good example is the saturated calomel electrode (SCE)

$$Hg/Hg_2Cl_2/KCl(saturated, aqueous)$$

which has a mercury pool in contact with solid mercurous chloride and potassium chloride. The aqueous electrolyte is therefore saturated with Hg_2^{2+}, K^+, and Cl^-, and the concentrations of all three species are controlled by the solubility equilibria. Passage of current in one direction or the other will not affect the activities of Hg or Hg_2^{2+}, hence the interfacial potential difference remains close to the equilibrium (zero-current) value dictated by the composition and related through the Nernst equation.

† See [4, 14, 17, 18, 22–24].

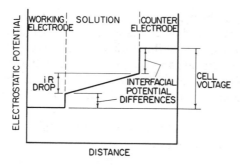

Fig. 4 Electrostatic potential profile along a path from the interior of the working electrode to the interior of the counter electrode of the cell in Fig. 2.

If the SCE were employed as the counter electrode in a cell, any change in cell voltage would be distributed between the working electrode potential and the iR drop. When the latter is negligible all of the voltage change goes to the working-electrode interface, so we have developed a basis for control. Now the counter electrode serves two functions: it is a charge compensator (the true counter electrode function), and it is an electrostatic standard. That is, it also acts as a *reference electrode*, and the operation of the working electrode can be referred to it unambiguously. With negligible iR drop, the overall cell voltage is the *potential* of the working electrode *with respect to the reference*.

The internationally accepted standard for the potential scale is the *standard hydrogen electrode* (SHE) or *normal hydrogen electrode* (NHE)

$$Pt/H_2(a = 1)/H^+(a = 1)$$

It is ideal device and cannot actually be constructed, but its properties can be extrapolated from measurements with real hydrogen electrodes. The SCE is by far the most widely used secondary reference. Its potential is 0.2415 V versus SHE at 25°C. Other reference electrodes may be more convenient in specific experimental situations, and many have been devised.

3. Potential at Equilibrium. The Nernst Equation†

Suppose we build a cell with say a Pt working electrode, an SCE as reference/counter electrode, and an electrolyte containing both ferri- and ferrocyanide. If we let it stand with no current flow for a while, we know that the potential of the working electrode with respect to the reference will come to a steady value that represents equilibrium between the electrode and electrolyte. The electrons on the electrode and those on ferrocyanide in

† See [1, 9, 14].

solution redistribute themselves among states on the electrode and on the electroactive solutes according to the usual principles of statistical thermodynamics. This process adjusts the state of charge on the electrode, and thus the potential of the electrode, to a value consistent with the energy levels and the populations of empty and filled states in solution, i.e., the E^0 for the redox couple and the concentrations of oxidized ("empty" states) and reduced ("filled" states) forms.

One can show by thermodynamic arguments that the Nernst equation describes the equilibrium condition

$$E = E^0 + \frac{RT}{nF} \ln \frac{a_O}{a_R} \tag{8}$$

where a_O and a_R are the activities of the oxidized species O and the reduced species R in the general process

$$O + ne \rightleftarrows R \tag{9}$$

The *standard potential* E^0 is an empirical constant (equal to the equilibrium potential when $a_O = a_R$) expressing the redox-active energy levels on O and R. Usually it is more convenient to work with concentrations, hence (8) is often rewritten as

$$E = E^0 + \frac{RT}{nF} \ln \frac{\gamma_O}{\gamma_R} + \frac{RT}{nF} \ln \frac{C_O^*}{C_R^*} \tag{10}$$

$$= E^{0\prime} + \frac{RT}{nF} \ln \frac{C_O^*}{C_R^*} \tag{11}$$

where the γs are activity coefficients, and C_O^* and C_R^* are the bulk concentrations of species O and R. The *formal potential* $E^{0\prime}$ includes the activity coefficients, and in more complicated systems, perhaps complexation constants, ligand concentrations, and proton concentrations; thus it is defined only for a specific medium. Large tabulations of standard and formal potentials exist [25–27].

4. *Free Energies and Sign Conventions*

It is well known that electrode potentials are related to the free energy changes in processes that go on inside cells. A large number of potentiometric measurements are made just to obtain thermodynamic information through that route. The principles are straightforward, fully discussed elsewhere [4, 9, 14], so we shall not deal with them here. Our interest is in standard potentials corresponding to the actual electrical measurements of an electrode in a cell. They are always found in tables where by convention the associated electrode reactions are written as *reductions*, e.g.,

$$Ce^{4+} + e \rightleftarrows Ce^{3+} \quad (1 \ M \ HClO_4) \qquad E^{0\prime} = 1.70 \text{ V versus NHE} \qquad (12)$$

B. Potentials and Rates

Heterogeneous electron transfer reactions are kinetic processes and are characterized by rate constants, just as homogeneous reactions are. Consider the general case

$$O + ne \underset{k_b}{\overset{k_f}{\rightleftarrows}} R \qquad (13)$$

but now let us add the two rate constants k_f and k_b. The rate of reduction must be proportional to the concentration of O, and the rate of oxidation similarly depends on the concentration of R. The net reaction rate is the difference between these two rates, which is also i/nFA, as we learned in Section II.

$$\frac{i}{nFA} = k_f C_O(0, t) - k_b C_R(0, t) \qquad (14)$$

This expression involves the concentrations of O and R at the electrode surface, because those concentrations are the only ones that the electrode can experience. In general, we have to consider concentration as a function of time and distance from the surface, so we write $C_O(x, t)$ and $C_R(x, t)$. By varying the potential, one can reduce or oxidize at faster or slower rates; thus k_f and k_b are clearly functions of potential. We need to determine the nature of this dependence.

1. Plotting Conventions

By writing the net reaction rate as in Eq. (14), we imply that a cathodic process (reduction) produces positive currents, whereas an anodic reaction gives negative currents. This is the most widely used definition, and we shall follow it here. It is unfortunately not ohmic, because negative changes in potential cause positive changes in current; and for that reason the IUPAC Commission on Electrochemistry discourages its use [28], recommending the complementary choice. For the present, anyone approaching the literature must determine on a case-by-case basis the current convention in use.

In plotting current–potential curves, we shall draw cathodic currents up, anodic currents down, negative potentials *right*, and positive potentials *left*. This arrangement is the most widely used in the literature, but every conceivable arrangement can be found.

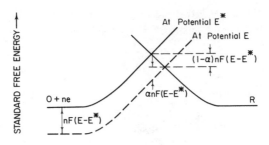

Fig. 5 Free energy changes during a reaction. At potential E^*, where $k_f = k_b$, the solid curves apply. The dashed curve describes $O + ne$ at some other potential E. Below we find that E^* is the same as $E^{0\prime}$.

2. Potential Dependence of Rate Constants†

The basic reason for the potential dependences of heterogeneous rate constants is that the relative energy of the electrons on the electrode moves linearly with potential. Negative potential changes destabilize the electrons, enhance k_f, and lower k_b. Positive potential shifts have the opposite effects.

Consider the solid curves in Fig. 5, which describe free energy changes during a reaction, as measured along an axis indicating reaction progress. A real system might involve sodium amalgam in contact with sodium ion in acetonitrile. The reaction coordinate then might be essentially the position of the sodium nucleus with respect to the interface. During an oxidation, the free energy rises as the sodium atom in mercury is desolvated, then it falls again as the ion is resolvated by acetonitrile. In Fig. 5 the solid curves represent the energetics of the general system at the particular potential E^* where k_f and k_b happen to be equal. Their value is k^0, the standard rate constant for the system.

Now suppose the potential is changed to a value E, which is more positive than E^*. This alteration stabilizes the electrons on the electrode, so it lowers the whole curve for $O + ne$ by $nF(E - E^*)$. Close study will show that part of this energy change goes to raise the activation barrier for reduction. The amount is $\alpha nF(E - E^*)$, where α is the *transfer coefficient*, a fraction with a value between zero and unity. It describes the symmetry of the barrier and in real systems it usually lies between 0.3 and 0.7. The remaining part of the energy change, $(1 - \alpha)nF(E - E^*)$, is the extent to which the barrier for oxidation is lowered. In a diagram like Fig. 5, α is determined by the slopes of the intersecting curves for $O + ne$ and for R. If the slopes are equal but opposite (as in Fig. 5), then $\alpha = 0.5$. For different slope magnitudes, the intersection becomes unsymmetrical and α becomes larger or smaller than 0.5.

† See [1, 2, 6, 9, 12, 16, 29–31].

If the rate constants k_f and k_b are of the Arrhenius form, then the changes in activation energy can be registered through exponential multipliers to k^0,

$$k_f = k^0 \exp\left[-\alpha nF(E - E^*)/RT\right] \tag{15}$$

$$k_b = k^0 \exp\left[(1 - \alpha)nF(E - E^*)/RT\right] \tag{16}$$

Now we see that there is an exponential dependence of the rate constants on potential, and the kinetic behavior of the system can be fully described by three parameters, k^0, α, and E^*. We can identify E^* by considering the zero-current condition.

3. At Equilibrium

When the net current is zero, the electrode is at equilibrium with the solution, and the surface concentrations are the same as the bulk values. From Eq. (14), then, $k_f C_O^* = k_b C_R^*$, and by substitution from Eqs. (15) and (16) we have,

$$\exp\frac{nF(E - E^*)}{RT} = \frac{C_O^*}{C_R^*} \tag{17}$$

or

$$E = E^* + \frac{RT}{nF} \ln \frac{C_O^*}{C_R^*} \tag{18}$$

In other words, the kinetic relations collapse, in the event of equilibrium, to an expression of the known thermodynamic form, which is the Nernst equation. Any valid kinetic model must do that, because it must describe not only the approach of the system to equilibrium, but also the maintenance of equilibrium.

Equation (18) resulted from a kinetic treatment in which E^* was identified as the potential where $k_f = k_b$. Since we know that this equation must be the same as the Nernst equation, we now know that $E^* = E^{0\prime}$. Thus the full kinetic relation describing current flow is

$$i = nFAk^0\{C_O(0, t) \exp\left[-\alpha nf(E - E^{0\prime})\right]$$
$$- C_R(0, t) \exp\left[(1 - \alpha)nf(E - E^{0\prime})\right]\} \tag{19}$$

where $f = F/RT$.

C. Coming to Equilibrium

It is useful to think about a mechanism by which an electrode comes to its equilibrium potential. How, for example, would a platinum working electrode at open circuit, suddenly immersed in a solution of ferri- and ferrocyanide adjust itself to the nernstian condition? With our kinetic model, we can supply an answer. Suppose the electrode, upon immersion, is at a relatively positive value. The solutes interact dynamically with it; but because the initial potential is positive, the rate of oxidation of ferrocyanide exceeds the rate of reduction of ferricyanide. Thus there is a net charge transfer resulting in electrons being deposited on the electrode. This charge is not removed by an external circuit, so the potential shifts negative. The shift slows the oxidation rate and speeds the reduction rate; hence the buildup of negative charge slows down. Eventually, the potential will approach the point where the rates of reduction and oxidation are equal, and there is no further change in the state of charge or the potential. This is equilibrium. The time required to reach it depends on the intrinsic facility of the kinetics. Note that the approach to equilibrium requires a net transfer of electrons from the solution to the electrode, or vice versa; therefore it involves chemical change. However, the amounts involved are very small. A 1-cm^2 electrode would typically require $\sim 10^{-10}$ equivalents per volt of shift.

In thinking about electrode reactions as dynamic processes with competitive forward and backward components, it is sometimes useful to translate the component rates into component currents. The forward reaction is a reduction, and we can regard it as causing a cathodic current flow i_c across the interface. Likewise, backward reaction generates an anodic current i_a. The net current is the difference between these two:

$$i = i_c - i_a \tag{20}$$

At equilibrium $i = 0$, and the rates of oxidation and reduction are equal. Thus i_a and i_c are both equal to a common value i_0, called the *exchange current* [1, 2, 6, 9, 12, 29–31].

Since Eqs. (19) and (20) are equivalent, i_c and i_a are given by the first and second terms of Eq. (19), respectively. When $i = 0$, either of them is the exchange current, and by manipulation one can easily show that

$$i_0 = nFAk^0 C_O^{*(1-\alpha)} C_R^{*\alpha} \tag{21}$$

Thus the exchange current depends on the bulk concentrations of electroactive species, as we would expect, and on the standard rate constant k^0.

The values of k^0 and i_0 are indices of a system's kinetic facility and its ability to deliver or accept current. Standard rate constants range from 1–50 cm/sec for very fast reactions to 10^{-9} cm/sec for very sluggish ones. The corresponding exchange currents range from tens of amperes to picoamperes per square centimeter. Kinetic facility is generally associated with very simple reactions, perhaps involving only solvation changes (such as aromatic/radical ion and metal ion/amalgam couples). Reactions involving major structural changes, such as alterations in an inner coordination sphere, are slower. Complicated reactions, such as the evolution of H_2 or O_2, are multistep processes requiring several rate parameters for description. Usually one of them is rate limiting, and the overall process can be very slow.

Note that the exchange current and the equilibrium potential are concepts that are defined only when both redox forms of a couple are present. In a case such as 1 mM $K_3Fe(CN)_6$ in 1 M KCl in contact with Pt, neither has meaning. However, there will be a *rest potential* where no current flows; and it will lie at some ill-defined value more positive than the region where $Fe(CN)_6^{3-}$ can be reduced, but more negative than the region where the medium can be oxidized (to O_2 or Cl_2) or the electrode can be covered with an anodic oxide film.

D. Moving Away from Equilibrium: Current Flow

Consider now our general system in a situation where both O and R are present in the bulk of solution, so that a true equilibrium potential exists and is defined by Eq. (11). At that potential, of course, no current flows. If we want to reduce O or oxidize R, then the electrode must be placed in a circuit where current can flow, and forced away from equilibrium. We can expect the speed and direction of the resulting reaction (i.e., the sign and magnitude of the current) to be controlled by the direction and the extent of polarization from the equilibrium potential.

1. Activation of Reactions. Overpotential†

Figure 6 quantifies these ideas, as dictated by Eq. (19). For simplicity, we have assumed that the currents are low and there is good stirring, so that the concentrations of O and R at the surface are essentially unchanged from their bulk values. It is useful now to define the *overpotential* η of an electrode as the difference between its actual potential and its equilibrium position. Thus net reductions occur at negative overpotentials and net oxidations take place at positive values. At any given current, a unique overpotential is required; and larger overpotentials are needed to produce larger currents. One can accurately regard η as the driving force required to make the electrode reaction operate at the rate needed to deliver a given current. It actually represents an activation energy, as we can understand by recalling how the potential dependence arose in the kinetic expressions above. In the case at hand, the current is limited by the rates of heterogeneous electron transfer, and η is called an *activation overpotential*.

† See [1, 2, 6, 9, 29–31].

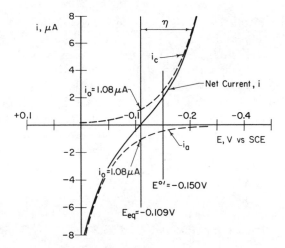

Fig. 6 Potential dependence of current flow for the reaction $O + ne \rightleftarrows R$ with $n = 1$, $A = 0.1$ cm^2, $E^{0\prime} = -0.15$ V versus SCE, $k^0 = 5 \times 10^{-5}$ cm/sec, $T = 298$ K, $\alpha = 0.5$, $C_o^* = 5$ mM, and $C_R^* = 1$ mM.

The amount of activation required to deliver any given current depends on the value of k^0; that is, on the intrinsic kinetic facility. If k^0 is large, the kinetics are facile and may not require activation. On the other hand, for very low values of k^0, one may require overpotentials of *volts* to observe any appreciable current.

2. Effects of Mass Transfer

The exponentials in Eq. (19) imply that the current rises very quickly with η, once one reaches the potential region where the net reaction takes place. On the other hand, as the current rises, the surface concentrations tend away from the bulk values. This effect has important implications as shown in Fig. 7.

For example, a net reduction depletes the concentration of O near the surface and causes an elevated concentration of species R. As the current rises, the value of $C_O(0, t)$ drops and tempers the effect of its exponential coefficient. The current is then limited in part by the activated kinetics of charge transfer (manifested in the exponentials) and in part by the supply of reactants reaching the surface by mass transfer [manifested by $C_O(0, t)$]. If large overpotentials are used, so that k_f becomes very large, then the reduction rate is no longer limited by charge transfer kinetics, and is wholly controlled by mass transfer. When this condition applies, species O cannot coexist with the electrode for very long, and its concentration at the surface drops to a very small value compared to C_O^*. Since the rate of mass transfer

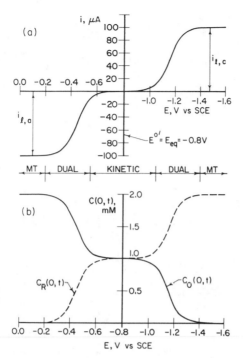

Fig. 7 Effects of mass transfer on current flow for a system involving O + $ne \rightleftarrows$ R with $n = 1$, $E^{0\prime} = -0.8$ V versus SCE, $\alpha = 0.5$, $C_O^* = C_R^* = 1\,\text{m}M$, and steady stirring so that limiting currents $i_{1,c} = -i_{1,a} = 100\ \mu\text{A}$ are reached. For this case $i_0/i_{1,c} = 0.001$, hence appreciable activation is needed for current flow. (a) Current–potential curve. (b) Surface concentrations versus potential. Between ~ -0.55 V and -1.05 V the current is kinetically controlled, and in the regions more positive than -0.2 V and more negative than -1.4 V it is mass-transfer controlled. Between -0.2 V and -0.55 V and between -1.05 V and -1.2 V, kinetics and mass transfer both influence the current.

independent of potential, the current is flat with respect to η. Its value is $i_{1,c}$, the *cathodic-limiting current*. This current is directly proportional to C_O^* and is independent of C_R^*, thus it is convenient for quantitative determinations.

Exactly parallel arguments can be made for the anodic branch, where $C_R(0, t)$ falls and the transport of R affects the current. The *anodic-limiting current*, $i_{1,a}$, is then proportional to C_R^* and is independent of C_O^*.

3. *Effects of k^0 and α†*

Figure 8 is a display of current–potential curves expected for different combinations of k^0 and α for a stirred solution with $C_O^* = 2\,\text{m}M$ and

† See [1, 2, 6, 9, 29–31].

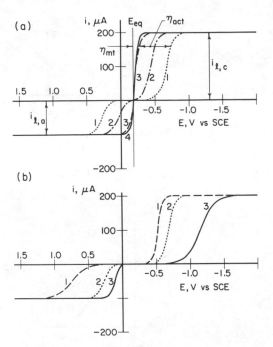

Fig. 8 (a) Effect of k^0 on the $i-E$ curve for $O + ne \rightleftarrows R$ with steady-state stirring and $n = 1$, $E^{0'} = -0.2$ V versus SCE, A = 0.1 cm^2, $C_O^* = 2$ mM, $C_R^* = 1$ mM, $\alpha = 0.5$, $i_{1,c} = 200$ µA, $i_{1,a} = -100$ µA, and $T = 298$ K. Curves 1–3 are for $k^0 = 10^{-6}, 10^{-4}$, and 10^{-2} cm/sec. Curve 4 is the reversible limit for $k^0 \to \infty$. (b) Effect of α on the current potential curve for the system described above with $k^0 = 10^{-6}$ cm/sec. Curves 1–3 are for $\alpha = 0.75, 0.50$, and 0.25.

$C_R^* = 1$ mM. In Fig. 8a, one sees the effect of k^0 with constant α. Each curve is really two waves, anodic and cathodic, reflecting two opposite chemical transformations. The effect of k^0 is to fix the splitting between them. With very large k^0 one obtains a limiting case in which the two waves merge without an intervening inflection. In this situation, no overpotential is needed to drive charge transfer kinetics; hence the actual overpotential required for any current just pushes mass transfer at the rate needed to support the current. With smaller k^0 one must also drive the electron transfer kinetics, and greater overpotentials are required. Thus the reduction wave moves to more negative potentials and the oxidation wave to more positive ones. Smaller and smaller values of k^0 cause greater and greater shifts.

Note that the limiting currents are the same in all cases shown in Fig. 8a. Even though the kinetics are very sluggish in several of the instances considered there, they can still be speeded by an applied overpotential to the point where they are no longer rate-limiting. Then the current is controlled

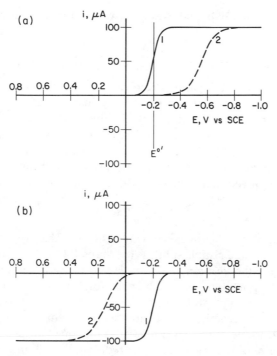

Fig. 9   Current potential curves for $O + ne \rightleftarrows R$ with steady-state stirring, but with only one form present in bulk solution. (a) For $C_O^* = 1$ mM and $C_R^* = 0$. (b) For $C_R^* = 1$ mM and $C_O^* = 0$. Curves 1 and 2 correspond to the reversible limit ($k^0 \to \infty$) and $k^0 = 10^{-5}$ cm/sec, $\alpha = 0.5$, respectively. The remaining parameters are $n = 1$, $E^{0\prime} = -0.2$ V versus SCE, $A = 0.1$ cm^2, $T = 298$ K and $i_{1,c}/C_O^* = -i_{1,a}/C_R^* = 100$ μA/mM.

by mass transfer independently of the kinetics. This ability to tune kinetics is a distinctive feature of electrochemical systems.

In Fig. 8b one can see the effect of α at constant k^0. We saw above that α controls the apportionment of the energy change represented by a shift in E between the two activation barriers for oxidation and reduction. Thus it controls the symmetry of the splitting between anodic and cathodic branches with respect to the equilibrium potential.

Very often we deal with systems in which only O or only R is present in the bulk solution. Then the equilibrium potential is not defined and both branches of the $i-E$ curve will not be seen. If only O is present, one just observes the cathodic branch (Fig. 9a); and if only R is available, the electrochemistry is restricted to the anodic branch (Fig. 9b). With fast kinetics, the electrochemistry is essentially centered on $E^{0\prime}$, but for smaller values of k^0 one must work harder to make the processes go; hence reduction waves are at more negative potentials, and oxidation waves at more positive potentials, than $E^{0'}$.

4. Reversible and Irreversible Systems

The manner in which the anodic and cathodic component currents combine to form the net current can be seen by inspection of Fig. 6 again. It shows the important point that, at sufficiently large overpotentials, the net current is the same as one of the two components. For example, at overpotentials beyond -100 mV, $i \cong i_c$ and i_a is negligible. In that region, the kinetics of the oxidation reaction have no impact, and one can regard the reaction as a unidirectional reduction with the single rate constant k_f:

$$O + ne \overset{k_f}{\to} R \tag{22}$$

By following along this line, one can easily see that when k^0 becomes sufficiently small that a sizable splitting (perhaps 100 mV or more) separates the anodic and cathodic branches, then i_a and i_c are never simultaneously important in the potential ranges where significant currents flow. Then the reduction and oxidation processes are said to be *irreversible*, because the kinetics of the back reaction do not influence the current response in any experimentally usable region of potential. Thus one can use, for example, only the first term of Eq. (19) to describe reductions, and only the second to describe oxidations. The resulting simplifications in mathematical treatments can be considerable.

If instead we consider the limit of very fast kinetics (large k^0), then the branches are not split, and one must everywhere consider both i_a and i_c as significant contributions to the net current. In other words, the reaction is fully *reversible* at all points (except at very high overpotentials, where mass transfer controls the current anyway). A very important relation applies at this limit. Let us rearrange Eq. (19) to a form in which we can test the behavior as k^0 becomes large:

$$\frac{i}{nFAk^0C_O^*} = \frac{1}{C_O^*}\left[C_O(0, t)e^{-\alpha n f(E - E^{0'})} - C_R(0, t)e^{(1-\alpha)n f(E - E^{0'})}\right] \tag{23}$$

With very facile kinetics (i.e., $k^0 \to \infty$), any measurable i will be small compared to $nFAk^0C_O^*$, hence the bracketed term must always tend to zero. It can be rearranged to

$$E = E^{0'} + \frac{RT}{nF} \ln \frac{C_O(0, t)}{C_R(0, t)} \tag{24}$$

This equation has a nernstian form, and it can be viewed as expressing a kind of equilibrium between the surface concentrations and the potential. In other words, the kinetics are so fast that the electrode potential and the surface concentrations adjust themselves to thermodynamic equilibrium regardless of how fast E is varied. Of course, the surface concentrations may differ from those in the bulk, and that disequilibrium elicits mass transfer that requires a continuous current, so that condition (24) is maintained. Fast charge transfer reactions are often called *nernstian* in consequence of these ideas. Note that Eq. (24), which is the essential descriptor of such a system, does not contain k^0 and α. We could have expected that, because this is the case in which kinetics are so fast as to be invisible. Obviously, reversible electrochemical responses will not allow us to measure kinetic parameters.

The terms reversible and irreversible describe limiting cases of kinetics— "very facile" and "very sluggish." Now we must recognize that these limits are relative, and we must consider the time scale of any perturbation to which we ask the kinetics to respond. For example, at slow rates of potential change, the system might be able to maintain condition (24) and would then give a reversible response. At somewhat faster sweeps the charge transfer kinetics might not be fast enough to adjust the surface concentrations to the nernstian condition, and one would see a response manifesting some activation of those kinetics. This circumstance is called quasi-reversibility and leads to the most difficult treatments of electrochemical responses, because none of the simplifications available at the two limits apply. At very fast sweeps, one might tax the kinetics so badly that the irreversible limit would be reached. Thus a single system may appear reversible, quasi-reversible, or irreversible in different time domains.

In reading the electrochemical literature, one must be wary of the ambiguous use of the word irreversible. It may have the meaning outlined above, or it may refer only to the chemistry following a charge transfer reaction. For example, reaction sequence (1)–(4) would be called irreversible (or chemically irreversible), because one cannot faradaically reverse the transformation of DPA to $DPAH_2$. However, the initial charge transfer reaction (1) is reversible in all senses.

5.   Current–Potential Curves†

In all electrochemical systems more than one electrode reaction can occur, depending on the potential; and in some potential ranges two or more reactions will take place simultaneously. Then the current describes the sum of the rates of the processes. This point is illustrated in Fig. 10, which shows current–potential data [32] for rubrene (a polycyclic aromatic hydrocarbon, A) in N,N-dimethylformamide containing tetra-n-butylammonium perchlorate (TBAP). The working electrode is a rotating Pt disk (Fig. 10, inset),

† See [7–10, 13–16].

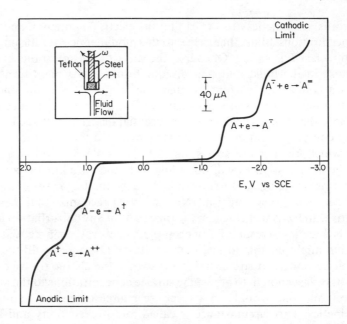

Fig. 10 Current potential curve for 1.0 mM rubrene (A) in N,N-dimethylformamide (DMF) containing 0.1 M tetra-n-butylammonium perchlorate (TBAP). The working electrode was a rotating Pt disk like that shown in the inset. The angular rotation rate is shown as ω. [Adapted with permission from J. T. Maloy and A. J. Bard, *J. Am. Chem. Soc.* **93**, 5968 (1971). Copyright 1971, American Chemical Society.]

which relies on the movement about its axis to throw the electrolysis products outward, so that fresh solution arrives continuously. Only a very small amount of electrolysis occurs, so we do not have to worry about the products building up in solution. The solution appears to the electrode as though it contains only A, the solvent, and TBAP. The curve in Fig. 10 is obtained by scanning the potential slowly while recording the current. At any given potential, the current is constant with time, because there is a steady rate of mass transfer to and from the surface. This curve shows the full range of electrochemical reactions in the system.

In the region between 0.8 V and −1.2 V versus SCE all species are stable at the electrode, and no current flows. If one moves to potentials more negative than −1.2 V, then reduction currents flow. First rubrene is reduced to its anion radical A$^{\bar{}}$, then near −2.1 V this species begins to be reduced to the dianion A^{2-}. The current for transformation of A$^{\bar{}}$ to A^{2-} is superimposed on that for A to A$^{\bar{}}$, because the latter continues at the mass-transfer-limited rate while the former takes place. At potentials beyond −2.8 V, there is another current rise caused by the addition of a third process, namely, reduction of the solvent/supporting electrolyte system. Since DMF and TBAP are present at very high concentrations, this current rises sharply and obliterates any electrochemistry at more extreme potentials. The *cathodic background limit* at −2.8 V is the negative end of the *working range* in this particular system. The location of the limit depends on the solvent, supporting electrolyte, electrode material, and surface preparation. See Section V.C for more details.

In a similar manner we could proceed in a positive direction from the intermediate region, and we would observe the first oxidation of rubrene to the cation radical A^+, then the oxidation of A^+ to A^{2+}, and finally the background oxidation; i.e., solvent or supporting electrolyte. This last process sets the *anodic background limit* at ~ 1.8 V versus SCE.

E.   *Potential as a Controlled Variable*

In Section III.A we covered the fundamentals of potential control. Here we take up some refinements that are usually employed in actual experimental situations. When we left the subject before, we had evolved two-electrode cells in which one of the electrodes served a reference/counter function, and a power supply imposed a voltage across the cell. That voltage was the quantity we wished to control, the potential of the working electrode with respect to the reference, *plus* the iR drop in solution. There are two major problems with this scheme.

First, the reference electrode must absorb the current flow in its dual role as the counter electrode. Now we understand that this feature will force the reference away from its equilibrium potential by the overpotential needed to accommodate the current. This overpotential is an error in the assumed potential of the working electrode. Real reference electrodes are usually designed with large concentrations of electroactive species, and they feature electrode processes with large k^0, so that the electrode can handle substantial currents with negligible error [17, 18, 22–24]. Problems arise in large-scale electrolysis or fast experiments, where large currents flow.

A solution is found by creating a *three-electrode cell* [8, 9, 15, 17, 18], in which the reference and counter electrode functions are separated, as shown in Fig. 11. Current flows between the working and counter electrodes, but one observes the voltage difference between the working and reference electrodes. A high impedance circuit can measure that voltage; hence only negligible currents are ever drawn through the reference. Potential control is achieved by a *potentiostat* [8, 9, 17, 18, 33–35], which applies power to the working and counter electrodes as necessary to maintain the voltage between the reference and working electrodes at the desired value. The potentiostat

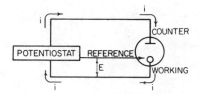

Fig. 11   Three-Electrode system for potential control.

is a feedback-controlled device, operating on the working-electrode/counter-electrode circuit and observing the effect on the working-electrode/reference-electrode circuit. Sections V.D and V.E provide more details.

A second problem with two-electrode cells is the inclusion of the iR drop in the observed voltage difference. The resulting potential control error can be quite large in high-current situations or in resistive media. The three-electrode configuration can alleviate the error to a considerable extent. The full iR drop in the cell then occurs between the counter and working electrodes, and the tip of the reference electrode makes contact with the solution at a third point. The voltage difference between the working and reference electrodes now is the true "potential" of the working electrode versus the reference plus the solution iR drop between the face of the working electrode and the tip of the reference probe. It is not feasible to eliminate the drop completely, but it can be minimized by placing the reference probe as close as possible to the working electrode.

The remaining iR drop is "uncompensated," and is the product of the current and the so-called *uncompensated resistance*, R_u [9, 15, 17, 18, 34–39]. One can profitably view the three-electrode system as a three-terminal potentiometer in which the full solution resistance extends between the reference and counter electrodes, and the reference electrode is the movable wiper, which can be positioned to include any fraction of the total resistance between the working electrode and itself. The included resistance is R_u. Most of the time iR_u is negligible, and the voltage difference beween the working and reference electrodes is the desired potential; however there are situations in which the uncompensated resistance causes experimental distortions and artifacts [17, 18, 34–39].

F.   *Mass Transport*

The rates at which electrode reactions occur are often limited by the rates at which reactants and products are carried to and from the surface. If we hope to understand current flow in a quantitative way we must therefore be able to treat mass transport.

1. Categories†

There are three ways to move a species from the bulk to the surface. Obviously, one can stir the solution and thereby achieve *convective flow*. Convection is a bulk process in which segments of solution can be regarded as moving, largely intact, from one location in the container to another. The flow can be *laminar* or *turbulent* and it can be carefully designed (as at a rotating disk) or it can be unpredictable (as might be caused by vibrations or thermally produced density gradients). *Diffusion* is, in contrast, a microscopic process based on the brownian motion of individual molecules and ions. It tends to randomize the distribution of molecules in a system, so its result is to transport species from regions of high concentration to regions of low concentration. In an electric field, ions can also be moved by coulombic forces in a process called *migration*. Since the ions respond individually, this is also a microscopic process. Electrolytic conduction in the bulk of the solution, where concentrations are uniform, is carried out strictly by migration. The electric field driving it is provided by the iR drop.

In general, we want to be able to describe the movement of mass on a fundamental basis, so we are often limited in our choices of transport processes for real experimental systems. Hydrodynamic flow patterns are difficult to derive with rigor, except in a few cases (such as laminar flow about a rotating disk or through a tubular electrode). Thus theory tends to concentrate on unstirred systems in which diffusion is the only mode of mass transport. It can be predicted straightforwardly through well-defined laws and boundary conditions. Migration is the least tractable mode, because it depends on the magnitude and shapes of the potential profiles in solution. They, in turn, depend on the shapes and relative dispositions of the electrodes and on the resistance of the solution. Steps are always taken to minimize any influence of migration on the flow of current.

That job is accomplished by adding a *supporting electrolyte* that is not electroactive in the potential range of interest. It serves to reduce the resistance to the point where electric fields within the solution (i.e., the iR drops) are negligible. As a rule of thumb, the supporting electrolyte must be present at a concentration at least 100 times that of the species under study [44]. Often the supporting electrolyte has important additional roles in the control of pH (e.g., as a buffer component) or in complexing the electroactive species.

---

† See [5–7, 8, 10, 12, 15, 16, 40–43].

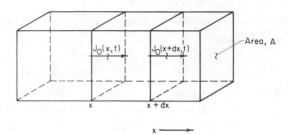

Fig. 12   A segment of a linear diffusion field taken along the axis of diffusion.

2. *Flux at the Electrode Surface†*

Usually we consider electrochemistry at planar electrodes immersed in unstirred solutions, so that we only have to consider the movement of species normal to the surface by diffusion. The distance away from the surface is x, as shown in Fig. 12. In general the concentration is a function of distance, and a net diffusion of material takes place at all points in an attempt to homogenize the system. Let us concentrate on the plane depicted in Fig. 12 at position x. The net number of moles of species O crossing that plane per square centimeter per second is called the *flux* of the solute, $J_O(x, t)$. One can show experimentally and theoretically that the flux is proportional to the concentration gradient at x (9, 43), thus we have *Fick's first law*,

$$J_O(x, t) = -D_O \frac{\partial C_O(x, t)}{\partial x} \tag{25}$$

where D_O, the constant of proportionality, is the *diffusion coefficient* of the species. The minus sign appears because a flux in the positive direction is associated with a negative slope in the concentration profile.

Now consider the situation in which species O undergoes reduction at the electrode, so that its concentration at the surface is depleted with respect to the bulk. This effect gives a positive slope to the concentration profile, and there is a flux of O formally crossing the interface, $J_O(0, t)$. This is the rate at which O disappears from the solution, thus it is the rate of reduction of O. An equal, but opposite flux of R enters the solution at the surface. In Section II we found that the rate of reaction is i/nFA; thus we can write from Eq. (25)

$$\frac{i}{nFA} = D_O \frac{\partial C_O(x, t)}{\partial t}\bigg|_{x=0} = -D_R \frac{\partial C_R(x, t)}{\partial t}\bigg|_{x=0} \tag{26}$$

† See [5, 9, 12, 15, 16, 41, 42].

This is an important result, because it tells us that we can describe the current as a function of time and concentration from the way in which the concentration profiles near the electrode evolve with time.

3. Fick's Second Law†

The time dependence of a concentration at any particular value of x can be considered with reference to Fig. 12. Suppose the concentration in the box between x and $x + dx$ is $C_O(x, t)$. At a later time $t + dt$, the concentration will have been changed by the difference between the inbound and outbound fluxes:

$$dC_O(x, t) = - \frac{[J_O(x + dx, t) - J_O(x, t)] \cdot A \cdot dt}{dx \cdot A} \quad \frac{\text{net moles}}{\text{box volume}} \quad (27)$$

or more formally

$$\frac{\partial C_O(x, t)}{\partial t} = - \frac{\partial J_O(x, t)}{\partial x} \quad (28)$$

By substitution from Eq. (25) we have

$$\frac{\partial C_O(x, t)}{\partial t} = D_O \frac{\partial^2 C_O(x, t)}{\partial x^2} \quad (29)$$

which is *Fick's second law* for linear diffusion. Other relations apply to different geometries.

In treating a typical electrochemical mass-transfer problem, the strategy is to use Eq. (29) to obtain the concentration profiles as functions of distance and time, then to use Eq. (26) to give the current stimulated by an applied potential program. Usually we must solve diffusion equations for both O and R

$$\frac{\partial C_O(x, t)}{\partial t} = D_O \frac{\partial^2 C_O(x, t)}{\partial x^2}, \quad \frac{\partial C_R(x, t)}{\partial t} = D_R \frac{\partial^2 C_R(x, t)}{\partial x^2} \quad (30)$$

thus we require six boundary conditions. Five of them are the same for almost all experimental situations. Typically only one of the species is present at the start of the experiment, and it is distributed uniformly; thus the *initial conditions* might be

$$C_O(x, 0) = C_O^*; \quad C_R(x, 0) = 0 \quad (31)$$

† See [5, 9, 12, 15, 16, 41, 42].

At distances far from the electrode (even millimeters), the solution composition is not changed by the experiment, so we also have the *semi-infinite conditions*

$$\lim_{x \to \infty} C_O(x, t) = C_O^*, \qquad \lim_{x \to \infty} C_R(x, t) = 0 \tag{32}$$

In addition, the *flux balance* at the surface holds [see Eq. (26)] so that

$$D_O \left. \frac{\partial C_O(x, t)}{\partial x} \right|_{x=0} + D_R \left. \frac{\partial C_R(x, t)}{\partial x} \right|_{x=0} = 0 \tag{33}$$

The sixth boundary condition is peculiar to the experiment and it determines the manner by which the potential program affects the concentration profiles, and thus the current. For example, a step to potential E would require a reversible system to adhere to Eq. (24) at all times after the step. We shall consider some specific examples below.

G. Double-Layer Structure

Electrochemical experiments are affected to a great degree by the arrangement of charges at the interface between an electrode and its adjacent solution, and in this section we consider some of the most important ideas about the details.

1. Models for the Double Layer†

One changes the potential of an electrode by changing the excess charge on it. Now we ask where that charge resides. Suppose no current flows through the electrode, so that no iR drops exist inside it. In that situation an excess charge cannot be found in the interior of the metal, because any such excess would generate fields that would rearrange the conduction electrons so as to eliminate the excess. Thus all of the excess charge q^M must reside on the surfaces of the metal. If the electrode contacts an electrolyte, the *charge densities* (charge per unit area) on the metal and in solution at the interface are equal and opposite,

$$\sigma^M = -\sigma^S \tag{34}$$

On both sides of the interface there is a competition between the electrostatic forces, which tend to hold the excess carriers (ions, electrons, or holes)

† See [1, 2, 9, 29, 45–49].

at the surface, and the thermal motion of the carriers, which causes diffusion and tends to eliminate the excesses. Thus the charged zones are three-dimensional in character and are sometimes called *space charge regions*. Their thicknesses depend on temperature and on the bulk concentrations of the carriers from which the excess charges must be accumulated. In a metal, there is a very large concentration of mobile electrons, and σ^M represents a small relative change, so the space charge thickness is negligible and σ^M can be regarded as a true surface charge. The ion concentrations in solution are typically much lower, and the assembly of σ^S affects the composition of the solution at significant distances (10–1000 Å) from the surface. In a semiconductor electrode the concentrations of charge carriers may be even smaller, hence the space charge on the electrode has appreciable thickness, perhaps approaching 1 μm in size [50–53].

The charge densities σ^M and σ^S are directly related to potential, and we can easily imagine that σ^M might be negative at very negative potentials and that it might be positive at the opposite extreme. Thus we can anticipate a *potential of zero charge* (PZC), E_z, where the charge excesses change sign. This is an important reference point for discussions of double layer structure, because it divides regions that are qualitatively different. On the positive side of the PZC there is an anionic excess in solution, but on the negative side cations predominate.

Most of our knowledge of double-layer structure comes from studies of surface tension (sensitive to σ^M) and capacitance (see below) at interfaces between mercury and aqueous solutions, whose behavior can be described and predicted in some detail by the Gouy–Chapman–Stern (GCS) theory [54]. The elements of the model are shown in Fig. 13. The electrode surface is viewed as being covered by water molecules polarized by the strong interfacial field (10^5–10^7 V/cm). Usually the bulk of σ^S is found in the *diffuse layer* of solvated ions distributed by thermal processes in solution. In the example shown here, the electrode is negative, so cations are in excess in the diffuse layer. These ions are said to be *nonspecifically adsorbed*, although they do not interact intimately with the electrode. In fact, they can only approach the electrode to a distance of one or two solvent diameters without desolvating. This distance, x_2, defines the *outer Helmholtz plane* (OHP), which is the inner boundary of the diffuse layer. One may also have species specifically adsorbed to the electrode by particularly intimate bonding, which may actually be strong enough to survive a coulombic force in opposition. For example, many anions are known to bind to mercury surfaces even at potentials more negative than the PZC. The locus of centers of these species defines the *inner Helmholtz plane* (IHP) at distance x_1. The layer itself is called the *Helmholtz, Stern*, or *compact layer*.

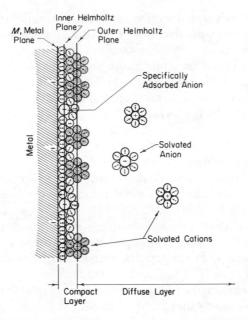

Fig. 13 Model for double-layer structure at a negatively charged electrode.

An array of charge like that in Fig. 13 generally produces a complex potential profile through the double layer, and its precise nature can affect the rates of electrode reactions. In addition, the adsorption of a species on a surface may block surface sites needed by an electrode process that involves adsorbed intermediates, or the adsorbate may catalyze an electrode reaction that would otherwise proceed slowly or not at all. Deliberate modification of electrode surfaces to promote designed *electrocatalysis* or *inhibition* is actually a major topic of current interest.† The range of influence of inter-facial structure is enormous and far exceeds our scope, so the interested reader is referred to the review literature for more details [1, 2, 9, 29, 45–54].

Much less is known about interfaces at solid electrodes such as Pt, Au, Ag, and C, than at mercury, but the general structure of the solution side is presumed to be similar to that at mercury. Solid electrodes have additional peculiarities associated with their usual crystalline structure, their roughness, and their tendencies to form oxides or other chemical modifications of the surface [1, 2, 6, 7, 56, 57]. These features also affect the faradaic chemistry that occurs at such electrodes, so work with them often involves careful

† See, for example, the articles listed in [55]. This list is intended to provide an entry into the literature. It is not exhaustive. The field has been quite active and has not been reviewed except in the last two references of [55].

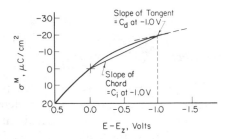

Fig. 14 Definitions of double-layer capacitance. [From A. J. Bard and L. R. Faulkner, "Electrochemical Methods." Wiley, New York, 1980, with permission.]

efforts to put the surface into a reproducible condition [2, 7, 17, 18]. (See also Section V.A.)

2. Interfacial Capacitance†

The double layer stores a charge that is monotonically related to potential, as shown in Fig. 14; thus the interface has a significant capacitance. It is not ideal, in the sense that the charge–voltage relation is nonlinear, but it offers an essentially ideal impedance small signals.

Because the σ^M–E relation is not a straight line (whose slope would be the capacitance), it is useful to think of the double layer capacity in two ways. For many experiments, we are interested in the small increment of charge density $d\sigma^M$ added by a small change in potential dE. The ratio is, in the limit, $d\sigma^M/dE$, which is the slope of σ^M versus E at any point. This is the *differential capacitance* C_d (usually in microfarads per square centimeter). At other times we may be interested in the total charge density σ^M divided by the total voltage inducing it, $E - E_z$. This ratio is the *integral capacitance* C_i (also usually in microfarads per square centimeter). The curvature in σ^M versus E implies that C_i is generally not equal to C_d and that both are functions of potential. Typical values are 10–50 $\mu F/cm^2$.

Whenever the area of the electrode, the potential, or the capacitance changes with time the charge on the double layer changes and current must flow to or from the electrode without causing actual electrooxidations or reductions. This current, in most experiments, is an interference to the observation of the faradaic current of interest, and very often it sets the limit of detection or the upper limit in speed with which an experiment can be carried out. Such *charging current* is of general concern in the design of electrochemical methods, and we shall have to think about it often.

† See [1, 2, 9, 29, 45–49].

IV. Methods

A. Purposes of Electrochemical Measurements

Electrochemical experiments are usually performed toward one of two ends: one wishes to *diagnose* the chemistry of the system, or one seeks to characterize the composition of a sample in the usual sense of *analysis*. For diagnosis, the points of concern are the identities of chemical species participating in electrode processes and the kinetics and thermodynamics of reactions that they undergo. Charge-transfer rates, homogeneous mechanisms, free energies of reaction, adsorption, and other such fundamental matters are the objects of the research. Techniques suited to them must provide for qualitative identifications of reactants and products, a general means for differentiating various types of chemical processes, a wide range of experimental time scales, and a good theoretical basis for interpretation. For analysis one is interested partly in qualitative identification, but primarily in quantitative determinations of concentrations; thus suitable techniques must be selective, sensitive, and precise. Flexibility in time scale and comprehensive exact theory are much less important. Since the experimental requirements of diagnosis and analysis are largely mutually exclusive, it is convenient to classify techniques along these lines. We follow such an approach below.

In this short chapter, it is impossible to cover the dozens of electrochemical methods available for characterizing chemical systems, so we limit our view in the next two sections to the most widely practiced diagnostic and analytical tools. Then we take up the growing body of hybrid methods featuring coordinated spectrometric and electrochemical aspects. Finally, we end with brief sketches of some techniques that deserve mention, but cannot be covered in detail. Generally we stress capabilities, rather than recipes for implementing specific measurements.

B. Diagnostic Techniques

1. Hydrodynamic Voltammetry: An Introduction to Concepts of Diagnosis†

Much of our discussion to this point has been based upon concepts of hydrodynamic voltammetry, which is the recording of current–potential curves at an electrode immersed in a stirred solution. Usually the stirring is steady, so that there is a constant flow of material past the electrode, which is

† See [5, 7, 9, 15, 40, 56–58].

a *microelectrode* of small area, so that electrolysis probes the solution but does not change its composition. The most popular approach is to use a rotating disk electrode (RDE) (see Section III.D.5; and inset of Fig. 10) [7, 9, 15, 40, 56–58]. Hydrodynamic voltammetry almost always involves a constant or very slowly scanned potential, so that the observed currents are steady-state values reflecting steady mass transfer. These aspects are important from an experimental standpoint, because they imply that charging currents are negligible and that long-term signal averaging can be applied to improve precision. Soluble electroactive species and products produce waves like those shown in Figs. 7–10, and the heights, positions, and shapes of the waves are the primary experimental information.

(a) *Wave Height*   For a reaction adhering to our usual simple O/R scheme [Eq. (13)], such as the reduction of rubrene to its anion radical (Fig. 10), Levich [40] has shown that the current at the rotating disk adheres rigorously to

$$i = nFAD_O^{2/3}v^{-1/6}\omega^{1/2}[C_O^* - C_O(0, t)] \tag{35}$$

where v is the *kinematic viscosity* of the solution (viscosity/density) and ω is the *angular rotation rate* ($2\pi \times$ rate of revolution).

Potential exerts influence through its effect on the surface concentration $C_O(0, t)$. When E is very positive with respect to $E^{0\prime}$, then species O is stable at the electrode. If only O is present in the bulk solution, then the surface concentration is the same as the bulk value, no electrolysis occurs, and $i = 0$. This is true for rubrene at potentials more positive than -1.3 V versus SCE (Fig. 10). On the other hand, negative potentials relative to $E^{0\prime}$ cause reduction of O, and sufficiently extreme values can cause $C_O(0, t)$ to be quite small, so that O is reduced as fast as it arrives. Then we have a limiting cathodic current, given by the *Levich equation*

$$i_{l,c} = nFAD_O^{2/3}v^{-1/6}\omega^{1/2}C_O^* \tag{36}$$

Likewise, for species R,

$$i = -nFAD_R^{2/3}v^{-1/6}\omega^{1/2}[C_R^* - C_R(0, t)] \tag{37}$$

$$i_{l,a} = -nFAD_R^{2/3}v^{-1/6}\omega^{1/2}C_R^* \tag{38}$$

where the minus signs arise from our current convention. The limiting anodic current $i_{l,a}$ would apply at positive potentials, where $C_R(0, t)$ is negligibly small.

In the example of Fig. 10, rubrene (A) serves as species O for the process $A + e \rightleftarrows A^{\bar{\cdot}}$, hence there is a clear $i_{l,c}$. However, species R *for this couple* (i.e., $A^{\bar{\cdot}}$) is not present in the bulk, therefore, anodic currents are not seen near $E^{0\prime}$ for the couple (~ -1.4 V). On the other hand, rubrene is species R for the process $A - e \rightleftarrows A^{\ddagger}$, and there is a clear $i_{l,a}$ at potentials positive relative to $E^{0\prime}$ for that couple (~ 1.0 V). Since the corresponding species O (i.e., $A^{\ddagger}$) is not present in the bulk, there is no $i_{l,c}$ for the reaction, $A^{\ddagger} + e \rightarrow A$. If both oxidized and reduced forms exist in the bulk, then one sees a composite response like the curves in Fig. 8.

(b) *Wave Position and Shape* The discussion in Section III.D, which concerned the general couple O/R, both stable and soluble, was based on just the sort of steady-state mass transfer that one finds at a rotating disk. Thus the conclusion that we reached there, and Figs. 6, 7, and 8, describe our expectations for equivalent conditions at the RDE.

If the charge-transfer kinetics are very facile, we can define a limiting case easily, because we have the nernstian relationship (24) between surface concentrations and potential. By rearranging and combining (35)–(38), one can express those surface concentrations in terms of the currents i, $i_{l,c}$, and $i_{l,a}$. Then by substitution into (24), the equation describing the current–potential curve is obtained

$$E = E^{0\prime} + \frac{RT}{nF} \ln \frac{D_R^{2/3}}{D_O^{2/3}} + \frac{RT}{nF} \ln \frac{i_{l,c} - i}{i - i_{l,a}} \tag{39}$$

Curve 4 in Fig. 8 follows this form.

The potential where the current is halfway between $i_{l,c}$ and $i_{l,a}$ is of special interest, because the third term of (39) goes to zero. This *half-wave potential* $E_{1/2}$ is then

$$E_{1/2} = E^{0\prime} + \frac{RT}{nF} \ln \frac{D_R^{2/3}}{D_O^{2/3}} \tag{40}$$

which is independent of the reagent concentrations. Since the diffusion coefficients of most electroactive species are restricted to a fairly narrow range (near 10^{-5} cm^2/sec in the usual solvents), the second term in (40) is generally no more than a few millivolts, and $E_{1/2}$ is a good approximation to $E^{0\prime}$.

If one member of the couple is not present in bulk solution, then a simple form of (39) arises. For example, with $C_R^* = 0$, $i_{l,a} = 0$, and we have,

$$E = E^{0\prime} + \frac{RT}{nF} \ln \frac{D_R^{2/3}}{D_O^{2/3}} + \frac{RT}{nF} \ln \frac{i_{l,c} - i}{i} \tag{41}$$

Curve 1 in Fig. 9 adheres to this relation, and the potential halfway up the cathodic wave is $E_{1/2}$, as defined by (41).

When the system shows sluggish charge transfer kinetics, an activation overpotential will be needed to drive current flow. The result is that the cathodic wave will be seen at a more negative values and the anodic part will be at more positive potentials relative to $E^{0\prime}$. Figures 8 and 9 show the effect.

Note that the wave is also more drawn out than in the reversible (facile kinetics) case. From the shifts in position relative to $E^{0\prime}$ and the shapes of the waves, one can extract the kinetic parameters k^0 and α [9, 15, 57].

Now let us suppose that we have just recorded a reduction wave for some species of interest, e.g., $Ru(bpy)_3^{2+}$ (where bpy = 2,2'-bipyridine) in acetonitrile. How do we diagnose the behavior? In other words, is $E_{1/2}$ close to $E^{0\prime}$ and are the kinetics facile, or not? The best bet is to plot E versus $\log[(i_{l,c} - i)/i]$ and check the slope. Equation (41) says that it will be $59.1/n$ mV (i.e., $2.3 \, RT/nF$) at $25°C$ for the reversible case. However, for quasi-reversible or irreversible behavior a larger figure (often $80/n$ to $150/n$ mV), reflecting the more drawn-out shape, is expected. If one knows the n value independently, this test is very useful and unambiguous for establishing the degree of reversibility. Note, however, that it cannot be used to establish *both* the reversibility and the n value. Prior knowledge of n might be available from coulometry (Sections IV.B.5 and IV.B.6) or from $i_{l,c}$ if something is known about D_R.

(c) Coupled Homogeneous Chemistry Very often the species O and R participate in other reactions in solution. The mechanism of Eqs. (1)–(4) is such as case. The existence of reaction steps involving O and R, other than the charge transfer reaction, quite generally perturbs the electrochemical response; thus we can use that response to diagnose the homogeneous chemistry to a certain extent [7, 9, 15, 57, 58]. A vast body of literature deals with the effects of various mechanisms and their diagnosis by electrochemical methods [9, 13, 15], including hydrodynamic techniques. We obviously cannot explore all mechanisms in any significant detail, but we can establish some of the basic ideas.

To put the problem in practical terms, consider the situation used above. We have dissolved $Ru(bpy)_3^{2+}$ in acetonitrile (with perhaps 0.1 M tetra-*n*-butylammonium fluoborate as supporting electrolyte) and recorded a reduction wave at the RDE. What does it tell us? We considered a diagnosis of charge-transfer kinetics above—under the assumption that O = $Ru(bpy)_3^{2+}$ and that both O and R were stable and soluble. Suppose O is not $Ru(bpy)_3^{2+}$, but instead is either something produced by a prior reaction of $Ru(bpy)_3^{2+}$ or something in equilibrium with that species. Or perhaps O and R adsorb, or R loses ligands or otherwise degrades. A great number of possibilities exist. Some might be ruled out on the basis of other knowledge. In this case, we know from its chemistry that $Ru(bpy)_3^{2+}$ is a very stable complex. Complications prior to the electrode reaction are unlikely. However we still need to identify R, test for its stability, and test for adsorption. Knowledge of the n value is extremely important in diagnostics because it provides the most important chemical connection between starting material and product, and because it is often important in numeric functions. These aspects were seen in part (b) just above. From hydrodynamic experiments it is hard to obtain n without ambiguity because it is usually coupled (as in $i_{l,c}$) with a diffusion coefficient, which generally is not known precisely. Thus we already see that we probably shall have to attack an unknown electrode process—the usual case in research—with several methods. For example, coulometry (Section IV.B.5) will tell us that $n = 1$, but it does not tell us whether the reduction might be

$$\text{Ru(bpy)}_3^{2+} + e \rightleftarrows \text{Ru(bpy)}_3^{+} \tag{42}$$

or

$$\text{Ru(bpy)}_3^{2+} + 2\text{BF}_4^{-} + e \rightleftarrows \text{Ru(bpy)}_2(\text{BF}_4)_2^{-} + \text{bpy} \tag{43}$$

or some other process.

The plot of E versus $\log[(i_{l,c} - i)/i]$ for the hydrodynamic voltammogram is then very useful. If, knowing n, we find a linear curve with a slope of $2.3\,RT/nF$ (59/n mV at 25°C), then the overall electrode process is reversible in a *comprehensive* sense. That is, charge transfer kinetics are rapid, and the species participating in the electrode reaction are either stable on the time scale of the experiment, or they exist in *rapidly established* and *fully maintained equilibria* with other species. In this particular example, reversibility holds, so we know that (42) is a possibility, but so is (43) if the ligand substitution process is sufficiently fast that all species are maintained at equilibrium.

Processes involving equilibria are quite common in electrochemistry [9, 10, 13, 14, 21, 26]. The most frequently encountered cases involve the coordination chemistry of transition metals, proton-based acid–base processes, solubility equilibria, dimerization, and keto–enol interconversion. The presence of an equilibrium generally affects the *position* of the wave, as measured by $E_{1/2}$, along the potential axis. For example, Ag^{+} can be plated onto an Ag RDE at about 0.5 V versus SCE, but in the presence of excess CN^{-}, the plating occurs at significantly more negative potentials. The shift occurs because silver in solution is present as Ag(CN)_2^{-}. The complexation stabilizes the oxidized form of the Ag(I)/Ag couple by lowering the free energy of that form. It is therefore energetically more difficult to reduce Ag(I), and $E_{1/2}$ shifts negatively. The extent of the shift is related to the free energy of stabilization, and thus to the stability constant for complexation. Likewise, protons stabilize the reduction product of p-benzoquinone. This molecule takes on two electrons *and* two protons. Protonation lowers the free energy of the product; hence reduction becomes easier as pH falls, and $E_{1/2}$ shifts positively. This general idea of interpreting wave shifts of reversible electrochemical systems in terms of stabilization or destabilization of the oxidized or reduced forms of a couple is extremely useful for understanding phenomena involving complexation, protonation, aggregation, and adsorption. Moreover, wave shifts in reversible systems in response to pH changes, concentration changes, or alterations in ligand concentration can provide a good deal of quantitative information about an electrode process, once the general chemical nature of the process is understood. In a reversible case, *kinetics are by definition so fast as to be unobservable*, so the available quantitative data are *thermodynamic* or *stoichiometric*. Stability constants, acid dissociation constants, ligand numbers, n values, and numbers of protons involved are typically derived from such measurements.

Most electrode processes are not reversible, either because the heterogeneous charge-transfer kinetics are slow, or because species in the electrode reaction engage in unidirectional chemistry in solution or on the electrode surface. The mechanism of Eqs. (1)–(4) is an example. An enormous variety of mechanisms exists in real chemical examples that have been observed electrochemically, and a good deal of theoretical work has gone into predictions of their effects on electrochemical responses [5, 7–18, 20, 21, 59, 60]. Many systems fit into a few general classes:

(*i*) *Preceding reaction (CE)* In this case a homogeneous *chemical* reaction supplies the species that undergoes the actual *electron transfer*:

$$Y \rightarrow O \tag{44}$$

$$O + ne \rightarrow R \tag{45}$$

Thus the letters CE are commonly used in the electrochemical literature to designate this scheme. Subsets include combinations in which (44) and (45) are reversible or not. A classic example involves formaldehyde in aqueous media [61, 62] where reducible formaldehyde $CH_2{=}O$ exists in equilibrium with an electroinactive hydrated form $CH_2(OH)_2$. The rate of reduction can be limited under some conditions by the rate of conversion of $CH_2(OH)_2$ to $CH_2{=}O$.

(*ii*) *Following reaction (EC)* In this scheme, the product of electron transfer is unstable and decays in a chemical step:

$$O + ne \rightleftarrows R \tag{46}$$

$$R \rightarrow X \tag{47}$$

where X is not electroactive. A typical example is the pattern of halide loss upon reduction [20, 21]

$$\tag{48}$$

In this particular case the benzyl radicals dimerize to form bibenzyl [63].

(*iii*) *The ECE and ECEC schemes* In a great many systems, the products of chemistry following the initial charge transfer are themselves electroactive at the potentials used for the initial step. A good example is the reduction of benzoyl peroxide to benzoate [64]

$$
\underset{\substack{\| \\ \text{PhCOOCPh}}}{\overset{O \quad O}{\| \quad \|}} + e \rightarrow [\text{PhCOOCPh}]^{\overline{\cdot}} \tag{49}
$$

$$
[\overset{O \quad O}{\underset{\text{PhCOOCPh}}{\| \quad \|}}]^{\overline{\cdot}} \rightarrow \overset{O}{\underset{\text{PhCO}}{\|}}\cdot + \overset{O}{\underset{\text{PhCO}}{\|}}^{-} \tag{50}
$$

$$
\overset{O}{\underset{\text{PhCO}}{\|}}\cdot + e \rightarrow \overset{O}{\underset{\text{PhCO}}{\|}}^{-} \tag{51}
$$

Process (51) occurs because benzoyl radical is easier to reduce than benzoyl peroxide, thus it is automatically eligible for reduction at the electrode at the potentials needed to initiate process (49). A general form of the ECE scheme can be written

$$ O + ne \rightleftarrows R \tag{52} $$

$$ R \rightarrow O' \tag{53} $$

$$ O' + n'e \rightleftarrows R' \tag{54} $$

Since O' is easier to reduce than O, one may have to contend with the homogeneous reduction of O' by R,

$$ R + O' \rightarrow O + R' \tag{55} $$

In the case of benzoyl peroxide, the importance of such a step would depend on the lifetime of $[\text{Ph(C=O)OO(C=O)Ph}]^{\overline{\cdot}}$, which probably is quite short.

Still more common than the ECE scheme is the ECEC case, in which R' in (54) undergoes decay to electroinactive product(s). The reduction of diphenylanthracene as outlined in steps (1)–(4) is a classic example [19–21].

(*iv*) *Catalytic schemes* Some following chemistry regenerates the original electroreactant. For example, Kuwana *et al.* [65] studied the reduction of molecular oxygen, as catalyzed by iron(III) tetra(4-N-methylpyridyl)-porphyrin [abbreviated Fe(III)TMP^{5+}]. In 0.05 M H_2SO_4 this ion can be reduced reversibly to a ferrous complex with $E^{0'}$ near -0.05 V versus SCE,

$$ \text{Fe(III)TMP}^{5+} + e \rightleftarrows \text{Fe(II)TMP}^{4+} \tag{56} $$

In the presence of dissolved oxygen, however, the ferrous species reacts to regenerate the ferric porphyrin and hydrogen peroxide,

$$ \text{Fe(II)TMP}^{4+} + \tfrac{1}{2}\text{O}_2 + \text{H}^+ \rightleftarrows \text{Fe(III)TMP}^{5+} + \tfrac{1}{2}\text{H}_2\text{O}_2 \tag{57} $$

The ferric species can then return to the electrode to gain another electron, which in turn is passed along to O_2. The porphyrin couple is acting as a true catalyst in this situation. It facilitates the *net* reduction of O_2 without being changed itself.

This type of scheme cannot be observed unless the follow-up reaction (57) is spontaneous. That in turn requires that O_2 be easier to reduce than $Fe(III)TMP^{5+}$, so one might ask why O_2 does not just become reduced directly at the electrode without the bother of shuttling electrons through the porphyrin. The answer is that O_2 is indeed eligible for reduction at less extreme potentials (more positive ones) than $Fe(III)TMP^{5+}$, but the kinetics of direct electroreduction are slow on the carbon electrode used in this work. Without the porphyrin, oxygen is not reduced significantly until the potential reaches ~ -0.4 V. The homogeneous reaction (57) is more facile. Thus the reduction of O_2 becomes possible when the reduction of $Fe(III)TMP^{5+}$ occurs, namely, at ~ -0.05 V. In other words the *homogeneous electro-catalyst* has effectively lowered the overpotential for the reaction by about 350 mV.

In general terms, an EC (catalytic) reduction can be written

$$O + ne \rightleftarrows R \tag{58}$$

$$R + Z \rightarrow O + X \tag{59}$$

where Z is the substrate and X is the net product. Since catalytic schemes facilitate the reduction of difficult substrates, there is a great deal of interest in them. The overpotentials required in the conventional reduction of these substrates are often very large and represent wasted energy; thus electro-catalysis afforts energy savings. Of particular interest is the catalysis of the reduction of O_2 to H_2O [65, 66], which is a very important process for the development of novel power sources.

(*d*) *General Strategy for Diagnosis: Ideas about Time*   Mechanisms all differ in their impact on the electrochemical response, (e.g., in hydrodynamic voltammetry at an RDE) and their differing characteristics can be used to diagnose them. For example, the most important feature of the EC scheme is that the immediate product of electrolysis is lost by reaction in the vicinity of the electrode, hence its operation is most easily detected through features of response that depend strongly on the concentration of that product. We shall point out some of them shortly. In contrast, the ECE and ECEC schemes are distinctive in that the following chemical reactions lead to additional electron transfer. The apparent n value of the process therefore depends on the extent to which the chemistry can proceed. Critical responses in detecting these mechanisms are those reflecting changes in n value (e.g., wave height) with changes in conditions, such as the concentration of a reagent inducing the following chemistry. Catalytic schemes are similar in that the following chemistry increases the apparent n, but very large n values can be observed because a molecule of the redox couple can shuttle through many catalytic cycles while it resides near the electrode. On the other hand, ECE and ECEC schemes go to a limiting value of $n + n'$ [in terms of (52)–(54)] as the chemistry is allowed to go to completion. The CE scheme differs from all these cases in controlling the electrochemical response by controlling the supply of the electroreactant. If the preceding chemistry is allowed to occur extensively during the time allotted to the electrochemical measurement, then much reactant is available and a large response (e.g., limiting current) is seen. Less extensive reaction leads to smaller response.

The important idea that results from these thoughts is that if we want to be able to make an active diagnosis of a mechanism, we must have some means for controlling the *extent* of any homogeneous chemistry associated with the electrode process. Our diagnostic basis must be the manner in which the extent of reaction affects various response characteristics. There are two general ways to control the extent of reaction. In many cases, there is the freedom to vary the concentrations of electroactive species and nonelectroactive agents. In addition, one can control the *time scale* of observation.

Understanding the experimental control of time is among the most important keys to electrochemical methods. By giving the various substances in a system a longer time to communicate with an electrode, one increases the available time for reaction, hence the extent of reaction and the effect of the reaction on the electrochemical response. We have already seen that current flow is a reflection of a reaction rate. The observation of current is a probe to the kinetics, and the length of time we make available for observing current controls the number and kinds of reactions that we can detect as part of an electrode process. All electrochemical techniques based on current flow have a controllable variable that manifests the time scale of observation, and it is extremely important to understand the effects of that time variable in each case. As we discuss several different approaches below, considerable stress will be laid upon this point.

In the case of the rotating disk electrode, the observational time scale is determined by the rotation rate. The hydrodynamic flow pattern shown in Fig. 10 shows that molecules are swept into the region of the disk from bulk solution and then outward, so that they have only transient communication with the disk. The effect of this limited time of communication is particularly easy to see for an ECE or ECEC system. Suppose a molecule of O is reduced to R. If that product reacts quickly to produce O′ before it is swept away, then additional charge transfer takes place, and the observed n value manifested in the wave height is $n + n'$. On the other hand, a slow conversion of R will not produce the O′ soon enough for it to interact with the electrode and the n' electrons will not be transferred. Now, for a given system, we might observe complete reaction and $n + n'$ electrons at low rotation rates, no reaction at all and n electrons at high rotation rates, and significant, but incomplete reaction with intermediate n values at rotation rates between the limits. From a plot of apparent n versus ω, we could even evaluate the rate constant for the homogeneous chemical step, once the mechanism had been diagnosed.

Practical rotating electrodes can be operated from a few rotations per second ($\omega \sim 30/\text{sec}$) to about 150 rps ($\omega \sim 1000/\text{sec}$). The time during which a molecule remains in communication with the electrode is on the order of $1/\omega$, thus this system has a variable time scale ranging from about 1 msec to several tens of milliseconds.

(e) *Reversal Experiments: The RRDE*† We have seen that much of the interesting chemistry associated with electrode reactions involves follow-up reactions of the initial product of charge transfer. The EC, ECE, ECEC, and EC (catalytic) schemes are examples. In order to detect the existence of this chemistry and differentiate among the various mechanisms, we need an experimental technique that offers direct monitoring of the initial product. Actually, electrochemistry offers a battery of such methods that as a group are called *reversal experiments* [7, 9, 11–13, 15, 17, 18]. Each experiment is divided into two phases. During the first, one initiates electrolysis of the sample in the direction of interest; for example, in the EC case this phase might involve a generation of reactive R.

$$O + ne \rightarrow R \tag{60}$$

$$R \rightarrow X \tag{61}$$

In the second phase the conditions are changed, so that the direction of electrolysis is reversed. That is, one seeks to convert R back to O by a reverse process (60). The current that flows as a result gives a direct measure of the amount of R that has survived the first phase, and that amount is in turn related to the extent to which reaction (61) has proceeded during the initial phase. The basic strategy, then, is to use a *forward phase* to electrogenerate a reagent, then to make a direct observation on that reagent during a *reversal phase*. The way in which the phases are sequenced and implemented varies with the type electrochemical approach being employed. We shall see several examples below.

In hydrodynamic voltammetry, one can accomplish reversal experiments at *a rotating ring-disk electrode* (RRDE), as depicted in Fig. 15 [5, 7, 9, 15, 56–58]. The flow pattern guarantees that a molecule interacting with the disk

† See [5, 7, 9, 15, 56–58].

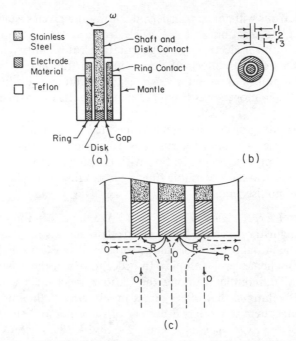

Fig. 15 Structure and operation of an RRDE. (a) Side cross section. (b) Face view. (c) Flow pattern in a collection experiment, where species O (present in the bulk) is reduced to R, which is collected at the disk. Note that some R escapes collection. Solid traces show paths for R, dashed ones for O.

will be swept past the ring, so we can plan to do the forward phase of the experiment at the disk [e.g., reaction (60)], then do the reversal at the ring. In order to accomplish this goal the ring and the disk must be operated as separate working electrodes. In the context of scheme (60) and (61), the disk would be set at a more negative potential than $E^{0'}$ (O/R) and the ring would be at a more positive value. Two potentiostats would be used to control the two respective electrodes.

This type of experiment is called a *collection experiment*, because a reagent is generated at the disk and collected at the ring. Since the current at the disk i_D is proportional to the rate of generation, and the current at the ring i_R is similarly proportional to the rate of collection, we can define a measurable collection efficiency

$$N = -i_R/i_D \tag{62}$$

One can show both theoretically and experimentally that for a simple electrode process like (60) with a stable product [i.e., a negligible rate con-

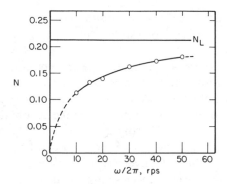

Fig. 16 Collection efficiency versus rotation rate for electrogenerated bromine in the presence of 8.69×10^{-4} M anisole. N_L is the limiting collection efficiency for stable species, which is 21.4% for the electrode used. The dashed extrapolation to $\omega/2\pi = 0$ shows that N tends toward zero at slow rotation speeds. In the limit of very large ω, N_L is approached. [Data from W. J. Albery, M. L. Hitchman, and J. Ulstrup, *Trans. Faraday Soc.* **64**, 2831 (1968).]

stant for (61)], the collection efficiency depends only on the geometry of the electrode. It can be calculated from knowing r_1, r_2, and r_3, as defined in Fig. 15. Typical values are 5–50%. This limiting N, which we call N_L, for a stable primary product, is an important characteristic of the system.

Now consider the effect of the EC scheme (60) and (61) on the measured value of N. If the following reaction (61) occurs negligibly, then R is effectively stable over the transit time required to move it from disk to ring, and N will approach the geometrically defined limiting value N_L. On the other hand, significant loss of R in transit by reaction (61) implies a correspondingly lower N. The transit time, which is the available reaction time, can be varied by changing the rotation rate ω, thus one would expect N to be small at low rotation rates and rise monotonically toward N_L as rotation rate is increased. The shape of this function can be predicted theoretically for various mechanisms involving following chemistry [57, 58], and it can be used to diagnose the mechanism and to determine a rate constant of a limiting reaction.

Figure 16 illustrates these points with some data reported by Albery *et al.* [67]. The chemistry involves electrogeneration of Br_2 at the disk and homogeneous reaction with anisole,

$$Br^- \rightarrow \tfrac{1}{2}Br_2 + e \tag{63}$$

$$Br_2 + CH_3OPh \rightarrow HBr + CH_3OC_6H_4Br \tag{64}$$

The ring collects the surviving Br_2 by reducing it back to Br^-. The expected trends in N with ω are nicely illustrated in the figure. The functional form can be used in conjunction with theory to evaluate the rate constant for (64) as 3.2×10^4 M^{-1} sec^{-1}. Taking the anisole concentration into account, one finds that this rate constant corresponds to a lifetime for Br_2 on the order of 30 msec.

Another useful aspect of reversal experiments is the ability to obtain qualitative information about species collected at the ring [57, 58]. For example, one can hold the disk at a potential where electrolysis occurs and scan the ring over a wide potential range to obtain a voltammetric overview of the products of electrolysis at the disk. The number, positions, and sizes of waves in the ring voltammogram can help in establishing the identities of those products. Conversely, the ring can be held at a constant potential, to monitor a single species, and the disk can be scanned. A plot of *ring* current versus *disk* potential shows the potential ranges where the collectible product is created at the disk. These kinds of experiments tend to highlight the chemical connections between various species that show up in the electrochemistry, and they are extremely useful for gaining an overall picture of the chemistry of a system.

This section on hydrodynamic methods is lengthy, because we have used our discussion as a vehicle for gaining a comprehensive view of the basic ideas involved in electrochemical experimentation. This body of concepts—about such things as kinetic reversibility, homogeneous coupled chemistry, reaction time scale, and forward and reversal phases—underlies *all* electrochemical investigation. Different investigative methods differ mainly in the style of perturbation applied to the system. They are all spotlights illuminating the same stage. The action that we see is the same regardless of which light is on, even though one of them may give us a clearer view than any of the others at a particular moment in the show. Thus the concepts that we have developed here must be carried over to our thinking about other methods below, even though they will not be discussed in such detail.

The strengths of hydrodynamic methods include (a) the ability to make measurements at steady state, so that superior signal-to-noise ratios can be obtained and double-layer charging currents are eliminated, (b) continuous supply of unelectrolyzed solution to the electrode surface, so that the voltammetry at the disk provides an accurate picture of the composition of the bulk, unmodified by prior electrolysis, and (c) simultaneous observations in the forward and reverse phases (at ring and disk), which makes the linkages between chemical species easy to delineate. The drawbacks are (a) a relatively bulky and complex mechanical system that is sometimes difficult to use with nonaqueous solvents, (b) the need for a dual potentiostat, (c) a requirement for critical machining in the fabrication of electrodes, especially RRDEs, and (d) a limited range of rotation rates, implying a limited range of time scales.

2. Chronoamperometry and Chronocoulometry†

To this point, we have considered only steady-state systems in which fresh electrolyte is continuously introduced to the electrode. Most electrochemical methods, on the other hand, involve *stationary electrodes*. Two important modifications in our thinking must now appear in these cases. (a) We must remember that diffusion is usually the only important mode of mass transport in properly designed systems. (b) We must understand that there is a tendency for electrolysis to exhaust the diffusion layer of electroreagents and to promote the buildup of electroproducts near the electrode. The composition of the diffusion layer is the product of the whole prior history of the system. Renewal can be accomplished by disconnecting the electrode or by returning its potential to a value where the bulk is not electroactive, then stirring for a moment, and finally allowing the system to settle down again.

(a) A Potential Step Consider a system involving a planar electrode, such as a Pt disk, that has been renewed as described above. Suppose species O is present in the bulk at concentration C_O^* and species R is not present. Then, as long as we hold the electrode at a potential much more positive than $E^{0\prime}(O/R)$, the solution coexists with the electrode without electrolysis and current does not flow. Now at some time t that we call zero, let the potential change abruptly to a value much more negative than $E^{0\prime}(O/R)$. What is the effect of this *forward step*?

The essential results are summarized in Fig. 17. The step has adjusted E to a point where O cannot coexist with the electrode, and those molecules of O present at the surface are reduced to R. A gradient in the concentration of O develops, and O tends to diffuse from the bulk toward the surface. When it arrives there, reduction occurs immediately, hence the current flow is controlled by the rate of arrival, which in turn is controlled by the gradient in concentration at the interface (i.e., the slope of the concentration profile shown in Fig. 17b). (See also Section III.F.)

As the process continues, the region near the electrode becomes depleted of O, as shown by successive profiles in Fig. 17b and the rate of diffusion to the electrode slows down, because the slope of the profile at the interface becomes shallower. Thus the current falls with time. Exact treatment of the problem [9, 13, 15] gives the *Cottrell equation* [68] as a description of the current–time relationship

$$i = nFAD_O^{1/2}C_O^*/\pi^{1/2}t^{1/2} \tag{65}$$

† See [7, 9, 12, 13, 15].

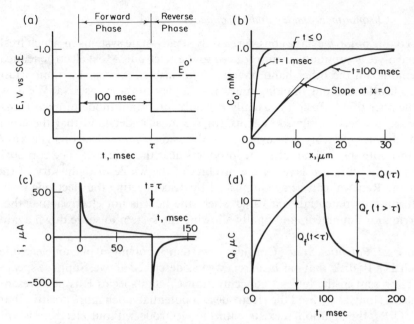

Fig. 17 Various aspects of double-step chronoamperometry and chronocoulometry performed on a 1-cm² electrode in a solution of 1 mM species O having a diffusion coefficient 1×10^{-5} cm²/sec and undergoing reduction to R with $n = 1$. The formal potential is at -0.57 V versus SCE. (a) Potential program. Initial and final potentials at 0.0 V; step potential at -1.0 V; $\tau = 100$ msec. (b) Concentration profiles for species O. (c) Chronoamperometric transients. (d) Chronocoulometric transients. This curve includes $Q_{dl} = 2\mu C$, corresponding to an interfacial capacitance of 20 μF/cm.

Likewise, the concentration profiles for O and R can be described exactly

$$C_O(x, t) = C_O^* \, \mathrm{erf}\left[x/2(D_O t)^{1/2}\right] \tag{66}$$

$$C_R(x, t) = C_O^*(D_O/D_R)^{1/2} \, \mathrm{erfc}\left[x/2(D_R t)^{1/2}\right] \tag{67}$$

where the *error function*, erf, and the *error function complement*, erfc, are standard mathematical functions frequently encountered in diffusion problems. Figure 17b shows clearly that the profiles thicken as electrolysis proceeds. There is no well-defined boundary between the *diffusion layer* [where $C_O(x, t) \neq C_O^*$] and the bulk [where $C_O(x, t) = C_O^*$], but the thickness is always of the order of $(D_O t)^{1/2}$ and the whole diffusion layer can, for practical purposes, be regarded as contained within a distance $6(D_O t)^{1/2}$ from the interface. For typical $D_O = 10^{-5}$ cm²/sec, this distance is $6 \times 10^{-4}, 6 \times 10^{-3}$, and 6×10^{-2} cm at $t = 10^{-3}, 10^{-1}$, and 10 sec, respectively. Thus the experiment does not affect the solution at a very large distance from the electrode.

Chronoamperometry is the recording of current–time curves as responses to steps of this sort. Almost always the step is of such magnitude that one begins at a potential where the system is electroinactive and moves to a step potential where the concentration of the electroactive species is driven effectively to zero at the electrode surface. *Chronocoulometry* [9, 15, 69] involves the recording of the integral of the current with time. Thus we predict the response in the stable (O/R) system by integrating (65)

$$Q = 2nFAD_O^{1/2}C_O^* t^{1/2}/\pi^{1/2} \tag{68}$$

(b) *Reversal Experiments* [9, 15] The most useful applications of chronoamperometry and chronocoulometry involve the double-step mode outlined in Fig. 17, frames (a), (c), and (d). The forward step begins at $t = 0$ and lasts for a period τ. Then for the reversal phase the potential is returned to the starting point. For stable O and R the forward part of the experiment proceeds exactly as we discussed above. The diffusion layer that exists at $t = \tau$ is then the basis on which the reversal operates. Note that R exists then at the surface in high concentration. An abrupt return to the initial potential, where O is stable at the electrode, causes reoxidation of R at the surface. This event introduces a concentration gradient that causes R to diffuse back to the electrode for reoxidation upon arrival. Thus the anodic current shown in Fig. 17c for $t > \tau$ reflects this recollection of R. An exact treatment of the problem is readily obtained [9, 15, 70].

The shapes of the i–t or Q–t curves are sensitive to coupled homogeneous chemistry [9, 15, 71]. For example, the size of the reversal current is an indicator of the persistence of the primary product of electrolysis and is useful for characterizing follow-up chemistry. On the other hand, the current in the forward phase can be useful for distinguishing among mechanisms, such as ECE, ECEC, or EC catalytic, in which follow-up chemistry leads to additional electron transfer.

The time scale of the experiment is τ, the width of the forward step, which defines the period over which any coupled chemistry can manifest itself. One of the strengths of chronoamperometry and chronocoulometry is the flexibility provided by an enormous range of time scales. The value of τ may be as short as 10 μsec or as long as several seconds. Very fast experiments ($\tau < 1$ msec) are difficult for reasons discussed below, but a four-order span in τ is readily available.

In general, these methods are excellent for the evaluation of rate constants, once the mechanism of an electrode process is known, because a wide span of τ's can be employed and because the perturbation is virtually the simplest possible one, so that theoretical connections between i–t curves and rate parameters are about as straightforward as they can be in electrochemistry. Chronoamperometry and chronocoulometry are not particularly useful for qualitative diagnosis of a mechanism, because one gains no view of different species that might become active in various potential regions. That is, there is no means for gaining an overview of the system like that afforded by a voltammetric method, such as hydrodynamic voltammetry, cyclic voltammetry, or polarography. However, the precision of measurements by these methods is often quite good, and they can be useful for distinguishing closely-related mechanisms [71, 72], such as the various types of ECE processes. Since large oversteps in potential are used (in the interest of simplifying the mass transfer problem by forcing the surface concentration of the electro-reagent to zero), charge transfer kinetics are automatically driven by a sufficient overpotential to make them fast, and therefore nonlimiting. Other methods, with smaller step amplitudes, have been devised for measurements of charge-transfer parameters [9, 11–13, 15].

(c) *Adsorbates Investigated by Chronocoulometry*† The discussion of the current–time curve above is only partly complete, because we have considered only current flow involved in oxidizing or reducing species diffusing to the electrode. At least two sinks for current must be added.

In any experiment one must supply the charge needed to change the double layer structure from the configuration at the initial potential to that at the step value. Current must flow as the step rises when this charge is injected. The double layer charge is the means by which potential is determined, so potential rise and double-layer charging are congruent acts. A charge Q_{dl} is required in the forward step, and that same charge is withdrawn in the reversal [69, 73, 74]. Since the potential rise is instantaneous, Q_{dl} is injected or withdrawn very quickly, then the time-dependent charge from diffusing species, e.g., as in (68), is added to this constant value.

An additional component of charge is that devoted to the oxidation or reduction of species that may be adsorbed on the electrode. Since they are electrolyzed in the first instant of a step, they contribute a second constant charge that adds to Q_{dl} in forming a base from which the charge from diffusing species rises [69, 73, 74].

We therefore write that the total charge in the forward step of the simple experiment discussed above is

$$Q_f = Q_{dl} + nFA\Gamma_O + (2nFAD_O^{1/2}C_O^*/\pi^{1/2})t^{1/2} \tag{69}$$

If R is stable, the reversal charge $Q_r(t > \tau) = Q(\tau) - Q(t > \tau)$ can be shown to be approximately [9, 15, 69]‡

$$Q_r = Q_{dl} + nFA\Gamma_R + (2nFAD_O^{1/2}C_O^*/\pi^{1/2})[\tau^{1/2} + (t - \tau)^{1/2} - t^{1/2}] \tag{70}$$

---

† See [9, 15, 69, 73, 74].

‡ An exact treatment takes into account modifications in the diffusion layer caused by adsorption and desorption of O and R [73].

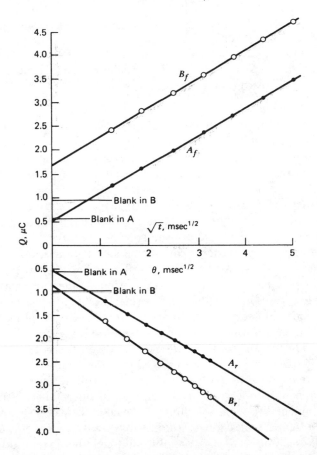

Fig. 18 Plots of data from chronocoulometric experiments. Potential initially at -0.200 V versus SCE, then stepped to -0.900 V and returned to -0.200 V. A: 1 mM Cd(II) in 1 M NaNO$_3$. B: 1 mM Cd(II) in 0.2 M NaSCN and 0.8 M NaNO$_3$. Blanks are values of Q_{dl} in equivalent solutions free of Cd(II). A_f and B_f are Q_f versus t; A_r and B_r are Q $(t > \tau)$ versus $\theta = \tau^{1/2} + (t - \tau)^{1/2} - t^{1/2}$. [From F. C. Anson, J. H. Christie, and R. A. Osteryoung, *J. Electroanal. Chem.* **13**, 343 (1967), with permission.]

In these two relations Γ_O and Γ_R are the *surface excesses* describing the extent of adsorption in mol/per square centimeter. Figure 17d shows the definitions of the various Q's in a graphical form.

These two equations suggest that if we plot Q_f versus $t^{1/2}$ and Q_r versus $[\tau^{1/2} + (t - \tau)^{1/2} - t^{1/2}]$, we should obtain two straight lines of equal slope. Their intercepts should differ by $nFA(\Gamma_O - \Gamma_R)$. In any given system we may have knowledge that one of the Γ's is zero, hence this method affords a convenient means for studying adsorption by measuring surface excesses. Figure 18 provides some data reported by Anson and co-workers [75] that clearly demonstrates that Cd(II) in $NaNO_3$ is not adsorbed at Hg, but that thiocyanate induces adsorption.

Chronocoulometry has become a rather powerful tool for the examination of surface-confined species, and its advent made possible rapid advances in understanding of adsorption mechanisms [49].

(*d*) *Experimental Limitations*† It is appropriate now to consider the limitations on the speed of electrochemical experimentation. In most cases the fundamental problem is that no potential change can be manifested without charging the double layer, and it takes a finite time to do that job. In electrical terms, one must charge a capacitance of perhaps 20 μF/cm^2 through the uncompensated solution resistance between the reference electrode and the working electrode. For a typical working electrode area of 0.1 cm^2 and an uncompensated resistance of 100 Ω, the cell displays an RC time constant of 200 μsec. Several time constants are required to complete any change in potential, hence measurements on a time scale shorter than about 1 msec would have little or no chemical meaning. Even worse situations can easily exist in resistive media. Fast experiments therefore require small electrodes, low resistance media, and careful cell design. Sometimes electronic compensation schemes can reduce the effective uncompensated resistance and usefully improve the time constant [17, 18, 39].

One also must recognize the power demands on the potentiostat. If the electrode cited above were to be changed in potential by 1 V in 1 μsec, an average current of 2 A would have to flow. Controlling such large currents in fast instruments is not trivial. These simple considerations illustrate some of the experimental problems in doing fast electrochemistry. In general, experiments on time scales longer than 1 msec are not so difficult. Work between 10 μsec and 1 msec is possible, but requires careful attention to apparatus. Work below 10 μsec is extremely difficult and is only possible in special situations.

Chronoamperometry and chronocoulometry are basically two faces of the same experiment. Making a choice between them comes down to experimental matters. Usually chronocoulometry holds the advantage because it preserves early-time events through integration, and because the act of integration improves the signal-to-noise ratio.

3. *Polarography*‡

An object of great importance to the development of electrochemistry is the dropping mercury electrode (DME). Heyrovsky's [77] application of this

† See [9, 17, 18].
‡ See [8–16, 62, 76].

device was ultimately recognized by a Nobel Prize. Its design is simple [62, 76]. A pressure head of mercury is used to drive the liquid metal through a fine capillary into a test solution. A droplet forms at the orifice and grows as mercury flows continuously into it. It remains in place as long as surface tension can support its weight, but eventually it becomes too heavy and falls, to be replaced with a new drop. The lifetime of the drop, called the drop time t_{max}, is dependent on the flow rate of mercury, the radius of the capillary, and the surface tension of the mercury/solution interface. Typical values are 2–4 sec. The growing drop is the working electrode.

Standard polarographic experiments involve recording current flow at the drop as a function of time. Usually the potential is swept at a low rate (1–3 mV/sec) from a positive limit to a negative one, and a signal proportional to current flow is fed continuously to a recorder. The term *polarography* has been reserved to mean voltammetry at the DME.

A typical polarogram is shown in Fig. 19. A characteristic feature is the oscillatory behavior of the current, which is derived from the growth and fall of the drop. Since the rate of potential change is slow, one can regard each drop's lifetime as a period of constant-potential control. Moreover the solution near the electrode is stirred when the drop falls so that the next drop is born into a freshly renewed solution. It is convenient to think about each drop as undergoing a potential step at the moment of its birth to the essentially constant potential that it experiences during its lifetime. In this way we consider each drop's duration to contain an individual forward-step experiment like that discussed in Section IV.B.2. As electrolysis takes place, we can expect a diffusion layer to build up and the electroreactant to become depleted. At the stationary electrode, this effect led to a continuous decrease

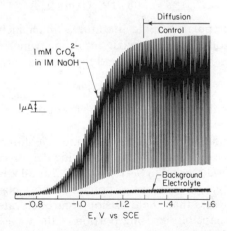

Fig. 19 Polarogram for the chromate system.

in current with $t^{1/2}$. However, the DME is not stationary. The drop grows continuously, and its expansion in area overrides the depletion effect, so that the current *rises* slowly with t until t_{max} is reached, then it falls because there is a sharp decrease in area [8–16, 62, 76]. This pattern of slow growth and abrupt fall is repeated again and again in the polarogram.

When the potential becomes sufficiently negative, the reducible species cannot coexist with the electrode, and its concentration is driven essentially to zero at the surface, as discussed above in Section IV.B.2. The current is then controlled by the rate of diffusion of the electroreactant to the surface. This is not a potential-dependent phenomenon; therefore essentially identical current growth curves are observed, regardless of potential, as long as the value is negative enough to enforce a zero surface concentration. This condition holds in Fig. 19 at potentials more negative than −1.3 V versus SCE. Since the same current oscillations are reproduced regardless of potential in this range, the polarogram shows a plateau (having a gentle slope from the underlying background currents). Currents there are said to be *diffusion-limited*.

The Ilkovic equation [8–16, 62, 76, 78] gives a good approximation to the current–time oscillations when diffusion control is achieved,

$$i_d = 708 n D_O^{1/2} C_O^* m^{2/3} t^{1/6} \tag{71}$$

where m is the mercury flow rate and t is measured from the birth of the drop. The constant 708 is a combination of factors evaluated for 25°C, and it implies that i is in amperes when the following units are used: $D_O^{1/2}$ (square centimeters per second), C_O^* (moles per cubic centimeter), m (milligrams per second), t (seconds). The current at the end of the drop's life is therefore

$$(i_d)_{max} = 708 \, n D_O^{1/2} C_O^* m^{2/3} t_{max}^{1/6} \tag{72}$$

Much of the older polarographic literature, which includes many useful measurements, involved galvanometric recording, which did not follow the current oscillations, but rather registered the average current,

$$\bar{i}_d = 607 \, n D_O^{1/2} C_O^* m^{2/3} t_{max}^{1/6} \tag{73}$$

In either case the *wave height*, $(i_d)_{max}$ or $\bar{i}_d$, is proportional to C_O^*, which can make it useful for analysis.

When the potential is insufficiently extreme to enforce diffusion control then the reducible species can coexist to varying degrees with the electrode, in much the manner discussed in Sections III.D.2 and IV.B.1. In the example of Fig. 19, chromate is not reduced at potentials more positive than −0.7 V and it is reduced at less than the diffusion controlled rate between −0.7 V and −1.3 V. The shape of the rising portion of the wave can be used diagnostically in exactly the manner discussed for hydrodynamic voltammetry in Section IV.B.1. [8–16, 67, 76]. In fact, for a reversible system, both techniques yield the same wave shape. In polarographic terms for $C_R^* = 0$

$$E = E_{1/2} + \frac{RT}{nF} \ln \frac{i_d - i}{i} \tag{74}$$

where $E_{1/2}$ differs only slightly between the two approaches. Here

$$E_{1/2} = E^{0'} + \frac{RT}{nF} \ln \frac{D_R^{1/2}}{D_O^{1/2}} \tag{75}$$

Plots of E versus $\log[(i_d - i)/i]$ are useful in polarography for essentially the same purposes as in hydrodynamic voltammetry.

A large volume of literature [8–16, 62, 76] describes the application of polarographic methods to systems with kinetic complications, and many data are available on actual chemical systems. Diagnosis is carried out by examining wave shape and wave height as functions of reagent concentrations and experimental time scale. Essentially, the time scale is the drop time, which is varied through the height of the mercury column supplying the pressure head.

A simple datum that can prove quite useful, especially in estimating n values, is the diffusion current constant [62, 76, 79], which is defined by rearranging either (72) or (73). For example,

$$(I)_{\max} = (i_d)_{\max}/m^{2/3} t_{\max}^{1/6} C_O^* = 708 n D_O^{1/2} \tag{76}$$

All the variables that ordinarily might change significantly from experiment to experiment have been used to normalize $(i_d)_{\max}$ to produce a number $(I)_{\max}$ that depends only on n and D_O. Since D_O does not usually vary a great deal from species to species, the size of $(I)_{\max}$ tends to be a useful indicator of n. In aqueous solutions values are 1.5–2.0 for $n = 1$, 3.0–4.0 for $n = 2$, and 4.5–6.0 for $n = 3$. In diagnosing an unknown process, a reliable estimate of n can be most valuable.

The DME is also useful for detecting adsorption [9, 45–48]. Since this process alters the interfacial tension between the Hg and the solution, the drop time changes when an adsorbable species is added. Plots of drop time versus potential (*electrocapillary curves*) recorded in blank electrolyte and in a test solution are very helpful for defining regions of potential where components of the test solution adsorb. Much of our knowledge about double-layer structure (Section III.G) comes from such experiments.

Even though polarography once held a preeminent position among electrochemical methods, it has serious drawbacks and has gradually been supplanted by other methods. Its problems include (a) a very limited positive working range (dictated by the oxidation of Hg near 0 V versus SCE), so that very few interesting oxidations can be studied at the DME, (b) in-applicability to any other electrode material of much interest, (c) a very restricted range of time scales (i.e., drop times), (d) a very complex mass-transfer problem, which renders theory extremely difficult, (e) the physical bulk and complexity of the DME, and (f) the lack of a companion reversal technique. Its strong points are (a) an exceptionally wide range of negative potentials, made possible by the high overpotentials of common discharge reactions on Hg, (b) excellent reproducibility, made possible by the continuous renewal of the mercury surface through a precise mechanical action, and (c) automatic renewal of the diffusion layer after each drop, so that the voltammogram characterizes bulk concentrations undistorted by the prior history of electrolysis. The strengths of polarography lend themselves better to analysis than to diagnosis. Likewise, the drawbacks particularly the limited time scale and lack of reversal are more serious for diagnosis, so most current applications of polarography and related techniques are in the analytical area. (See Section IV.C.1.)

4. Cyclic Voltammetry†

An extremely useful means for gathering information about an electrochemical system is to use a stationary working electrode and to apply a potential sweep with reversals, like that shown in the inset to Fig. 20. Current

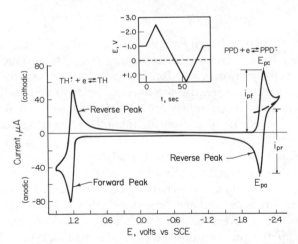

Fig. 20   Cyclic voltammogram of 1 mM thianthrene (TH) and 1 mM 2,5-diphenyl-1,3,4-oxadiazole (PPD) in CH_3CN/0.1 M TBAP. Inset shows potential program. Note that the reverse peak current (i_{pa} for $PPD^{\overline{\cdot}}$ or i_{pc} for $TH^{\dot{+}}$) is measured with respect to the forward current decay, which is extrapolated in *time* and therefore is folded at the switching potential.

† See [7, 9, 13, 15, 80].

is then recorded as a function of potential to produce a record like that in the main part of the figure. This experiment is an example of cyclic voltammetry, which is the most powerful and popular of electrochemical diagnostic tools.

The particular system used in Fig. 20 involved 1 mM thianthrene (TH) and 1 mM 2,5-diphenyl-1,3,4-oxidiazole (PPD) in CH_3CN with 0.1 mM tetra-n-butylammonium perchlorate as supporting electrolyte. The working electrode was a small Pt disk. As usual in cyclic voltammetry, the solution was not stirred, except to renew the diffusion layer between scans.

The response in Fig. 20 is interpreted as follows. As the potential moves from -1.0 V to -2.2 V, very little current flows because nothing in the solution is electroactive in this range. Beyond -2.2 V there is a cathodic current rise as PPD is reduced to $PPD^{\overline{\cdot}}$, an anion radical. As the scan continues, a peak is observed at -2.3 V, then the current falls because electrolysis has depleted the region near the electrode of PPD. Thus the rate of its diffusion to the electrode slows down, and the current reflects the corresponding decreased rate of reduction. Reversal of the scan at -2.5 V causes the potential to move positively again, and one soon returns to the region (~ -2.2 V) where the electrode became capable of reducing PPD to the anion radical. At still more positive values, the $PPD^{\overline{\cdot}}$ that was just generated is reoxidized by the electrode, so an anodic peak develops. The anodic current eventually decays to zero (i.e., as the scan continues from -2.2 V toward positive potentials), because all of the $PPD^{\overline{\cdot}}$ in the *diffusion layer* near the electrode is collected back again by diffusion.

As the scan in Fig. 20 continues between -1.8 V and 1.0 V versus SCE, there is again a region of no electroactivity, but at $+1.1$ V an *anodic* current appears as TH is oxidized to TH^{+}, its cation radical. A peak again appears as the diffusion layer is depleted of TH, and reversal of the scan produces a *cathodic* peak as TH^{+} is recollected from the diffusion layer and reduced. When the scan returns to the starting potential of -1.0 V, the region near the electrode has been restored essentially to its initial condition.

One of the most useful features of cyclic voltammetry is its ability to generate a potentially reactive species and then to examine it immediately by reversal. The responses in Fig. 20 are for two systems, TH and PPD that produce completely stable ion radicals. One can tell that by examining the ratio of reverse (r) and forward (f) peak currents. For example, for the PPD/$PPD^{\overline{\cdot}}$ peaks, i_{pf} is the forward peak current, and i_{pr} is the reverse. For a stable product $i_{pr}/i_{pf} = 1$ [80]. Note from Fig. 20 how the reverse peak current is measured. Consider the contrasting results if $PPD^{\overline{\cdot}}$ had not been completely stable, but decayed to an electroinactive product. Then, the anodic peak would be smaller because less $PPD^{\overline{\cdot}}$ would have survived in the diffusion layer until reoxidation could take place. If none had survived, because the following reaction was very fast, then there would be no current peak on reversal and $i_{pr}/i_{pf} = 0$. Even so, at a faster scan rate, one might still be able to catch a glimpse of the PPD$^{\bar{}}$ in the form of a measurable i_{pr}. At a very fast scan rate, one might even be able to see PPD$^{\bar{}}$ essentially as a stable species with $i_{pr}/i_{pf} = 1$, because one would then generate and re-oxidize the species before it had a chance to decay at all.

Scan rate is the important time-related variable in cyclic voltammetry [9, 13, 15, 80, 81], because it defines the elapsed time between generation of a product in the forward phase and collection of it in reversal. It can extend from ~ 10 mV/sec as high as 2000 V/sec with fast potentiostatic and recording apparatus. With x–y recorders, the practical range extends only to about 1 V sec.

The shape of the voltammogram depends strongly on the mechanism of the electrode process, and this aspect gives cyclic voltammetry its diagnostic utility. We cannot cover many cases here, but we shall consider the reversible case [80], in which the kinetics of electron transfer to and from the electrode are very fast and the oxidized and reduced forms are both stable. In this instance, the average of E_{pf} and E_{pr} is essentially $E^{0'}$, and E_{pf} and E_{pr} are separated by ~ 60 mV/n. Both peak potentials are independent of scan rate and concentration. The forward peak current (e.g., for PPD reduction or for TH oxidation) adheres to the relation

$$i_{pf} = (2.69 \times 10^5)n^{3/2}AD^{1/2}v^{1/2}C^* \tag{77}$$

where the current is in amperes when A is in square centimeters, D is in square centimeters per second, C^* is in moles per cubic centimeter, and the scan rate v is in volts per second. The constant is for $T = 25°C$. The important points are the linear dependences on $n^{3/2}$, C^*, and $v^{1/2}$. The inverse square root of time dependence embodied in the linearity with $v^{1/2}$ reflects control of the peak current by diffusion of reactants to the electrode.

The parameters E_{pf}, $|E_{pf} - E_{pr}|$, i_{pr}/i_{pf}, and $i_{pf}/(v^{1/2}C^*)$ are often evaluated as functions of scan rate for diagnostic purposes. Different patterns of behavior reflect different mechanisms [9, 13, 15, 80, 81]. For example, the current ratio i_{pr}/i_{pf} is very useful for detecting following chemistry, such as EC, ECE, or ECEC processes. The so-called "current function," $i_{pf}/(v^{1/2}C^*)$ tends to reflect the effective n value, so it is a sensitive indicator of following chemistry leading to further electron transfer, as in the ECE, ECEC, and EC catalytic cases. The peak separation, $\Delta E_p = |E_{pf} - E_{pr}|$, is sensitive to the kinetic facility of heterogeneous electron transfer [82]. Species that do not have to diffuse to the electrode, such as adsorbates or groups chemically bound to the surface, tend to show sharp peaks and $i_p \propto v$ [9, 13, 15, 83]. Extensive theory is available for the mechanistic cases frequently encountered in electrochemistry, and some of it can be used to evaluate parameters such as rate constants and α values [9, 13, 15, 80–83].

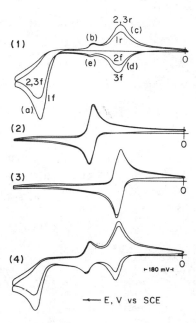

Fig. 21   Cyclic voltammograms of aniline and related compounds at pH 2.3. (1) aniline, (2) benzidine, (3) N-phenyl-p-phenylenediamine, (4) mixture of the three compounds. The working electrode was carbon paste. All scans start at 0.0 V and first move positively. The numbers on the traces refer to cycle numbers and the letters r and f label forward and reverse scans. [Adapted with permission from J. Bacon and R. N. Adams, *J. Am. Chem. Soc.* **90**, 6596 (1968). Copyright 1968, American Chemical Society.]

A very useful aspect of cyclic voltammetry is the electrochemical overview that it provides for a reaction system. This point becomes clear by considering Fig. 21, which deals with the oxidation of aniline in an aqueous system [84]. Part (1) shows that on the first scan toward positive potentials, one sees only a single irreversible oxidation peak, (a). Reversal, however, reveals two reduction waves (b) and (c) that probably can be assigned to the products of the reactions undergone by aniline after its electrooxidation. A second cyclic scan, carried out without stirring, confirms these ideas. The two reduction peaks of products turn out to be at least chemically reversible, in offering oxidation peaks (d) and (e). Continuing the positive scan brings the electrode back to peak (a), which is now smaller than before, because the chemistry following the first electrooxidation permanently depleted the diffusion layer of aniline. Reversal brings us back to peaks (b) and (c) which are *larger* than before because the products responsible for them are building up in the diffusion layer.

From this experiment, we can be fairly sure that the basic process follows the general scheme

$$\text{Aniline} - ne \rightarrow \text{Primary Product} \tag{78}$$

$$\text{Primary Product} \rightarrow \text{Product I} + \text{Product II} \tag{79}$$

where Products I and II are the species detected voltammetrically. The positions of the peaks corresponding to them can be useful in their identification. Parts (2) and (3) of Fig. 21 are voltammograms for benzidine and N-phenyl-p-phenylenediamine, both of which are oxidized reversibly at potentials corresponding closely to those characterizing Products I and II. These molecules are sensible products of the oxidation of aniline. One can invoke a process such as

$$2 \left[\underset{}{\overset{NH_2}{\bigcirc}} \right]^{+\cdot} \rightarrow \bigcirc -\underset{H}{N} - \bigcirc -NH_2 + 2H^+ \tag{80}$$

to produce N-phenyl-p-phenylenediamine, which is more easily oxidized than aniline. It therefore reacts at the electrode when the potential is in the region of wave (a),

$$\bigcirc -\underset{H}{N} - \bigcirc -NH_2 \rightarrow \bigcirc -N = \bigcirc = NH + 2H^+ + 2e \tag{81}$$

and the resulting diimine is then seen in reduction in the following negative scan as the larger product peak. The mechanism producing it is an ECE type, with aniline cation radical as a primary product. The benzidine is formed similarly, by tail-to-tail coupling of the aniline radical cations [84].

Cyclic voltammetry's chief strengths are (a) applicability to a wide range of electrode materials, including Hg (as a hanging drop), C, Au, Pt, and a host of other metals and semiconductors, (b) a range of five orders of magnitude in sweep rates, (c) great flexibility in setting up scan limits and reversal conditions, (d) an intrinsic facility for highlighting the chemical connections between various electroactive species present in the voltammogram, and (e) highly developed theory. Its weakness is mainly in the difficulty of making quantitative evaluations of kinetic parameters. Peak currents are often difficult to define with high precision, because baselines are always extrapolated and are almost never flat. The measurement of the reversal current, as illustrated in Fig. 20, is particularly prone to uncertainty. One typically can record trends in the parameters fairly easily, so that the qualitative power of the method is preserved, even though optimum precision for evaluation of parameters is unavailable. Once a mechanism is defined, perhaps by cyclic voltammetry, one may be better off doing quantitative work by step techniques or by hydrodynamic voltammetry.

5. Coulometry at Controlled Potential†

We have long since encountered the need to obtain the n value of an unknown electrode process, which is the most important stoichiometric connection between reactants and products. In all of the methods discussed so far, the current depends on n, but there is always a codependence on D. Any evaluation of n by these means requires knowledge of D or an assumption that the value for the compound under study is close to that for some other species whose electrochemical response has been previously measured under identical conditions. A much less ambiguous approach involves the electrolysis of the entire sample solution (*bulk electrolysis*), so that all electroreactant molecules are converted to products. If this job can be done with a *current efficiency* approaching 100% (meaning that virtually all of the current goes to the process of interest, with negligible amounts attributed to a background process), then the integral of the current gives the total charge passed Q, which in turn is linked to the number of moles of electroactive species present at the start by Faraday's law

$$Q = \int_0^\infty i \, dt = nFVC_O^*$$ (82)

where V is the volume of solution. Knowing C_O^* and V, we can straightforwardly obtain n from Q. Alternatively, C_O^* can be found if the electrode process (and therefore n) is known. Note that coulometry is an absolute technique that requires no calibration. Basically, one counts the electrons causing a known amount of chemical change.

The concept here is very different from the microelectrochemistry that we have encountered earlier. We have previously used small electrodes to *probe* a solution, with minimal changes in composition of the bulk solution being induced by electrolysis. Now we seek to convert the whole system from one oxidation state to another. Before, we used small electrodes and well-defined theoretically tractable mass transfer; now we must use large electrodes and efficient transport processes. The quantitative definition of the transport process is often unimportant. A typical cell may have a working electrode 5–100 cm² in area, such as a pool of mercury or a Pt gauze, immersed in a volume of 10–100 mL of solution. Stirring is implemented as efficiently as

† See [9, 14, 18, 85–87].

possible throughout the electrolysis. The working electrode is held potentiostatically at a potential where the process will occur at the mass-transfer-limited rate (i.e., where $C_O(0, t) = 0$). A choice of working potential is usually made from a voltammogram. For example, one might use the polarogram in Fig. 19 to design a bulk electrolysis of chromate at -1.3 to -1.5 V versus SCE.

Large-scale chemical conversions at the working electrode imply that equally sizable amounts of products will be generated at the counter electrode. For example, in the bulk electrolysis of chromate, the reduction at the working electrode is matched by an oxidation at the counter electrode. If a Pt gauze were used, the oxidation product might be O_2. It could then go into solution and either oxidize the reduction product of chromate or become reduced itself at the working electrode. In either case, its effect is to destroy the stoichiometric link between current flow and chromate electrolysis. Therefore, in all coulometric measurements, the counter electrode is either isolated in a separate compartment connected by an ion-permeable barrier (such as a frit), or it is designed to produce an insoluble product or an innocuous gas (e.g., AgCl or N_2).

One can show that for a simple process there is an exponential decay of the current with time

$$i = i(0)e^{-pt} \tag{83}$$

where $i(0)$ is the current at the beginning of electrolysis and p is a kind of rate constant for the overall process of electrolysis. Its reciprocal is the *time constant*. A consequence of this kind of relationship is that the same amount of time is required to complete electrolysis to a given degree under a single set of conditions, regardless of the starting concentration. For 99% completeness, that time is 4.6 time constants, whether the solution is initially 10^{-5} M or 10^{-1} M in electroactive species. The time constant is proportional to the efficiency of mass transfer (which is controlled by the quality of stirring) and the ratio of electrode area to solution volume. Common values are 2–10 min, which imply electrolysis times for 99% completeness of 10–50 min. Coulometry is therefore a slow technique.

If the electrode process involves slow homogeneous chemistry, e.g., follow-up reactions leading to subsequent charge transfer, then the deviation of the current decay from exponentiality can be useful for diagnosis and evaluation of rate constants [85, 86]. Also worthwhile is *reversal coulometry*, in which one completes a forward electrolysis of the sample, then reverses the process by an appropriate change in potential, with a second integration of charge. This procedure tests the stability of the initial product of electrolysis.

Coulometry usually involves the passage of substantial currents, frequently 100 mA and sometimes amperes. Potentiostats used for this purpose need high current capacity, but speed is not important. If resistive media are to be used (e.g., THF, CH_3CN, or CH_2Cl_2) then these currents will have to be pushed through appreciable solution resistance. A voltage drop as large as 10–100 V between the counter and working electrodes is not unusual. Designing cells with efficient mass transfer, effective separations of working and counter electrodes, adequate ability to handle heat from power dissipated in the cell, and minimal resistance is an important art [18, 85–87].

The advantage of coulometry is that it supplies information that is hard to obtain in other ways. Its drawbacks are its slowness and the special experimental problems associated with bulk electrolysis.

6. Thin-Layer Electrochemistry†

Throughout our discussion so far, we have generally regarded the diffusion layer as involving only a small part of the solution in contact with an electrode. In other words, the system is *semi-infinite*, in that the solution phase extends effectively indefinitely from the interface. On the other hand, suppose we built a *thin-layer cell*, in which the thickness of the solution is made less than about 50 μm, perhaps by the scheme in Fig. 22a, where only a small amount of solution is held between the electrode and a glass plate. In this situation, every electroactive species in solution would have access to the electrode, by diffusion alone, within a few seconds.

Thin layer experiments have gained a fairly prominent position because they afford a means for doing bulk electrolysis on a short time scale. Equation (82) is valid for electrolysis carried out under thin layer conditions, and it affords the n value of an unknown process in a straightforward fashion. One loads the thin-layer cell with solution and then holds the electrode potential where electrolysis does not occur, then a step is made to a potential in the mass-transfer-controlled region, and the resulting current is integrated. Only 1–30 sec are required to complete the process under normal conditions. From the integral charge Q, n can be calculated if C^* and V are known. The volume of the cell is usually calibrated by using a solution of a species that undergoes a well-characterized electrode reaction of known n value.

Cyclic voltammetry is also a useful method to study electrode reactions in thin-layer cells. The theory works from the assumption that the cell is sufficiently thin that diffusion can homogenize the solution continuously, so that concentration gradients do not exist and no explicit account of mass transfer need be maintained. To realize this condition in practical cells, one must not change the conditions at the surface more rapidly than the whole solution can be homogenized by diffusion; therefore the potential scan rate must not be too large. For cells with 50-μm thicknesses, v can be no larger

† See [9, 88–90].

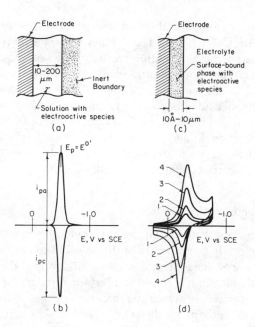

Fig. 22 (a) Construction of a thin layer system. An inert boundary traps a thin layer of solution against the working electrode. Usually this assembly is immersed in a larger volume of test solution containing the reference and counter electrodes. Electrical contact is maintained by ionic conduction through the thin layer of solution and its junction with the bulk. (b) Theoretical cyclic voltammetric response for $n = 1$, $T = 25°C$. (c) Schematic of a surface-bound phase showing the formal resemblance to a thin-layer cell. (d) Actual voltammograms for species of the $Ru(NH_3)_6^{2+}/Ru(NH_3)_6^{3+}$ couple bound electrostatically to crosslinked, sulfonated polystyrene on a Pt electrode surface. Curves 1, 2, 3, and 4 are for $v = 50$, 100, 200, and 500 mV/sec. Linearity of i with v and narrow peak splitting establish surface-bound condition. The broadness of the peaks and nonzero splitting [compare with (b)] indicate interactions between redox centers in the film or slow kinetics for homogenization of charge in the film. [Frame (d) courtesy of B. R. Shaw.]

than about 20 mV/sec. For thinner solution phases, higher rates can be tolerated.

When proper conditions apply, the relationship between current and potential is easy to work out [89, 90]. For example, in the reversible case, one assumes that the Nernst equation links the potential with the concentrations throughout the cell

$$E = E^{0\prime} + (RT/nF) \ln(C_O/C_R) \tag{84}$$

If we start with only species O present at concentration C_O^*, then $C_O + C_R = C_O^*$ at all times, hence

$$E = E^{0\prime} + (RT/nF) \ln[C_O/(C_O^* - C_O)] \tag{85}$$

The rate of change in C_O is related to the current,

$$-dC_O/dt = i/nFV \tag{86}$$

By combining (85) and (86) and doing some algebra, one obtains the expression for the current–potential curve,

$$i = \frac{n^2 F^2 v V C_O^*}{RT} \frac{\exp[nf(E - E^{0\prime})]}{\{1 + \exp[nf(E - E^{0\prime})]\}^2} \tag{87}$$

where $v = -dE/dt$. This is a symmetrical peak like that shown in Fig. 22b. A characteristic of reversible thin-layer behavior in cyclic voltammetry is that the forward and reverse scans show equal peak magnitudes and identical peak potentials, positioned at $E^{0\prime}$.

The peak current is

$$i_p = n^2 F^2 v V C_O^*/4RT \tag{88}$$

This relation highlights two other aspects of thin-layer behavior. The current response depends linearly on the scan rate v and on the square of n.

Thin-layer concepts apply to many electrochemical systems other than those physically constructed as thin-layer cells in the manner described here. Thus it is very important to understand the essential character, which is an ability to change redox state homogeneously throughout a phase on a time scale short by comparison to any perturbation. Thin-layer behavior is seen, not only for thin layers of solution held against the electrode, but also for adsorbates and for redox species bound to the electrode by chemical modification (Fig. 22c). In these cases, the system may become quite homogenized very quickly, so that thin-layer characteristics are maintained even on millisecond time scales. Diagnosis of surface-bound systems is often made by looking for very small peak separations in cyclic voltammetry and for currents that are proportional to v [9, 55, 83, 91, 92]. (See Fig. 22d.)

Conventional thin layer cells are most useful for n value determinations and for spectroelectrochemical measurements as described in Section IV.D.2. They are not very helpful for mechanistic diagnosis, because the time scale range is narrow and tends toward the slow end. The actual hardware requires more attention than equipment designed for more ordinary experiments, but there are good sources of experimental information in the literature [89, 90].

C. *Analytical Techniques*

1. *Polarographic Approaches*

Polarography is the classical method that was the first widely practiced electroanalytical technique. Using polarography, many useful procedures have been worked out for determinations of analyte concentrations. In recent years the newer methods of *normal pulse polarography* and *differential pulse polarography* have supplanted the classical method because they offer superior sensitivity. Before we can understand their design and operation, however, we must first comprehend the applicability and limitations of conventional "dc" polarography.

(*a*) *Conventional Polarography*† Large compilations and reviews give convenient access to the enormous body of literature on applications of polarography in analytical chemistry [26, 62, 76, 93, 94]. Because mercury is itself easily oxidized, the dropping mercury electrode offers a very limited positive working range. Thus reduction of electroactive species is stressed in polarographic analysis.

One can make direct measurements of most metal ions. The literature dealing with transition metals is especially rich, and most practical analytical polarographic measurements concern them. Amalgam-forming metals give especially pretty reversible reductions to the zero valence state, and these waves are the most widely exploited. Examples of species that can be determined in this way are Cu(II), Cd(II), Zn(II), Pb(II), Tl(I), and Bi(III). Most of these have become important as part of environmental monitoring in recent years.

Reducible organic species are also eligible for polarographic determination. Pharmaceuticals in particular have received stress in the literature.

One should consult the original publications dealing with a specific method, and care should be taken to follow the specified procedures. The development of a polarographic method requires special attention to the choice of medium. The solvent, complexing agents, and pH are generally selected to maximize sharpness in the wave and to reduce interferences. A good deal of clever chemistry is embodied in available procedures.

The half-wave potential is a useful qualitative indicator to identify an unknown constituent. In fact, the general strength of electrochemical methods is their ability to provide information, through such characteristic potentials, about the particular *species* present in a chemical system. For example, one can often distinguish oxidation states and complexes of a given metal. Many competitive methods (such as atomic spectroscopic methods) frequently provide only total elemental content.

Quantitative evaluation is based on the proportionality between the diffusion-limited current (plateau height) and concentration [14, 76]. Calibration is usually by standard solutions. Precisions on the order of 1% are attainable by a practiced worker, and careful attention to detail can result in 0.1% precision [95]. Standard addition and internal standard evaluations

† See [8, 9, 13, 14, 62, 76].

are other choices that offer precision to a few percent. Standard addition is particularly recommended for determinations in complex unknown media such as environmental samples. As usual in analytical chemistry, one must take care that the standards resemble the unknown as closely as possible.

Factors determining the detection limit deserve careful scrutiny here, because they bear on the design of the pulse voltammetric schemes discussed below. In conventional polarography, one is interested in a faradaic current that is always superimposed on a background of charging current. As the mercury flows from the capillary it must be charged [8, 9, 14, 62, 76] to the operating potential E from the potential of zero charge E_z. The charge required on the drop at any time is

$$q^M = C_i A(E - E_z) \tag{89}$$

where C_i is the integral double-layer capacitance. The charging current is the rate of accumulation of this charge, $-dq^M/dt$. (The minus appears because electron flow toward the interface from the external circuit is considered positive.) Since E is essentially constant during the life of a drop, the charging current is

$$i_c = C_i(E_z - E) \, dA/dt \tag{90}$$

Thus the charging current comes from *steady drop expansion* at the DME, not from a change in potential. It also increases approximately linearly as one moves away from E_z, and this feature is seen in the sloping backgrounds of polarograms, such as that of Fig. 19. The size of this current is typically comparable to the faradaic current for an electroactive species present at the 10^{-5} M level. Thus, the polarographic wave for any analyte present at concentrations much below this figure will be obscured by the background of charging current. This sort of detection limit corresponds to 0.1–1 ppm and is comparable to the figures attainable by many standard analytical methods, such as atomic absorption, UV–visible absorption, and gas chromatography with flame ionization detection. A number of variants in experimental technique have been advanced with the purpose of improving detection limits by reducing the capacitive background. The pulse methods that we shall now examine are the most widely practiced.

(b) *Normal Pulse Polarography*† To understand how the capacitive background can be reduced, we must first consider the time dependence of i_c. At a DME, each drop grows in a spherical form upon a steady inflow of mercury at the rate m mg/sec. From simple geometry [8, 9] one readily finds that dA/dt is proportional to $t^{-1/3}$, thus i_c follows the same time function. In other words, the charging current is largest at the instant of birth and

† See [8, 9, 96–99].

falls steadily through each drop's lifetime. If we set up our experiment so that current flow is not measured continuously, but instead is *sampled* at the moment just before drop fall, then the smallest possible charging current will be included in the measured total current.

Instruments designed for normal pulse polarography include special electronic circuitry to allow this kind of current sampling. A voltage proportional to the sampled current is presented steadily to the recorder until it is replaced by a value representing a new sampled current, flowing at the end of the next drop's lifetime. The sampling operations are synchronized with drop growth and fall by a mechanical *drop knocker*, which dislodges the drop upon command by timing circuits in the instrument.

Since no current measurements are to be made until the moment just before drop fall, there is no point in electrolyzing the sample during the whole lifetime of the drop. In fact, there is a considerable advantage in avoiding electrolysis altogether until a time shortly before the current sample is to be taken [8, 9, 97]. In that way, the analyte is less depleted near the electrode at the time of measurement than it would be if electrolysis had occurred continuously. Therefore one can expect a steeper concentration profile near the electrode at the sampling time, which leads in turn to higher faradaic currents (see Section IV.B.2.a).

This plan is accomplished by applying a waveform like that shown in Fig. 23a. During the long period of drop growth, the potential is held at a value (the *base potential*) where the analyte is not electrolyzed. At some point

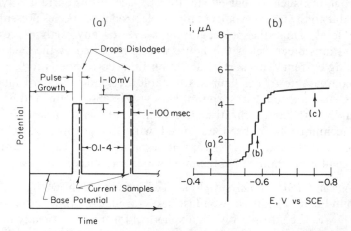

Fig. 23 (a) Timing diagram for normal pulse polarography. The sequence of events is shown for two successive drops. (b) Normal pulse polarogram for 10^{-4} M Cd^{2+} in 0.01 M HCl. The base potential was -0.41 V and each successive pulse was about 9 mV higher than its predecessor. Potential axis is potential of the pulse. Vertical dislocations occur on the current axis as new samples replace old ones. Points (a), (b), and (c) are discussed in the text.

just before sampling (usually about 50 msec), the potential is stepped to a new value. The current is sampled; the potential returns to the base value; and finally the drop is dislodged. The whole process is repeated for the next drop, but a little larger potential pulse is used. The record is a plot of sampled current versus step potential, and it typically involves samplings from a few hundred drops. Figure 23b is a typical polarogram.

In the pulse experiment, larger faradaic currents and smaller charging currents are observed than in conventional polarography. The overall effect is about an order of magnitude improvement in detection limit—to perhaps 10^{-6}–10^{-7} M. Moreover, the method is applicable in many cases to stationary electrodes, such as the hanging mercury drop or a Pt disk, as well as the DME. The theory of current response in normal pulse voltammetry is actually much simpler than for conventional polarography [9], because the effect of drop expansion can generally be ignored for the short time of electrolysis. There is a considerable, and growing, body of literature dealing with normal pulse voltammetry as a diagnostic tool. One can expect an expansion in its use for diagnosis.

(c)   *Differential Pulse Polarography*†   The normal pulse method, when used with a DME, minimizes the charging current from drop expansion by sampling at the latest possible moment. However, the charging current is still present to some extent, and in accord with Eq. (90), it increases linearly as one moves away from the potential of zero charge. If the concentration of the analyte is reduced below 10^{-6} M, it becomes progressively more difficult to see a faradaic wave on this sloping background; thus even for normal pulse polarography, the detection limit is set by the capacitive background current. Further improvement in sensitivity would require a new method featuring (a) a charging current of smaller magnitude, (b) a flatter background, and (c) perhaps a peak-shaped faradaic response that would be easier to detect and quantify than a wave. All of these improvements are achieved with *differential pulse polarography*.

The potential program is outlined in Fig. 24a. A potential E is imposed during most of a drop's lifetime, but for a brief period before the end, a pulse of height ΔE is added, so that the potential becomes $E + \Delta E$. The pulse width is typically about 50 msec. Two current samples are taken: one just before the pulse at τ' and one at the end of the pulse at τ. The response function is the difference between these currents $\delta i = i(\tau) - i(\tau')$.

Successive drops are treated similarly, except for a small shift in E with each new drop. The *pulse height* ΔE remains constant at a preselected value between 10 mV and 100 mV. As the experiment proceeds, then, E is scanned

† See [8, 9, 96–99].

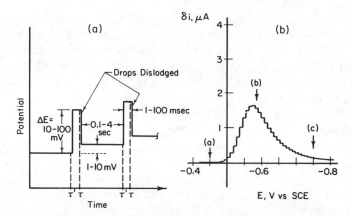

Fig. 24 (a) Timing diagram for differential pulse polarography. The sequence of events is shown for two successive drops. (b) Differential pulse polarogram of 10^{-4} M Cd^{2+} in 0.01 M HCl. The potential axis is the value of potential during the growth phase of each drop. A shift of about 9 mV occurs with each drop. The pulse height, ΔE, was -25 mV. This curve corresponds to the normal pulse polarogram in Fig. 23b.

from an initial value to a final value in increments of a few mV as the new drops appear. The polarogram is a record of δi versus E. Figure 24b is an example.

The timing sequence of this experiment is similar to that found in normal pulse polarography. Everything is synchronized with the birth of each mercury drop by the use of a mechanical drop knocker; and sample/hold circuitry is used to sample the current. Any commercial instrument capable of the normal pulse technique should also permit differential-pulse measurements.

The peak-shaped response in Fig. 24b is a direct result of the differential current measurement. One can easily understand the shape by considering δi values that would be recorded at potentials (a), (b), and (c) on the polarograms in Figs. 23b and 24b. First think about point (a). With E at a rather positive value with respect to $E^{0'}$, reduction of the analyte is not possible during the early part of drop growth, hence the faradaic part of $i(\tau')$ is essentially zero. The application of the pulse forces the potential to a slightly more negative value, but not sufficiently negative to begin the reduction; hence the faradaic part of $i(\tau)$ is also quite small, as must be the difference δi. At point (c), matters are quite different. There, E is sufficiently negative so that the analyte is reduced at a diffusion-controlled rate throughout the period of drop growth, hence $i(\tau')$ is sizable. On the other hand, application of the pulse does not change the rate of reduction, because the process is already mass transfer controlled. The value of $i(\tau)$, the second sample, is nearly the same as the first, so the difference δi is again quite small.

By now it should be clear that δi will only have a significant component representing electrolysis of the analyte when the application of the pulse causes a detectable change in the rate of that electrolysis. Such a condition is generally met at potentials near $E^{0\prime}$ (for a reversible system), because this is the region at which the electrochemical balance of oxidized and reduced forms is most sensitive to potential. Point (b) in Figures 23b and 24b is in this region. The method, in effect, measures the value of δi stimulated by ΔE, as a function of E. A differential pulse polarogram, therefore, is approximately the derivative of the corresponding normal pulse polarogram [8, 9, 97].

The height of the peak for a reversible analyte depends on ΔE. It can at most be the same as the wave height for the normal-pulse curve run under the same conditions (drop time and pulse width). Pulses of 100–250 mV amplitude are required to reach this maximum δi. Smaller values of ΔE produce smaller peaks. Thinking about the differential pulse method as a kind of derivative technique helps one to see that pulse width also affects the width of the peak at half height. The larger the value of ΔE, the coarser is one's approximation to the true derivative, and the broader is the peak in δi. In practical work, one sacrifices some sensitivity to achieve reasonably narrow peaks, so that resolution of peaks for a mixture of analytes is enhanced. Typical values of ΔE (10–100 mV) give peak heights that are 10–99% of the maximum possible value [8, 9, 97].

For a reversible system, the potential at the peak is $E_{1/2} - \Delta E/2$, where $E_{1/2}$ is the polarographic half-wave potential. In most applications, the scan is made from positive to negative potentials and ΔE is a negative quantity, so the peak potential is slightly more positive than $E_{1/2}$.

An extremely important feature of this method is the much flatter charging current background that it yields, by comparison to the normal pulse technique. The improvement comes about because a differential current measurement is made. The charging current at time τ is very similar to that at time τ', because the rate of drop expansion has slowed to a small value late in the drop's lifetime. From (90) one can show that the charging current difference δi_c is proportional to $C_i \Delta E m^{2/3} \tau^{-1/3}$. This is a constant level across the polarogram, insofar as C_i is constant with potential. Moreover it is generally smaller in magnitude than the background currents present in normal pulse polarography. The reduced background currents and improved readout lead to excellent sensitivity in measurements of concentrations by differential pulse polarography. Detection limits are often near 10^{-7} M and sometimes approach 10^{-8} M (about 1 ppb for MW = 100).

Differential pulse polarography, is theoretically complex. Treating the expected responses for various reaction mechanisms is possible in principle, but is difficult. Thus the method is not generally useful for mechanistic diagnosis. It has been designed for the measurement of concentrations at high sensitivity, a job it does superbly. Differential pulse polarography has become a very popular analytical tool.

This method, like normal pulse voltammetry, can also be applied to stationary electrodes such as C, Pt, and Au. Sensitivities are not as good as with a mercury electrode, because these other materials always feature sizable background currents, relating often to oxidation and reduction of surface structures on the electrode itself. Interference from these faradaic components limits detection of analytes in solution.

(*d*)   *Static Mercury Drop Electrode*†   We have seen that detection limits in polarographic methods are almost always set by charging currents needed to support drop growth at a DME. An improved dropping electrode could be imagined. One would prefer to have each drop grow to maturity very quickly after birth, then hang for awhile without growth, finally to be dislodged and replaced with a new drop. In this way normal and differential pulse polarography could be carried out on static drops *with no charging current background*. Then, detection limits would be set by residual faradaic currents from electrolysis of impurities or discharge of the solvent/supporting electrolyte system.

Princeton Applied Research Corporation, a division of EG and G, has devised and marketed such an electrode. It overcomes virtually all of the drawbacks of the DME. It is fairly compact, and the mechanisms for extrusion and dislodgment of drops are electronically controlled in a manner that is readily compatible with the timing circuitry essential for proper sequencing of events in pulse techniques. For analytical measurements this device is superior in concept and performance to a DME, so one can expect the static mercury drop electrode to dominate the field of electroanalysis by pulse polarography in the future.

2.   *Stripping*‡

When one is faced with determining traces of analytes, it is common to use a *preconcentration* step in the procedure. The idea is to gather analyte from a large volume of sample into a small space, where it can be readily detected and quantified by standard tools. Solvent extraction, ion exchange, and evaporation are common approaches to preconcentration, and they are as applicable to electrochemical determinations (e.g., by differential pulse polarography) as they are to spectrometric or chromatographic determinations.

† See [100, 101].
‡ See [99, 102–109].

For certain analytes, however, there is the option of using a particularly convenient electrochemical preconcentration step prior to a voltammetric determination. The whole procedure is called *stripping analysis.*

The best and most widely practiced procedures feature a mercury electrode, such as a hanging drop (HMDE), and involve amalgam-forming analytes, such as Cd, Zn, Pb, Tl, Bi, and Cu [102–107]. In the preconcentration stage, the electrode is held potentiostatically at a potential sufficiently negative to reduce any analyte ion of interest into the amalgam, e.g.,

$$Pb^{2+} + 2e \rightarrow Pb(Hg) \tag{91}$$

The solution is stirred in order to bring as many ions as possible up to the electrode surface, where they are reduced and collected into the mercury drop. By allowing adequate time for this deposition, one effectively concentrates the ions from a large volume of solution into the much smaller mercury droplet. Accuracy and precision demand that the same amount of material be collected into the drop during the preconcentration stages of replicate runs with a given analyte concentration in solution. Thus it is important that precise stirring be used and that the deposition time be carefully controlled. Usually no attempt is made to collect all analyte ions (or even a large fraction of them) out of the volume of test solution in contact with the electrode. It is only necessary to use sufficiently long deposition times to achieve measurable concentrations of atoms inside the drop. Deposition times ordinarily range from 1 to 30 min.

Once the deposition is complete, stirring is halted, a brief rest period is allowed so that the system becomes quiescent, and then the potential is swept toward positive values. As E approaches, then passes, $E^{o\prime}$ for an analyte's deposition reaction, then the atoms of that analyte are oxidatively stripped from the electrode, e.g.,

$$Pb(Hg) \rightarrow Pb^{2+} + 2e \tag{92}$$

The anodic current manifesting this process goes through a peak for exactly the reasons discussed in the section on cyclic voltammetry above. As the scan continues, additional peaks are seen for other strippable analytes. The peak potentials give clues to the identities of the metals present, and the heights of the peaks are proportional to the concentration of metals *within the mercury.* If a standardized deposition step is used, the peak heights are also directly proportional to analyte concentration in solution at the time of deposition. Figure 25a shows an example of a stripping voltammogram obtained in this manner.

An alternative to this technique is to use a mercury film electrode (MFE). With this method, one normally employs a rotating pyrolytic graphite disk or a rotating glassy carbon disk as the working surface [105–107, 109] and

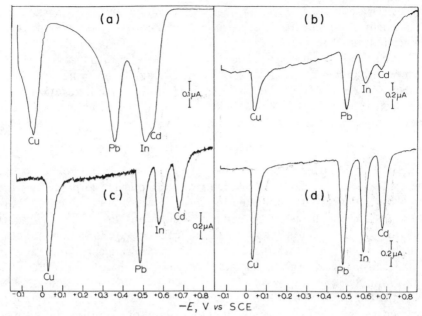

Fig. 25 Stripping voltammograms. Solutions contained 2×10^{-7} M Cd^{2+}, In^{3+}, Pb^{2+}, and Cu^{2+} in 0.1 M KNO_3. For (b)–(d) there was also 2×10^{-5} M Hg^{2+} Initial deposition at 0.85 V versus SCE. Scan rate of 5 mV/sec. (a) HMDE with 30 min deposition. (b) Pyrolytic graphite substrate. (c) Unpolished glassy carbon substrate. (d) Polished glassy carbon. Curves (b)–(d) involved deposition times of 5 min and electrode rotation rates of 2000 rpm. [Reprinted from T. M. Florence, *J. Electroanal. Chem.* **27**, 273 (1970), with permission.]

Hg^{2+} is added to the test solution in concentrations of 10^{-5}–10^{-4} M. Preconcentration is carried out as described above, in that the electrode is held at a sufficiently negative potential to deposit all analytes of interest. During this time, the added Hg^{2+} is also reduced so that a film of mercury is created, into which the analytes are codeposited. These films are typically a few hundred angstroms thick, and the preconcentration factor is enormous because there is a very large ratio of the film's area to its volume. The stripping step is then carried out as described above, although rotation is usually continued. Stripping voltammograms (b), (c), and (d) in Fig. 25 were obtained in this way.

The advantages of the MFE are in improved sensitivity, resulting from the high preconcentration factor, and superior resolution of peaks. Both advantages are apparent from a comparison of Figures 25a and 25d. The fundamental reason for the improvement in resolution is that, with respect to the stripping step, the MFE is like a thin layer system (Section IV.B.6). All deposited atoms of analyte are accessible to the electrode/solution interface without transport limitation, and they are all stripped from the film rapidly. This aspect leads to a sharp trailing edge to each peak (on the positive side), so that the succeeding peaks have a good chance to rise from a flat baseline. In contrast, an HMDE is not a thin-layer system, and the stripping step is controlled by diffusion of atoms within the droplet to the mercury/solution interface. Stripping does not exhaust the droplet quickly, hence the peaks are wider and the trailing edges are much more extended than with an MFE. Finally, the MFE tends to be more precise, because the hydrodynamics of deposition at the RDE are efficient and well-defined.

There are drawbacks to the MFE. They are chiefly related to interferences from analyte–substrate interactions and analyte–analyte interactions. Since the film is so thin, every analyte atom has ready access to the electrode substrate. There are specific interactions between some analytes and substrates that lead to nonlinear concentration relationships and other problems in evaluation. Analyte–analyte interferences arise from intermetallic complexes as a result of the high concentrations of atoms that can exist in the films. Many examples of this kind of interference are documented in the literature [103–107]. Problems of this sort are much less severe with an HMDE.

Improvements in sensitivity and resolution in stripping procedures at HMDEs can be gained by using differential pulse voltammetry for the stripping step [99, 105–107]. This modification makes procedures with the HMDE competitive with those involving MFEs. One therefore finds both approaches in widespread use.

Stripping methods can have quite low detection limits. Figures between 10^{-8} and 10^{-9} M are routine; values between 10^{-9} and 10^{-10} M are fairly common; and values below 10^{-10} M are occasionally reached. Actual limits for any type of sample depends on its interferences and the deposition time employed. By far the preferred evaluation method is standard addition. Trace determinations are subject to serious problems of contamination, sample loss through adsorption on utensils, and matrix effects (e.g., from complexation or adsorption on particles). It is therefore very difficult to create standards that accurately reflect conditions in an unknown. The only reasonable action is to add a comparable amount of standard to the unknown.

Stripping methods exist for some species other than amalgam-forming metals [104–109]. Most involve the electrolytic formation of insoluble films on an electrode during the deposition step. For example, one can anodically deposit Cl^-, Br^-, or I^- on mercury by

$$2Hg + 2X^- \rightarrow Hg_2X_2\downarrow + 2e \qquad (93)$$

After deposition, a negative scan causes cathodic stripping of the halide by reversal of the deposition process.

The procedures for determination of amalgam-forming metals are now standard and are widely practiced. They are selective, simple, accurate, and precise, so one can expect them to continue to hold a strong position in day-to-day analysis for these substances at low levels. Research in the field is aimed mainly at finding techniques for determining other species, usually through the formation of precipitate films. One can expect some useful new methods to appear from time to time.

3. *Coulometric Titrations*†

A great number of practical determinations of concentration involve titrations, and many of these, particularly with automated apparatus are done coulometrically. The fundamental idea behind this approach is that, in each situation, we must develop a strategy that will allow us to add electrons to (or subtract them from) a test solution through a working electrode until a *stoichiometric* number has been transferred, so that the analyte is chemically transformed and a precise endpoint is reached. In our scheme the electrons, rather than a reagent in a burette, serve as titrant. We shall always add them at a constant rate (i.e., at constant current) through a working electrode of large area into a stirred solution. The arrangement is shown schematically in Fig. 26a. The constant current is a convenience that allows us to use time as a measure of the amount of total charged passed, since $Q = it$. The remaining conditions are needed because this approach is a type of bulk electrolysis (Section IV.B.5), in which full chemical conversion is essential. As usual in bulk electrolysis, the counter electrode is isolated to prevent interference from electrolytic products generated there.

In the simplest coulometric titrations, the electrolysis generates a reagent in a straightforward manner, and this reagent reacts with the analyte rapidly and cleanly. A good example is the production of protons by oxidation of water at Pt [14],

$$2H_2O \rightarrow O_2 + 4H^+ + 4e \qquad\qquad (94)$$

For each electron withdrawn one proton is produced. It, of course, will react with any base present in the solution. Thus the passage of a constant oxidation current at the working electrode generates H^+ at a constant rate, which titrates base at a constant rate. At some well-defined time, an exact stoichiometric quantity of H^+ will have been generated, and an equivalence point is reached. It could be observed colorimetrically with an indicator, potentiometrically with a glass electrode, or by any other standard means of endpoint detection.

† See [9, 14, 16, 110]. See also the biennial reviews on electroanalysis that appear in *Analytical Chemistry* [111].

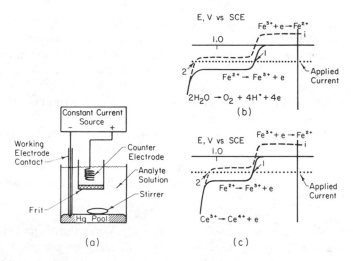

Fig. 26 (a) Typical arrangement for coulometric titrations with a mercury pool as a working cathode. The analyte is confined to the volume outside the counter electrode compartment. (b) Current–Potential curves at the working electrode for a solution of Fe^{2+} in sulfuric acid. The solid curve applies at the start. Since $|i| < |i_l|$, the electrode adopts a potential on the rising portion of the wave for oxidation of Fe^{2+}. Operation is at point 1 and current efficiency is 100%. Dashed curve applies after half the Fe^{2+} is converted to Fe^{3+}. Now $|i| > |i_l|$, so the potential has shifted to enable the evolution of O_2 to occur (point 2). (c) Situation as in (b) except that a large excess of Ce^{3+} has been added. Point 2 in this case is not so positive, and excess current produces Ce^{4+}.

If we call the generation time to the endpoint t_{ep}, then the total charge devoted to reagent generation is it_{ep}, which by Faraday's law gives rise to it_{ep}/F equivalents of acid, which in turn react with an equal number of equivalents of base. From the known current and the precisely measured time to the endpoint, we can therefore calculate the amount of base in the sample *without standardization*. This absoluteness of coulometric titrations is an important feature, because it implies that one can avoid standardizing titrant solutions and storing those solutions, and one does not have to suffer added imprecision from the standardization step. This feature and the intrinsically high reproducibility by which i can be controlled and t_{ep} can be measured, bring exceptional precision to coulometric titrations [14, 16]. One often expects repeatability to one part per thousand. Some titrations with considerably better precision have been devised [14, 26]. The accuracy of the method depends on the current efficiency of reagent generation (Section IV.B.5), the efficiency with which the reagent reacts with the analyte, and the match between the detected endpoint and the actual equivalence point. In well-designed procedures, accuracy comparable to precision is obtained.

Other advantages of coulometric titrations are their ready adaptability to automation, their suitability for remote operation, and their ability to deal with quite small *absolute quantities* ($\sim$ nanomole) of analyte. On the other hand, the *concentration* range of analytes for which they are suitable is the same as for conventional titrations, because the lower limit is set by the ability to detect an endpoint. It is difficult to do any titration on an analyte at a concentration below about 10^{-4} M.

Approaching perfect stoichiometry in converting electron flux to reagent flux is easy to achieve for the acid–base titration discussed above, because the *reagent precursor* is the solvent, which is in very large excess. An equivalently efficient process is the production of hydroxide by reduction of water,

$$2H_2O + 2e \rightarrow 2OH^- + H_2 \tag{95}$$

The OH^- obviously can be used to titrate acids. When the reagent precursor is a solute, the danger of less than 100% efficiency in the production of reagent increases, but it still can be quite satisfactory. An interesting example is the generation of bromine from concentrated (perhaps 0.5 M) bromide for titration of alkenes in alcoholic solvents [14]

$$2Br^- \rightleftharpoons Br_2 + 2e \tag{96}$$

$$\tag{97}$$

Probably the largest group of coulometric titrations involves redox reactions. The classic example [14, 16] involves the titration of Fe^{2+} by electrogenerated Ce^{4+}. The reagent precursor is Ce^{3+}, which is added to the test solution in large excess. Passage of an oxidation current then partly goes to oxidize Fe^{2+} directly

$$Fe^{2+} \rightarrow Fe^{3+} + e \tag{98}$$

but most goes to produce Ce^{4+}, which in turn oxidizes Fe^{2+} in solution

$$Ce^{3+} \rightarrow Ce^{4+} + e \tag{99}$$

$$Ce^{4+} + Fe^{2+} \rightarrow Fe^{3+} + Ce^{3+} \tag{100}$$

Note that *every electron passed at the working electrode has the same net effect on the analyte.* Thus 100% *titration efficiency* is achieved, even though neither electrode reaction, (98) nor (99), occurs with perfect currect efficiency.

As the equivalence point is passed the Fe^{2+} is used up and an excess of Ce^{4+} appears. Endpoint detection can be achieved, again by any standard means, such as potentiometry, amperometry, or a redox indicator.

One might ask why the reagent precursor is needed at all. Why not just titrate the Fe^{2+} by direct oxidation at the electrode? The answer is outlined in Fig. 26b. If this route is chosen, one is not free to use an arbitrary constant current, because the oxidation of iron can be carried out no faster than the mass-transfer-limited rate. Thus the applied i must be smaller in magnitude than i_l. Any larger current could not be supported by the direct oxidation of iron. Instead, a part i_l would go to that process, but the remainder would have to be adsorbed by the next eligible oxidation, which in this case would be the evolution of O_2. When we control the current, the potential adopts whatever value is required to support the current, so in this latter instance it would shift to a sufficiently positive value to allow the secondary process to fill in the excess current as needed. Of course, this action reduces the current efficiency for oxidation of Fe^{2+} to an indeterminate value below 100% and destroys our ability to quantify the amount present. For quantitative direct oxidation of Fe^{2+}, we therefore *must* have $|i| \leq |i_l|$. Yet $|i_l|$ decreases as we use up the Fe^{2+}, so we could only maintain high current efficiency by choosing $|i|$ much less than $|i_l|$ at the start. However, that choice entails a very slow titration rate.

By adding a titrant precursor we solve many problems. Figure 26c shows the effect. The titrant precursor, in the case of the titration of Fe^{2+}, exists just to absorb excess current in a controlled fashion. If we apply $|i| > |i_l|$, the amount of current greater than $|i_l|$ is taken up as before by the next eligible oxidation. The titrant precursor, such as Ce^{3+}, is deliberately added to be this next eligible electroactive species. Thus the excess current generates Ce^{4+}, which oxidizes Fe^{2+} homogeneously, so that the current ultimately goes to the desired reaction. Note that this would not be true if O_2 were generated in the backup process, because most O_2 would escape. The titrant precursor is always added in very large excess, so that it has the ability to handle quite sizable excess currents. In this way, one is free to apply currents much larger than $|i_l|$ and thereby shorten the titration time. Then most of the current is absorbed by the titrant precursor, yet the efficiency in converting the analyte is still preserved.

The freedom from standardization lends special flexibility to coulometric procedures. One is free to electrogenerate titrants such as Cr^{2+}, I_2, Br_2, Ag^{2+}, and Karl Fischer reagent [14, 26] that are difficult to store and standardize for conventional titrations.

4.   Ion-Selective Electrodes†

In considering selective electrodes, we return to the concept of measuring potentials at equilibrium, rather than under conditions of current flow. In

† See [9, 14, 24, 112–116].

Sections III.A–C, we saw how an equilibrium potential was established by creating a charge separation at a phase boundary and maintaining it by a dynamic exchange of charge carriers. Here the charge carriers of importance are ions rather than electrons.

(*a*) *Junction Potentials*† Perhaps the easiest way to understand the origin of the effects exploited in ion-selective electrodes is to consider a junction between two solutions of different composition. Suppose, for example we built the cell

$$Hg/Hg_2Cl_2/NaCl, 0.1\ M/HCl, 0.1\ M/Hg_2Cl_2/Hg$$

This could be a piece of glassware with two compartments separated by a frit, with Hg and $HgCl_2$ in both sides, and with NaCl and HCl solutions each in one chamber. What happens at the frit?

Of course Na^+ and H^+ diffuse across the barrier because of concentration gradients. Since H^+ diffuses much faster than Na^+, an excess positive charge develops on the NaCl side and an excess negative charge forms on the opposite side. The field tends to slow down the movement of H^+ and speed up Na^+, so that a steady state is soon reached. This steady charge separation across the boundary creates a potential difference, which is a component of the measured overall cell potential. Whenever two electrolytes of differing compositions meet, a liquid junction potential will exist. Its magnitude can be as high as 50 mV. Approximate equations are available for calculating junction potentials from the concentrations, charges, and mobilities of the ions present.

Most of the time junction potentials are nuisances and efforts are ordinarily made to minimize them. This is a major function of the salt bridges made of concentrated KCl and KNO_3 that are widely employed in electrochemistry. The basis for their operation is discussed fully in the literature [4, 9, 14, 24].

(*b*) *Selective Interfaces*‡ Ion-selective electrodes are based on the creative exploitation of junction potentials. The single idea underlying their technology is to seek an interface with a test solution where charge exchange between a sensor phase and the solution can be dominated by one species of ion. For example, an ion selective system would be created if we separated the NaCl and HCl solutions in the cell above by some barrier that would pass only H^+. Then the junction potential would be a function only of the ratio of activities of H^+ on the two sides of the barrier. By controlling the activity on one side, the value of the junction potential could be used to measure an unknown activity on the other side.

† See [4, 9, 117].
‡ See [9, 24, 112–117].

The most common embodiment of this idea is the glass electrode [24, 112, 118], which operates in a cell of the following design:

$$\text{Hg/Hg}_2\text{Cl}_2/\text{KCl(saturated)/test solution/glass/HCl(0.1 } M)/\text{AgCl/Ag}$$

$|$—— reference electrode →$|$ $|$—— glass electrode →$|$

The glass electrode is really a glass envelope containing 0.1 M HCl and a silver/silver chloride reference. A thin, specially formulated glass membrane is part of the envelope. It is a selective sensor that is immersed in the test solution. Paired with the glass electrode is a conventional reference electrode, such as an SCE. The pH meter is just a high-impedance voltmeter that measures the potential difference across the whole cell. As discussed in Section III.A. the overall cell potential is the sum of potential differences at the various phase boundaries. However, the boundaries inside the glass electrode and inside the reference electrode are sealed and never change from experiment to experiment. The test solution can only affect its junction with the glass and with the tip of the reference electrode. Since the reference electrode usually has a concentrated KCl electrolyte that is intended to minimize variations in the junction potential at its tip, the test solution's only real point of influence is the junction with the glass.

Even though we normally think of glass as an insulator, some formulations with high cation contents exhibit appreciable ionic conduction and can be regarded as complex electrolytes. The types used in glass electrodes are predominantly sodium ion conductors. When the glass is exposed to an aqueous solution, a thin outer zone, perhaps 100 Å in thickness, is hydrated and forms a rather open structure that can freely exchange cations with the solution. The rigid silicate lattice offers fixed compensating negative charges that have a tendency to bind univalent cations *selectively*. The glass used in pH electrodes is specially formulated to bind H^+ selectively. Thus immersion of the glass electrode unit into an acidic test solution tends to charge the glass surface positively. The field resulting from increased charge tends to discourage additional adsorption, and an equilibrium is reached. The extent of charge buildup reflects the concentration of H^+ in solution, and the charge is itself reflected in the junction potential between the glass phase and the test solution.

The measured cell potential contains a constant sum of interfacial potential differences, as discussed above, plus this junction potential, which depends logarithmically on the H^+ activity

$$E_{\text{cell}} = \text{const} + (RT/F)\ln a_{H^+} = \text{const} - 2.3\,(RT/F)\,\text{pH} \qquad (101)$$

The act of "standardizing" the system is just a way of evaluating the constant.

One simply uses a buffer of known pH and adjusts the zero on the meter until it reads the correct value.

When the buffer is replaced with a test solution whose pH is to be measured, one hopes that the constant does not change [9, 112, 116, 119]. Since almost all the contributors to it are sealed up, this hope is generally well founded. However, the junction potential at the reference electrode's tip can be expected to vary by a millivolt or two from one test solution to the next; and this uncertainty in the constant implies an uncertainty in pH of about 0.02 unit.

It is important to realize that we did not require a permselective membrane to achieve this overall selectivity in measurement. That is, we did not require selective transport of H^+ between the internal electrolyte within the electrode and the test solution. All that matters is that charge transport within the part of the membrane *affected by the test solution* be dominated by the ion of interest. Selectivity of binding gives us a chemical basis for a domination of charge transport by a single species. All ion-selective electrodes are built on this principle of finding a chemical basis for allowing one species to dominate charge transport within a small region of the current path that is directly affected by the test solution. We do not have space here to review the various chemical schemes for accomplishing this task, but several complete discussions of ion-selective electrode technology are available in the literature [112–116].

While some very good devices exist, no ion-selective electrode is immune to interference from other ions. For example, glass electrodes also tend to bind Na^+, K^+, Li^+, NH_4^+, Ag^+, and other species. The extent of preference for H^+ determines the degree of selectivity in a pH electrode. Other glass electrodes are formulated to have enhanced selectivity for Na^+, but they show interference from H^+, Ag^+, and K^+. In general, Eq. (101) must be modified to account for the effect of an interferent j on ion i of interest,

$$E_{cell} = \text{const} + (RT/F)\ln(a_i + k_{i,j}^{pot}a_j) \tag{102}$$

where the a's are activities and $k_{i,j}^{pot}$ is a *potentiometric selectivity coefficient*. The values of $k_{i,j}^{pot}$ are available for various interferents in the manufacturer's literature. In using these electrodes, it is obviously important to be aware of possible interferents and to control their concentration levels.

Most electrodes are effective for measurements of concentration only above the 10^{-5} M level. At lower figures various effects, including competition from other ions or slight contamination from the membrane material, tend to make measurements meaningless. Only pH electrodes operate effectively at very low concentrations, a fact related to the participation of the solvent itself in the control of pH.

(c)   *New Directions and Outlook*†   Ion-selective electrodes are important commercial devices, because they make selective measurements very simply and inexpensively. They are often very rugged and reliable. These features will keep them important for the foreseeable future.

New developments have mostly involved schemes for making existing electrodes respond selectively to other analytes. For example, gas-sensing electrodes for NH_3, SO_2, and CO_2 have been constructed by placing a gas-permeable membrane over a glass electrode and trapping between the membrane and the glass a solution that reacts selectively with the molecule of interest to change the local concentration of an ion (e.g., H^+ or NH_4^+) that can be detected by the electrode. A similar strategy is used in enzyme-coupled systems [116, 120–123]. In that case an enzyme is immobilized on an electrode that responds to a product of the enzyme's operation on a substrate. In this way the electrode's response becomes sensitive to the substrate concentration, while selectivity is preserved through the natural selectivity of enzyme catalysis. Commercialization of some of these systems has begun, and one can expect additional developments from research along this line.

Nieman and co-workers [124] have recently suggested an interesting alternative to the standard scheme of potentiometric measurement. Their approach is to apply a bipolar voltage pulse to the cell, so that current flows, and to measure an apparent resistance. The change of this impedance from one test solution to another reflects the *dynamics* of the selective charge transport processes, rather than equilibrium characteristics of the state of charge. Since the rates of charge transport are related to concentration, useful calibration curves of impedance versus concentration can be realized. An important advantage of this approach is that measurements can be completed very quickly, sometimes in milliseconds, so that ion-selective electrodes could be considered for monitoring concentrations on time scales that are too short for potentiometric methods, which generally require 2–30 sec for equilibration. In addition a conventional reference does not seem to be needed, and this aspect could lend considerable flexibility in the design of compact, special purpose sensors.

D.   *Spectroelectrochemical Methods*

Over the years investigators have found that purely electrochemical means for characterizing an electrode process may not allow clear-cut identification of intermediates or adequate distinction between mechanistic alternatives. There has been much effort spent on improving the availability

† See [114–116].

of information from electrochemical experiments by making various parallel spectrometric measurements. In the past fifteen years research in this area has been intense and productive. There are now several techniques that could be regarded as standard methods, readily available to investigators who might occasionally want to use them. There are other workable ones that are intricate in the experimental details and in theory. These generally require a large commitment of time, money, and interest to the method itself. In keeping with the spirit of this chapter, our stress must be on the first group.

1. *Absorption Measurements: Semi-infinite Cells*†

Probably the most straightforward spectroelectrochemical experiment involves the measurement of absorbance changes induced by an electrode reaction. Virtually all work of this type has featured *optically transparent electrodes* (OTE) that allow light to pass in a direction normal to the surface, as shown in Fig. 27a. The various types of OTEs have been reviewed by Kuwana and Heineman [127]. In general, they are thin semitransparent films of metals (Au, Pt, Ni), carbon, or semiconductors (SnO_2, In_2O_3) on glass substrates, or they are "minigrids," which are meshes of bulk metal (Au, Ni) with a few hundred wires per cm. As electrode products build up, or as electroreactants are depleted, in the diffusion layer adjacent to the electrode, corresponding absorbance changes are registered by the beam of light.

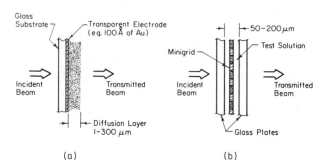

(a)   (b)

Fig. 27 (a) Spectroelectrochemistry in a semi-infinite system. Incident beam comes from outside the cell, passes through the electrode and its diffusion layer, then continues through a bulk of solution and on to the detector. The absorbance changes of interest come from alterations in the composition of the thin diffusion layer. (b) An optically transparent thin-layer electrode. Working electrode and test solution are trapped between two glass windows. The assembly is emptied by vacuum and filled by capillarity. A minigrid is shown here in cross section as the working electrode. The bottom edge of this assembly usually opens into a larger container of solution, in which the counter and reference electrodes are immersed.

† See [9, 125–129].

In the experiments that we are considering now, the cell wall is placed sufficiently far from the electrode surface not to interfere with the growth of the diffusion layer.

One important experimental mode is to hold the wavelength constant, often at an absorbance maximum for a product, and monitor absorbance $\mathscr{A}$ versus time, as a step experiment is performed. Since the light passes through the diffusion layer, it samples all of the product remaining there at the time of measurement. If the product is stable, the absorbance rises steadily as material accumulates. In this case the total amount of product at any time per unit area of a planar electrode is Q/nFA, where Q is the faradaic charge devoted to generation of the product. The trace of $\mathscr{A}$ versus t therefore runs parallel to that of Q versus t. The exact function is

$$\mathscr{A} = 2\,\epsilon\,C^*D^{1/2}t^{1/2}/\pi^{1/2} \tag{103}$$

where ϵ is the molar absorptivity of the product, and C^* and D apply to the electroreactant. Reversal of the potential step leads to an absorbance transient that also follows the chronocoulometric response.

In general, the information available from these absorption measurements is quite similar to that from corresponding chronocoulometric data. The shapes of the absorbance transients can be used to diagnose [71] the mechanism of an electrode process, perhaps more conveniently than with chronocoulometry, since there is no contribution from double-layer charging. Moreover, there is the added freedom to vary the wavelength, so that transients for more than one species can be followed.

Since absorbance is time-dependent, obtaining spectra at OTEs in semi-infinite cells requires care. One can repeat steps at various wavelengths and sample the absorbance at a fixed delay, or one can make sufficiently rapid scans over wavelength that the time dependence in $\mathscr{A}$ is negligible. Either of these procedures adds complexity. For spectral data it is usually simpler to use a thin-layer cell in the manner that we discuss next.

2.   Absorption Measurements: Thin-Layer Cells†

Some substantial benefits accrue when a thin-layer system like that in Fig. 27b is used for spectroelectrochemical measurements. From an electrochemical standpoint, the operation is the same as the kind of system discussed in Section IV.B.6. The major difference is that an OTE has been incorporated, permitting absorbance measurements on the trapped thin layer of solution. Typical cell thicknesses are 50–200 μm, hence seconds to tens of seconds are required to achieve homogeneity within the cell in response to a change in

---

† See [9, 127–129].

potential. These periods are long by comparison with those of thin layer cells used for purely electrochemical ends, but the added thickness that gives rise to slow response also improves the absorbance signal. The assembly of working electrode and cavity is often called an *optically transparent thin-layer electrode* (OTTLE).

These devices find their greatest utility with systems in which the electro-reactants and electroproducts are stable. Then, for reasons outlined in Section IV.B.6, the contents of the trapped layer will reach essentially a nernstian equilibrium with the electrode potential within a short time after the potential is imposed. Spectra are very easy to obtain on the thin layer at equilibrium, and they can provide most valuable clues to the identities of products of new electrode processes.

We have already seen that one of the most valuable services of a thin-layer cell is to provide n values by integration of the current transient upon application of a potential that will fully convert the electroreactant from one oxidation state to another. That procedure can also be valuable in OTTLE cells. However, there is also a very useful alternative [128]. By taking spectra from the equilibrated cell at intervals of a few mV over a range of potential where an electrode reaction occurs, one gathers data for the completely oxidized and completely reduced solutions, and for various mixtures of oxidized and reduced forms. From these spectra one can easily calculate the ratio of concentrations of oxidized and reduced forms present at any potential. Then from the Nernst equation,

$$E = E^{o'} + (RT/nF)\ln[O]/[R] \tag{104}$$

we see that a plot of E versus $\ln([O]/[R])$ should be linear with intercept $E^{o'}$ and slope RT/nF. This method allows determinations of $E^{o'}$ and n with no uncertainties from mass transfer coefficients and with little interference from background electrolysis.

Thin-layer spectroelectrochemical techniques have proven extremely useful in the study of redox couples of biological components. The data supplied by these methods, namely, spectra, n values, and formal potentials are often very helpful in understanding the chemistry of such species in vivo. Also valuable is the fact that an OTTLE requires only minute amounts of sample (as little as nanomoles in some cases), most of which can be recovered, if necessary.

Biologically active redox agents often show very poor charge transfer kinetics at electrodes. Thus it may be quite difficult to obtain reversible faradaic responses in conventional electrochemistry, even though the basic redox process is reversible in a chemical sense. Thin-layer cells are especially well adapted to the use of *mediators* [127, 128], which are chemical agents added to circumvent problems with slow heterogeneous kinetics. Their role is to shuttle electrons between the electrode and the components of real interest, hence they must be able to react with good facility both at the electrode and with the substrate in solution. In a thin-layer cell, where a reversible system comes to the equilibrium dictated by the electrode potential, one can expect the electrode itself to impose a nernstian balance of concentrations of oxidized and reduced forms on the mediator, then the mediator will impose the equilibrium balance on the substrate. By use of the mediator, then, the spectroelectrochemical determination of $E^{o'}$ and n, as outlined above can be applied. A group of mediators covering the whole potential range of biological interest has been developed by Kuwana, Heineman, and co-workers [128–130].

OTTLEs are simple to use, conserve the sample, and provide valuable information more readily than competing approaches. Methods based on them have already become quite popular and probably will grow to become the most widely used spectroelectrochemical approaches.

3.　Electron Spin Resonance†

Electrochemistry produces a host of odd-electron species, such as radical ions, that are often very conveniently characterized by electron spin resonance (ESR). A few studies have involved quantitative measurements of signal strength versus time, with an eye toward evaluation of kinetic parameters; but by far the majority of electrochemical–ESR experiments are qualitative in nature. The goal is either to detect and identify an intermediate or product of an electrode process, or to study unpaired spin distribution in a known odd-electron product.

Simple cells for generating radicals in situ at constant potential can be constructed, and a few are even commercially available from manufacturers of ESR spectrometers. Alternatively, one can generate the species of interest by bulk electrolysis outside the spectrometer and transfer it later. Techniques for this work are well developed and have been reviewed extensively [131, 132].

4.　Other Methods

We cannot take space to discuss all valuable spectroelectrochemical techniques, so the more extensive discussion above dealt only with those that can be adopted for use without major dedication to the method itself. Here we briefly outline other more elaborate approaches.

Internal reflection spectroelectrochemistry [9, 125, 126] involves the passage of a light beam by internal reflection through a substrate, such as a glass plate, on which an optically transparent film electrode adheres. At each

† See [9, 131, 132].

point of reflection on the side of the glass plate in contact with the electrode, the light beam is eligible for absorption by components in the electrolyte. However, only species within about 1000 Å from the surface can absorb photons, so this technique probes the near-surface region exclusively. It has been useful for measurements of homogeneous rate constants on time scales short enough to keep the diffusion layer thickness comparable to the sampling depth [133]. Potential steps no wider than about 500 μsec are required for this purpose.

Ellipsometry [9, 134, 135] *and reflectance spectroscopy* [9, 136] both involve the observation of the intensity and the state of polarization of light specularly reflected from a planar electrode. The theory behind these experiments is elaborate. Both methods have been used extensively for studies of film growth and adsorption in electrode processes.

Raman [137, 138] *and infrared spectroscopy* [139] could be extremely valuable for characterizing electrolytic products and adsorbates. Their spectra are rich with specific information about molecular structure, but there is, of course, very little material on or around most electrodes in operation. There are, therefore, serious problems with sensitivity. Even so, a good deal of experimental effort has been put into these areas recently, and worthwhile general tools could arise. Most attention has gone into Raman spectroscopy, because it appears that certain adsorbates interact with certain electrodes to enhance the cross section for Raman scattering by large factors (perhaps 10^6), and there is, of course, a corresponding dramatic increase in sensitivity. The origin of this *surface-enhanced Raman effect* is currently a topic of hot debate.

Auger electron spectrometry and x-ray photoelectron spectrometry are tools that can provide elemental composition and some information about oxidation states of elements present within 10–100 Å of the surface of a sample [140–143]. The sample is held in a vacuum chamber at 10^{-8}–10^{-10} Torr during observation, so these techniques cannot be applied to parts of operating cells. Even so, they can be used to examine working electrodes ex post facto, and they have become popular for characterizing the growth of surface films and for verifying deliberate chemical modifications of electrode surfaces. Their chief drawbacks are low sensitivity (>0.1–1% atomic in each element detected) and the uncertainty that the observed surface is actually that present in operations within a cell.

E. Additional Electrochemical Methods

The fundamental idea in all electrochemical techniques involving the passage of current is to perturb the working interface by some type of waveform and to record the response. Theory gives us a guide for interpretation. The shapes of waveforms that might be used are limited only by the imagination, and a great many have actually been explored. We have had space above only to examine those that have become the most widely exploited. Here we look more briefly at a few others.

An important group involves perturbation of the working interface by a sinusoidal signal of small amplitude. In one approach the cell is arranged with both oxidant and reductant present in the test solution, so that an equilibrium potential is defined. Then the potential is made to vary sinusoidally about the equilibrium value with an amplitude of 1–10 mV and a frequency of 10 Hz to 100 kHz. The cell is placed in one arm of an impedance bridge, and by balancing the bridge one obtains the real and imaginary components of the cell impedance [8, 9, 11, 12, 15, 144–147]. Theory is readily available for translating this information into chemical parameters, such as exchange currents and rate constants. This *ac impedance method* has a very wide range of time scales, and it produces data of excellent precision, because it is a null-balance steady-state approach. Its main use has been in the evaluation of $k°$ values of chemically reversible electrode processes. In general, it is an excellent quantitative tool but it does not yield data that readily lend themselves to diagnosis.

An alternative method [8, 9, 34, 36, 148] producing very similar information is *ac voltammetry* (or with a DME, *ac polarography*). The idea is to superimpose a small amplitude (again, 1–10 mV) sinusoidal signal (again, 10 Hz–100 kHz) on a slow linear ramp in potential, like that normally used in polarography or slow cyclic voltammetry. The ramp will generate the type of response that we have already discussed. Of interest here is the small alternating current that flows in response to the ac modulation in potential. This current is resolved electronically and its amplitude and phase angle are measured. Voltammograms of alternating current amplitude at a given phase angle are often plotted automatically. This information is much more extensive than that available from the bridge method and it has greater diagnostic value, but the method's strength still lies mainly in its quantitative precision over a wide frequency range. In recent years, Smith and co-workers have developed some novel instrumentation for multifrequency ac excitation and data analysis based on Fourier transformations [149].

Most electrochemical experiments involve control of potential, because one obtains the most selective control of the electrode process in that manner. An alternative is to control the current. The response is then the working electrode's potential observed as a function of time. *Chronopotentiometry* [9, 12, 14, 16, 150] is the name given to this method. In the simplest form, a constant current is applied, so that some species of interest is oxidized or reduced at a constant rate. As its concentration near the surface falls due to electrolytic depletion, a point is reached, called the *transition time*, when the analyte can no longer support the enforced current. Then, there is a shift in potential as the working electrode starts to oxidize or reduce the next most eligible species. The value of the transition time and the shape of the potential–time curve provide information about an electrode process similar to that obtained from voltammetry. Chronopotentiometry has fallen into disuse because it offers no real advantage, and suffers from some important drawbacks, with respect to voltammetry in studies of dissolved species. However, it has some positive features for studies of thin layers [89], and, given the current stress on surface films and modified electrodes, there could be a modest revival.

Finally, we need to consider the long-neglected variable of temperature in electrochemical studies. Van Duyne and Reilley [151] showed a few years ago that electrochemistry could be performed with simple apparatus at temperatures down to $-50°C$ and that such studies could be quite valuable in deciphering complex electrode processes involving homogeneous chemistry. The methods have been adopted rather slowly, but their adherents are extremely positive about them, and some nice results have been published. One can expect growth in this area over the next few years.

V. Experimental Aspects

Some excellent books on the practice of electrochemistry have been published in recent years by Meites [76], Gileadi *et al.* [17], Sawyer and Roberts [18], and Kissinger and Heineman [129]. A person seriously interested in taking up experimental work in this field is strongly advised to become acquainted with these sources. Our purpose in the last few sections is to sketch roughly the spectrum of choices that one can make in experimental apparatus and materials.

A. Working Electrodes

Upon embarking on an electrochemical investigation, immediate and careful selections of the working electrode and the medium must be made, because these two elements dictate the working range and the nature of the interface where all the electrochemistry will occur. The working electrode is probably the more important element, because its surface character can have profound effects on adsorption and charge transfer kinetics, and because its form and size determine, in part, the nature of the experimental approaches that can be tried with it. Moreover, one may be particularly interested in, e.g., oxide film growth on some particular metal, such as W or Ir, so that the substance of the electrode is a critical element in the chemical system. More usually, however, the electrode is intended as a probe of chemistry outside itself.

1. Shape and Size

Very early one must decide whether the goal is bulk electrolysis or characterization by a method dependent on well-defined mass transfer. In the first case a large area is needed. Shape may not be critical. In the second, a small area with a well-defined shape is essential. Large metal foils, pools of mercury, and porous beds of carbon are useful for bulk conversions, whereas polished disks of carbon or metals, hanging mercury drops, and DMEs are used in microelectrode measurements. Any microelectrode must be compatible with the theory used to treat the experiments that are planned. In general, that condition demands planar, cylindrical, or spherical electrodes. Moreover, it may require additional features, such as the ability to rotate without eccentricity (for hydrodynamic voltammetry) or a reproducible dropping action (for polarographic methods). After these requirements are understood, a choice can be made among several materials.

2. Mercury†

Mercury is always the working electrode of choice, if it is compatible with the experimental goals. It can be employed as a pool for bulk electrolysis, a hanging drop for voltammetry and chronocoulometry, a DME for polarography, or an electrodeposited film on C, Au, Ag, Ni, or Pt for hydrodynamic voltammetry, stripping, or spectroelectrochemical experiments. Its positive features are (a) excellent reproducibility, (b) an amorphous surface structure, (c) ready renewability, and (d) very high overpotential for cathodic background reactions, so that it yields a working limit at the most negative potentials possible in any solvent.

The drawbacks are (a) easy oxidation of the mercury itself, which makes operation more positive than ~ 0 V versus SCE impossible, (b) awkwardness in handling of the liquid metal in situations requiring sealed cells, e.g., with nonaqueous solvents, (c) a tendency to react with radicals produced in electrode reactions.

The new static mercury drop electrode, discussed in Section IV.C.1.d, is a particularly convenient device for using mercury in a dropping form or as a hanging drop.

† See [64, 76].

3. *Platinum*†

Usually electrochemists turn to platinum when they cannot use mercury. For bulk electrolysis this material is employed in the form of foils or gauzes, and for microelectrode measurements, as polished disks. The disks are generally fabricated by sealing wire into soft glass tubing and grinding a flat face perpendicular to the axis of the tubing, or by force-fitting a Pt cylinder into a bored-out Teflon mantle, which is then polished to give a flush face. Preparation of a Pt surface for actual measurements is an art that varies widely among laboratories. Section V.A.7 below deals with some of the important considerations.

Platinum is a very good general-purpose electrode. Although the overpotential for hydrogen evolution in aqueous media is quite small, it has a negative working range in nonaqueous solvents that is bettered only by mercury, and for most purposes it is as good as any other electrode material on the positive side. In aqueous media, it tends to form thin oxide films at potentials more positive than about 0.7 V versus SCE; thus the surface character of the electrode can change during the course of an experiment. These alterations can in turn affect electrochemical behavior. One must generally expect this type of problem for any solid electrode. Platinum is, of course, a metal that can show specific marked adsorption and catalysis.

4. *Gold*‡

Gold has a less catalytic surface than platinum, and it is used when inertness is desirable. It also is less subject to surface oxidation, hence one can perhaps more easily maintain the integrity of the metal/electrolyte interface throughout a given experimental series. The working range is similar to that found with Pt in any medium except aqueous systems containing species that tend to complex gold ions. For example, the halides tend to facilitate anodic oxidation of the gold and produce a drastically narrowed positive working range.

Only rare use of gold is encountered for bulk electrolysis. Virtually all applications of it are confined to microelectrode measurements, where it is used in the form of polished disks sealed in glass or mounted in Teflon. The availability of gold minigrids and the ease of evaporating it as thin films onto glass has brought gold prominence in optically transparent electrodes [125–129].

† See [17, 18, 152].
‡ See [18].

5. *Carbon*†

It is only a slight exaggeration to say that carbon is the electrode material of technology. Processes at carbon electrodes have been interesting to electrochemists for a long time, partly because a large number of industrially important electrochemical systems feature this material. In recent years it has also come to occupy an important position in fundamental research and analysis, because carbon is exceptionally versatile. One can purchase it in powders, fibers, rods, cloth, and pellets, and it can be used in amorphous or crystalline states. The technology of carbon is extremely important to the industrial establishment, hence numerous carbon products exist, and there is a large body of knowledge about its behavior in various forms. Much of this information is useful in the interpretation of its properties in electrochemical service.

Most of the forms in which carbon is found in electrodes are fundamentally graphitic. This is true of essentially all powders (including carbon black), fibers, spectroscopic carbon rods, pyrolytic graphite, and single crystals of graphite. In the graphite structure, carbon atoms are hybridized sp^2 and are bound to three neighbors in planar sheets of hexagonal rings. Sheets are stacked together to form crystals, but the forces holding the sheets together are van der Waals attractions only, and are weak. Dislocations along the intersheet plane are very easy, and this feature gives graphite its high lubricity. Differences between various graphites are in the sizes of the individual crystals and their degree of perfection.

Large single crystals are difficult to obtain and are expensive. They are used to some extent in electrochemistry, where one wishes to exploit or study variations in behavior of particular crystal faces.

Such differences are extreme with graphite. For example, electron mobility within a sheet is high, but from sheet to sheet, movement is slow. Thus single crystals show high conductivity in the plane parallel to the molecular sheets, but they are quite resistive along the axis perpendicular to this plane. Thus the *basal planes* of the crystal, parallel to the molecular planes, are much less active electrochemically than the *edge planes*, which cut across the sheets and therefore terminate them. On the other hand adsorption of aromatic molecules is favored on the basal planes, apparently from good π–π overlap. The lattice termination at edge planes leaves dangling bonds that are ultimately satisfied by reaction by adventitious contaminants, such as O_2 [153]. Various functional groups appear to develop, some of which are redox-active (such as quinoid groups) and contribute to the faradaic background. Much attention has been paid to the adsorptive power of basal-plane sites and the

† See [7, 18, 153].

reactivity of functional groups at edge planes during the past few years of intense research on surface modification in electrochemistry [55].

The next step down in order from single crystals is found in *stress-annealed pyrolytic graphite*, which is polycrystalline, but shows a high degree of alignment of the basal planes of the crystal. This material is now quite popular for research on surface modification because it offers many of the features of single crystals at moderate cost.

Still less ordered is ordinary *pyrolytic graphite*, which nonetheless retains considerable alignment of basal planes in a polycrystalline sample. This substance is quite inexpensive and is popular as a general-purpose carbon for voltammetry. It is also studied extensively in research aimed at surface modification. Wide use of pyrolytic graphite is made in stripping analysis as a substrate for mercury film electrodes. In this work the graphite is usually impregnated with wax to improve reproducibility and to minimize sample carryover in the pores between crystal grains. Carbon fiber ultramicroelectrodes recently developed by Wightman and co-workers for electroanalysis in vivo [154], are very similar to pyrolytic graphite.

Spectroscopic carbon is usually sold as rods, which are essentially compressed random graphite crystals. This material is quite porous and requires wax impregnation before use as an electrode. It has been employed in stripping voltammetry, although it is not as widely utilized as wax-impregnated pyrolytic graphite.

Powdered graphite has been used in porous beds for flow-through bulk electrolyzers, and it has been frequently dispersed into inert matrices and fabricated into microelectrodes. *Carbon paste* [7], which is graphite powder in mineral oil, makes a surprisingly well-behaved electrode for electrooxidations in aqueous media. The paste is simply packed into a shallow well that is drilled into a plastic mantle, then trimmed flush with the mantle's face. Contact is made by a metallic lead mounted in the interior of the well. More recently, machinable carbon dispersions in Kel-F and in polypropylene have been developed [155]. They offer many of the positive features of carbon paste, but are readily usable with nonaqueous solvents.

Carbon black is the most finely divided, least perfect form of graphite. The particle size is very small and the surface area is extremely high, thus carbon black has been of interest as a conducting support for electrocatalysts. Porous electrodes are fabricated from it by dispersing it with a binder, such as Teflon powder, under heat and pressure treatment. Such structures are employed in some industrial applications, including fuel cells.

A very important form of carbon is a comparatively new material called *glassy* or *vitreous carbon* [156]. It is noncrystalline and can be viewed as an amorphous state between graphite and diamond. It is impermeable and quite hard. Glassy carbon is available mostly in the form of rods, and it has become quite popular in electrochemistry for use in stationary or rotating disk electrodes.

Carbon electrodes in general offer a narrower working range than Pt in most media, and their background currents are larger, because they exhibit extensive surface electrochemistry. Interest in them derives from their particular chemical properties.

6. Semiconductors

Numerous semiconductors have been studied as working electrodes [9, 50–53], but interest has mainly been in the specific properties of each semiconductor, rather than in their development as general working electrodes. These materials have the special property of being able to undergo photoactivated electrochemistry upon absorption of light within the semiconductor [9, 50–53]. In recent years, they have received enormous attention because they might prove useful for solar energy conversion [52, 157, 158]. The details are beyond our scope here.

Two such materials, SnO_2 and In_2O_3, have, nonetheless, been brought into fairly routine use in electrochemical characterization. Both are deposited as thin films on glass substrates, and they are ordinarily so heavily doped as to show practically metallic behavior. Because they are transparent in the visible region, they have been used mainly as OTEs [125, 127, 128].

7. Preparation of Surfaces†

A great advantage of mercury is that it requires no preparation of the surface. All other electrode materials will show irreproducibilities in behavior if steps are not taken to create the same type of surface in successive experiments. Any electrode, of a solid material or of mercury, tends to change as it is exposed to a solution, because it picks up adventitious adsorbates, or it becomes oxidized, reduced, or film-covered. The gradual change in an electrode surface may be inconsequential or crucial, depending on the chemistry of the experiment.

For bulk electrolysis, electrodes are not usually pretreated because of the inconvenience. However, care should be given to microelectrodes, particularly in quantitative work, so that they start each experiment in approximately the same state. Attention is focused mostly on the *roughness*, state of *oxidation*, and *chemical cleanliness*. Solid microelectrodes are generally polished with metallographic abrasive grits in a sequence starting at perhaps 200 mesh and finishing with 0.3–0.05 μm Al_2O_3 or diamond. Various methods

† See [17, 18].

are used to control the state of oxidation. One may dip the electrode into an oxidant or reductant, or one may cycle the electrode in an electrolyte between oxidizing and reducing potential. For example, a popular procedure for treating Pt disks is to carry the electrode through several cycles of alternate steps into the anodic and cathodic background discharges. This sequence builds up an oxide film and is thought to destroy adsorbed organics during the positive phases, then it is believed to reduce the oxide to clean, active platinum during the negative phases. The sequence ends with a negative step, and the electrode is ready for the planned experiment. In contrast with this electrochemical pretreatment sequence, pyrolytic graphite can be renewed simply by cleaving it on a basal plane, removing thin sections.

In any series of experiments, one should at least be aware of possible effects of irreproducibilities or alterations in surface character; and have a rational plan to deal with such problems. A workable approach will usually arise from a little experimentation based on thoughts about the probable chemistry of the interface.

8. *Modified Electrodes*†

Deliberate attention to the preparation of an electrode surface in the manner described above is a kind of surface modification designed to ensure a reproducible starting point for experiments. A more sophisticated idea is to make chemical alterations on the surface, not for cleaning, but rather to create an interface tailored to a particular end, such as catalysis of an electrode reaction of interest. Modified electrodes involve the deliberate binding of chemical moieties generally quite different from the electrode material itself.

The past five years have seen extremely active research in this area, and several useful approaches to modification have been developed. Desired functional groups can be attached to the electrode by adsorption, by covalent binding, or by coating the electrode with a polymer that itself can bind electroactive centers. The references [55] listed provide a useful entry to the large, growing body of literature.

We deal with the subject here because it seems inevitable that particular modified surfaces will become standard, probably even commercially available, tools for electrochemists. A great deal is now being learned about the chemistry of electrode surfaces. Within a very short time these advances will change our now old-fashioned manner of thinking about the working electrode as a nearly unalterable element, whose properties just have to be accommodated.

† See [55].

B. Media

An electrochemical medium is generally a fairly complex solution, whose constituents together establish a reaction environment for an electrode process. In ordinary practice one has at least a *solvent* and a *supporting electrolyte*, the latter being added in high concentration (≥ 0.1 M) simply to assure an adequate ionic strength for conductivity. However, there may also be complexants, buffer components for pH control, secondary reagents intended to engage in follow-up reactions with a primary electroproduct, or species added to modify the interface by adsorption.

1. Solvents†

Any useful solvent must have sufficient polarity to dissociate ionic solutes at least partially into free ions, which are the charge carriers. In very nonpolar solvents, such as hexane or benzene, supporting electrolytes exist so prevalently in the form of ion pairs that their solutions are not usefully conductive. The dielectric constants of practical electrochemical solvents generally exceed five. Table I gives some of the properties of the most frequently used solvents.

Of course, a great deal of electrochemistry is done in aqueous media. *Water* [18, 76] is conveniently available in pure form‡; it has a high dielectric constant, so that its solutions of supporting electrolytes are quite conductive; and there is a great deal of knowledge about the chemistry that takes place in it. Acid–base chemistry is a dominant element in much work done in aqueous media. One must constantly be aware of the possible involvement of H^+ and OH^- in any electrode process. In quantitative measurements, whether for analysis or diagnosis, it is usually important to control the pH with high concentrations of acids or bases, or with buffers. Usually the additives that control pH do double duty as supporting electrolytes (e.g., 1 M HCl, 1 M NaOH, 0.5 M phosphate buffer), and sometimes they even serve a third purpose as complexants (e.g., 1 M HCl, ammoniacal and citrate buffers).

The drawbacks of water include a fairly narrow working range only 1.5–2.5 V wide, limited fundamentally by the evolution of hydrogen on the negative end and the evolution of oxygen at the positive limit. Of course, any particular system may have more restrictive limits imposed by an electrode reaction of some other constituent, such as the anodic dissolution of a DME

---

† See [18, 159, 160].

‡ Very pure water may be needed for detailed studies of interfacial structure or heterogeneous kinetics. The references provide an entry to the literature.

TABLE I

Properties of Electrochemical Solvents[a]

Solvent	Freezing point[b] (°C)	Boiling point[b] (°C)	Viscosity (cP)	Dielectric constant
Acetonitrile	−45.7	81.6	0.325_{30}	37.45_{20}
Ammonia	−77.7	−33.4		23.7_{-36}
Benzonitrile	−13.5	191.1	1.111_{30}	25.58_{20}
Dimethyl sulfoxide	18.55	189.0	2.003_{30}	46.7_{25}
1,2-Dimethoxyethane	−58	85		3.49_{20}
N,N-Dimethylformamide	−61	153.0	0.796_{25}	36.7_{25}
Methanol	−97.49	64.51	0.445_{25}	32.6_{25}
Methylene chloride	−96.7	39.95	0.393_{30}	8.93_{25}
Pyridine	−41.6	115.5	8.78	13.24
Tetrahydrofuran	−108.5	64		7.39_{25}
Water	0	100	0.8937_{25}	78.54_{25}

[a] Data for all solvents except water from C. K. Mann, *Electroanal. Chem.* **3,** 57 (1969). Data for water from C. D. Hodgman, ed., "Handbook of Chemistry and Physics," 42nd ed. Chem. Rubber Publ. Co., Cleveland, Ohio, 1960.

[b] At 1 atm.

or the reduction of a nitrate used as supporting electrolyte. In addition, aqueous media often are worse than others in promoting the formation of anodic oxide films on solid electrodes.

Other protic solvents, such as ethanol or methanol, generally give behavior similar to that which would be seen in a corresponding aqueous medium. They are used mostly to improve the solubility of organic solutes.

Getting away from the drawbacks of aqueous systems generally requires the use of an aprotic solvent, i.e., a solvent that does not ordinarily engage in protic acid–base chemistry. Examples are acetonitrile and methylene chloride. Advantages of aprotic media are considerable: they offer working ranges as wide as 6 V, and the removal of protic acid–base chemistry from the picture often allows reversible study of rather reactive species, such as radicals and radical ions. Their drawbacks are mainly in the difficulty of purification and in the relatively high resistance of their solutions. A great many solvents have been evaluated [159], but most work nowadays is being done in one of four.

(a)   *Acetonitrile* [18, 159, 160] is probably the most widely used aprotic solvent. Its dielectric constant is high enough to assure reasonable conductivity, and it gives the widest available working range (+ 3 V to − 2.8 V in a solution of 0.1 M tetra-n-butylammonium fluoborate at Pt). It is as inert a solvent as one can find, except for its tendency to form complexes with some metals. Acetonitrile is, however, not a very good solvent, in the sense of dissolving compounds readily, and one sometimes runs into complications from limited solubility of electroreactants or products. Its volatility makes it easily removed in procedures intended to isolate products. The background limits are controlled by discharge of supporting electrolytes or water in the solvent. The quality of any batch is controlled in large part by the water content, which is difficult to reduce below 1 mM. Commercial spectrograde acetonitrile will prove adequate for a good deal of interesting electrochemistry, but for the widest operating range and longest lifetimes of primary electroproducts, further purification of the solvent just before usage is necessary.

(b) *Methylene chloride* [18, 159, 160] is an excellent solvent for stabilizing oxidants, such as cation radicals. It has a positive working range (as wide as acetonitrile's) that is limited by the discharge of supporting electrolyte. Its negative limit, between -1.5 V and -2.0 V versus SCE, is set by the solvent's tendency to take up electrons and eliminate Cl$^-$. This process also limits the stability of strong reductants in methylene chloride. It is superior at dissolving things, and will produce concentrated solutions of many interesting nonpolar compounds, including polymers. Methylene chloride is also readily evaporated in work-up procedures. Commercial solvent of high quality (spectrograde or even reagent quality) is rather good, and since water is not appreciably soluble in methylene chloride, its quality does not rapidly deteriorate with exposure to the atmosphere. Some improvement in performance can, nonetheless, be made by distillation. A problem with this solvent is its low dielectric constant, which leads to quite high resistant electrolyte solutions.

(c) *N,N-Dimethylformamide* (DMF) is another solvent with superior dissolving power [18, 159, 160]. Reduced species often have excellent stability in it, and it displays very low background currents in the negative range. The negative background limit is as extreme as can be achieved in any solvent (almost -3.0 V versus SCE at Hg or Pt). On the positive side, it tends to show a background curve of slowly rising sizable currents from $+1.0$ V to a limit at about $+1.6$ V versus SCE. Strong oxidants tend not to be very stable in it. DMF is a fairly involatile substance and is generally hard to remove during the workup of products. It does possess a high dielectric constant and gives solutions of good conductivity. Since DMF tends to decompose with time and has an affinity for water, the commercial solvent is generally not very satisfactory for electrochemistry. Purification by distillation is essential for good performance.

(d) *Tetrahydrofuran* (THF) has the lowest dielectric constant of the four aprotic solvents considered here [18, 159, 160]; consequently its solutions of electrolytes are the least conductive. Its virtue is in its excellent properties at negative potentials. Strong reductants often show high stability in THF, and the cathodic background limit is about as negative as with DMF. At positive potentials, its properties are more limiting. Again, a range comparable to that of DMF is available. THF is very volatile and can be removed easily in workup procedures. The commercial solvent is stabilized against peroxide formation by addition of radical scavengers, hence it must be distilled before electrochemical use.

2. *Supporting Electrolytes*†

In aqueous media, any ionic solute can be used as a supporting electrolyte, although one must obviously choose something that is not electroactive in the potential range of interest. Usually, the concentration should be 0.1 M or greater. As we have seen, pH control is very important in aqueous media, and the supporting electrolyte often serves that end. Favorite strong acids are HCl, H_2SO_4, and $HClO_4$. Strongly basic media include solutions of NaOH, KOH, and tetraethylammonium hydroxide. Among the more popular buffered media are those based on acetate, phosphate, and ammonia. If the electrode process is known not to involve H^+ or OH^-, neutral unbuffered media may be worthwhile. Solutions of KCl are most frequently used.

Nonaqueous media require specialized supporting electrolytes with much larger ions, so that pairing is minimized and the greatest possible conductivity is obtained. Only three find extensive use: tetra-*n*-butylammonium perchlorate (TBAP), tetra-*n*-butylammonium fluoborate ($TBABF_4$), and tetraethylammonium perchlorate (TEAP). All three are commercially available in satisfactory purity, although performance is improved by drying them in vacuo at an elevated temperature. In general, TEAP is less satisfactory than the other two, because it has a smaller cation and suffers more from ion pairing. At negative potentials, in any solvent, all three supporting electrolytes will perform about the same; but at positive potentials, perchlorate starts to undergo electrooxidation near 1.6 V versus SCE, and this process establishes the background limit in acetonitrile [159] or methylene chloride. Moreover, the perchlorates are harder to dry and are more dangerous to handle than the fluoborate. $TBABF_4$ is the strongly preferred supporting electrolyte for all these reasons. It is not electroactive in any solvent between $+3$ V and -2.9 V versus SCE.

† See [18, 159, 160].

3. Deaeration†

For reductions at potentials more negative than about 0 V versus SCE, one must deaerate the solution to avoid interference from the reduction of molecular oxygen, which is present in air-saturated solutions at concentrations up to 2 mM. The cell must contain some means for sealing out the atmosphere and for maintaining an inert blanket of gas over the solution.

Traditionally, the job of deaeration has been accomplished by bubbling a purified stream of N_2 or Ar through the solution for 3–20 min, then switching the stream, so that it purges the space above the solution continuously during the experiment. This approach is simple, effective, and inexpensive. One need only take care that the gas stream is both free of O_2 and saturated with solvent before it enters the cell. Various purification trains have been described in the literature for treating bottled N_2 or Ar for electrochemical use.

With aprotic solvents, many workers have chosen to prepare samples on vacuum lines instead. Since many of the solvents need to be protected from exposure to the atmosphere, it can be helpful to store them in sealed containers that are easily connected to vacuum lines. The solvent can then be distilled through the line by a bulb-to-bulb process, and in this manner one can achieve perhaps the best exclusion of water from the test solution. If deaeration is still necessary, it is convenient to carry out that operation by freezing the solution with liquid nitrogen, pumping out the space above the solution, then thawing. Repeating this freeze–pump–thaw cycle two or three times gives extremely low levels of oxygen. Additional information about the use of vacuum lines is available in the literature [129, 159].

It is worth remembering that deaeration is usually unnecessary if the intended operations are restricted to potentials more positive than 0 V versus SCE.

C. *Reference Electrodes*‡

Most electrochemical instrumentation senses the reference electrode's potential through high impedance circuitry that draws extremely small currents; hence it is no longer important for that electrode to have an ability to deliver sizable currents with small overpotentials. The important attributes are stability, compactness of construction, intercomparability among laboratories, and chemical compatibility with the system of interest.

Stability and intercomparability require a *poised* system within the reference electrode, which means that a well-defined kinetically facile

† See [18, 76, 160].
‡ See [17, 18, 22–24, 76, 129].

electrochemical couple controls the potential there, and that both oxidized and reduced forms are present so that an equilibrium potential, predictable from the Nernst equation, exists. One could make reference electrodes, for example, of Pt/Fe^{3+} (1 mM), Fe^{2+} (1 mM) in 1 M HCl or Ag/Ag^+ (0.01 M in CH_3CN), based on the processes,

$$Fe^{3+} + e \rightleftarrows Fe^{2+} \tag{105}$$

$$Ag^+ + e \rightleftarrows Ag \tag{106}$$

It is important to specify and control the concentrations of each active component. Since there are several well-behaved couples, a great many different electrodes can be constructed and used validly. Sometimes unusual experimental situations make it convenient to devise a special reference based on these principles, but the great majority of work is done with three or four standard systems.

The most popular is the *aqueous saturated calomel electrode* (SCE), which is based on

$$Hg_2Cl_2 + 2e \rightleftarrows 2Hg^0 + 2Cl^- \tag{107}$$

It is generally a glass envelope containing a pool of Hg overlaid by a mixture of solid Hg_2Cl_2 (calomel) and KCl, all of which is in equilibrium with a solution saturated in Hg_2Cl_2 and KCl. A wire extends into the mercury without contacting the solution, so that external electrical connection can be made. Figure 28a contains a common design. At 25°C this reference electrode has a potential of 0.2415 V versus SHE.†

SCEs are easy to construct, and they operate faithfully. Their main drawbacks are in having (a) components that often lead to an intrinsically bulky design, (b) a high temperature coefficient of potential, and (c) an aqueous electrolyte. This last point is only a problem for work with non-aqueous solvents. Then, there must exist at the tip of the reference electrode a liquid junction (Section IV.C.4.a) between the aqueous KCl solution (within the reference electrode) and the nonaqueous test solution. Uncertain junction potentials can arise, and one may suffer contamination of the test solution by water diffusing from the reference electrode.

The *silver/silver chloride* electrode is similar to the SCE in concept. A typical design is available in Figure 28b. The electrode process is,

$$AgCl + e \rightleftarrows Ag + Cl^- \tag{108}$$

---

† *Hydrogen electrodes* such as $Pt/H_2/H^+$ are occasionally used for reference purposes, but they are not very convenient. The *standard hydrogen electrode*, where H^+ and H_2 exist at unit activity, is the thermodynamic and electrostatic reference for the potential scale, but it cannot be constructed experimentally. See Section III.A.2.

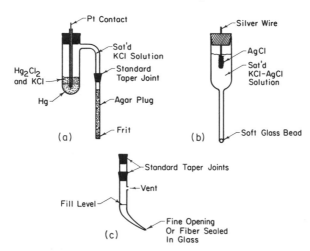

Fig. 28   (a) A typical saturated calomel electrode. (b) A type of Ag/AgCl electrode. The silver wire is first anodized in chloride medium to create a coating of AgCl. (c) Luggin capillary. The upper standard taper joint is meant to receive a matched one on the reference electrode [like that in (a)]. The lower one is a male joint for mounting the capillary into the cell. In operation, the test solution fills the capillary to the level shown, and the working electrode's tip makes contact within the capillary.

and solubility equilibria of AgCl and KCl are used to control the concentrations of potential-determining constituents. This system is supplanting the SCE to some degree, because it is simpler to construct and lends itself to very compact designs. Its potential is 0.197 V versus SHE, or -0.045 V versus SCE. The drawbacks are those cited as (b) and (c) for the SCE.

In aqueous systems, it is sometimes necessary to avoid contamination by Cl^-. Then the SCE and the Ag/AgCl electrodes are unsatisfactory. Alternates based on Hg/HgO or Hg/Hg_2SO_4 are commonly used.

To get around the problem of contamination of nonaqueous test solutions by water, one might go to a reference electrode with a nonaqueous solvent. A good bit of development along this line has been carried out, but the only device with much currency is the silver/silver-ion electrode, Ag/Ag^+ (0.01 M). The solvent in which Ag^+ is dissolved (usually as the perchlorate) can vary, depending on the composition of the test solution. A change in the solvent causes a corresponding change in potential, hence the potential of this type of reference should always be measured against an SCE, in the interest of standardizing the interchange of information among laboratories. The main drawback to the silver/silver-ion electrode is having to maintain the concentration of Ag^+ to compensate for leakage and evaporation of solvent.

Most workers prefer to use an SCE or an Ag/AgCl electrode for all work and deal with leakage of water in other ways. One of them is to interpose a

Luggin capillary (Fig. 28c) between the tip of the reference and the test solution. Water contaminating the solution within the buffer chamber will not affect the part of the test solution examined by the working electrode. Alternatively, one can put a tip on the reference electrode that has a high mass transfer resistance, so that ionic contact between the test solution and the reference electrode's electrolyte is maintained, but only a very low rate of cross diffusion can occur. Favorites [18] are fine frits, fine asbestos fibers sealed in glass, porous ("thirsty") Vycor, and Pyrex/soft-glass junctions, which develop fissures on cooling. These tips are usually rather resistive in an electrical sense, but for most work this feature is inconsequential. The main problem is eventual loss of a good reference standard by plugging of the tip with precipitates.

Finally, we should examine quasi-reference electrodes (QRE) [160, 161], which are nothing more than silver or Pt wires immersed directly in test solutions. These are not true references at all, because they are not poised. However, they are buffered against changes in potential by their own double-layer capacitances and by any faradaic surface processes peculiar to them (e.g., oxide formation), and when they are sensed by high-impedance circuitry they may be quite stable over a period of hours in a given matrix. They are useful because they have extremely simple construction, they are compatible with the sealed cells often used with nonaqueous solvents, and they do not contaminate the test solution. Their drawback is in not providing a stable reference from day to day. In doing voltammetry with them, one frequently adds a species with known electrochemistry to the solution to act as an internal reference. Then the waves or peaks for the solutes of real interest can be referred to those of the standard. Of course, the standard must be innocuous.

D. Cells

The design of cells is electrochemistry's highest art form. The literature contains innumerable figures of devices of exquisite compactness, with marvelously blown glass and intricate, difficult metal-to-glass seals. These cells have been developed for many specific purposes, such as spectroelectrochemistry, bulk electrolysis, or operations with a vacuum line; hence it is difficult to generalize the principles of design. A few discussions of important aspects are available in the literature [18, 76, 87, 129, 160].

In its simplest form, an electrochemical cell can be a beaker with a Teflon top that has a few holes drilled to allow insertion of the three electrodes. Elaborations might permit the bubbling of nitrogen through the test solution and subsequent diversion of the gas stream after deaeration to maintain a blanket above the solution. Sophisticated cells generally feature standard-taper joints in the top, so that vacuum tight seals around the electrodes are obtained. This approach is essential if the cell is to be filled or deaerated on a vacuum line.

The relative positions of the electrodes can be important. In general, it is a good idea to place the tip of the reference electrode as close as possible to the working electrode without obstructing mass transfer. This step reduces *uncompensated resistance*, which is the part of the iR drop in solution (Section III.E) that falls between the face of the working electrode and the tip of the reference. A *Luggin capillary* (Section V.C) often helps with this placement. In work with microelectrodes, the currents are so small that relative placement of the counter electrode with respect to the working electrode is not critical. Moreover, it is usually unnecessary to isolate the counter electrode by placing it in a separate compartment, because only small amounts of electrolytic products will be generated there. For bulk electrolysis, on the other hand, isolation is often essential, and placement becomes very important. A useful rule is that most of the current in a cell passing large currents will flow between points on the working and counter electrodes *in closest proximity*. To obtain uniform current density at the working electrode and thereby utilize its full area, one must employ parallel or concentric working and counter electrodes.

Beyond these generalities, the matter of cell design is determined by the details of the particular purposes for which the cell is intended. It is not profitable for us to delve further here. For investigators thinking about a general-purpose cell for occasional diagnostic and analytical measurements at microelectrodes, perhaps the best policy is to avoid worry over design and buy a ready-made cell and electrodes from one of the prominent manufacturers of electrochemical equipment.

E. Instrumentation

Electrochemists are peculiarly independent, and in their papers they often describe custom-built equipment that seems inaccessible to someone seeking occasional use of electrochemical measurements. In fact, the equipment can often be quite simple, and can be constructed quickly and inexpensively. However, most new users prefer to buy ready-made facilities, so it is fortunate that several excellent manufacturers have put good equipment into the field at prices that are inexpensive by comparison to those of other routine items like spectrophotometers and chromatographs.

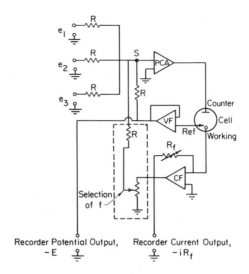

Fig. 29 An adder potentiostat. CF = current follower, VF = voltage follower, PCA = potential control amplifier, S = summing point. Dashed box shows circuitry for compensation of resistance by positive feedback.

1. *Basic Design*

All *potentiostats* are now based on *operational amplifiers* [9, 15, 18, 33–35, 129, 162], which are high-gain stages that are manipulated as single circuit elements. In practice they may be monolithic integrated circuits or they may be packages of discrete transistor circuits. The use of operational amplifiers is simple and elegant, but it is too much a side issue for us to cover here. Figure 29 is a diagram of an *adder potentiostat*, which is a practically universal basic design [9, 15, 18, 33–35, 129]. The working electrode is held virtually at circuit ground by a *current follower*, which also converts the current flow through the working electrode to a proportional voltage, which in turn is made available to a recorder. The *potential control amplifier* (or *summing amplifier*) is the central element. It forces enough current through the counter and working electrodes to maintain the potential of the working electrode at the desired value versus the reference. (Actually, it controls the voltage of the reference versus ground, in electronic terms.) The potential of the reference is sensed through a *voltage follower*, which has an extremely high input impedance. The desired potential of the working electrode is indicated to the potential control amplifier by applying one or more voltages to the inputs. If it has the necessary speed and power, the potentiostat will always maintain the working electrode at a potential equal to the sum of these voltages,

$$E_{\text{wk vs. ref}} = e_1 + e_2 + e_3 \qquad (109)$$

Complex time-dependent waveforms can be created by putting various signals at e_1, e_2, and e_3. The *summing point* S, at the input of the potential control amplifier, is the junction where the actual potential of the reference electrode versus ground (i.e., reference versus working) is compared against the expected value. Any discrepancy will cause the summing amplifier to react correctively.

To achieve a specific potential program, e.g., for cyclic voltammetry, one must make $e_1 + e_2 + e_3$ follow the desired waveform. This is done in practice by using another electronic box, a *function generator*, to feed the proper time-dependent voltages into the potentiostat's inputs. Often electrochemical instruments are constructed by putting the function generator and the potentiostat together in a single chassis, but one can also buy the units separately. Function generators can be made from a variety of analog and digital circuits.

2. Performance

The important characteristics of a potentiostat are speed, voltage compliance, and current capacity. Commercial equipment can establish control over a cell in characteristic times ranging from < 1 μsec to perhaps 1 msec. If fast experiments are contemplated, this figure may be important, but for most work it is not. A good general-purpose instrument for doing sweeps up to 100 V/sec and steps or pulses no shorter than 10 msec can do quite well with a 50 μsec rise time, which is a routine figure. For bulk electrolysis, speed is not at all critical.

Voltage compliance is the range of voltage between the counter and working electrodes that the potentiostat can impose. Figures between ± 10 V and ± 100 V are available in commercial equipment. Large compliance is needed when there is a big iR drop between the working and counter electrodes. This can happen when there are very high currents, as in bulk electrolysis. For example, passing 1 A through only 50 ohms of solution resistance requires 50 V. These figures could be found in an aqueous system with a frit separating the counter and working electrode compartments. Good compliance is also important where resistive media, such as nonaqueous solvents, are employed. In THF, for example, one can easily have 20 kΩ between the counter and working electrodes; hence 100 V would be needed to push even 5 mA through the cell.

The capacity is the limiting current that can be delivered by the amplifiers within the potentiostat; e.g., 10 mA to 10 A are often encountered. High current capacity may be essential for bulk electrolysis or for very fast experiments (Section IV.B.2.d).

Most routine electrochemical work with microelectrodes, and even with bulk electrolysis, can be done satisfactorily with a potentiostat having 50 μsec rise time, ± 15 V compliance, and 200 mA capacity. For fast experiments with microelectrodes in aqueous media one desires < 1 μsec rise time, ± 15 V compliance, and 1 A capacity. With nonaqueous media compliance up to ± 100 V is helpful. For bulk electrolysis at exceptionally large electrodes, capacities of 5–10 A, with the highest available compliances (usually ± 20 V) are necessary. In general, speed and power (compliance times capacity) cost money, so it is worthwhile in buying instruments to match them to one's real needs.

3. *Electronic Compensation of Resistance*†

In Section III.E, we saw that the measured difference in voltage between the working and reference electrodes in any cell is the true potential difference, which we desire to know and control, plus a part of the iR drop in solution, which just complicates the measurement. This iR drop is the product of cell current and the so-called uncompensated resistance R_u. The size of R_u determines the maximum rate of potential change that can be imposed on the working electrode (Section IV.B.2.d), and it controls the importance of certain artifacts that can complicate the interpretation of experimental results. Fundamentally, the problem is that the real potential of the working electrode is not, in terms of $e_1 + e_2 + e_3$ as shown in Fig. 29, but is instead‡

$$E_{true} = e_1 + e_2 + e_3 + iR_u \tag{110}$$

Thus, the actual potential program felt by the working electrode is not $e_1 + e_2 + e_3$, nor is it the one used by the theory underlying the interpretation of results. It is always in our best interest to minimize the discrepancy iR_u.

Experimentally, one can help matters by using small electrodes, high supporting electrolyte concentrations, and low analyte concentrations. However, in fast experiments or in resistive media iR_u may still be significant.

Quite a few workers have adopted an electronic measure for counteracting iR_u called *positive feedback compensation*. Facility for it is often available on commercial potentiostats. The dotted block in Fig. 29 shows how the scheme is implemented. The current follower, as we have seen, produces a voltage proportional to i. In fact, its output is $-iR_f$. The potentiometer gives us the ability to take a fraction f of that output and feed it back to the summing junction S. This voltage $-fiR_f$ is a kind of input, just like e_1, e_2, or e_3, hence the working electrode will not behave according to (110), but instead it will follow,

$$E_{true} = e_1 + e_2 + e_3 + i(R_u - fR_f) \tag{111}$$

† See [9, 15, 17, 18, 34–39, 129].

‡ The term iR_u is added, rather than subtracted, because we have taken cathodic i as positive.

By selecting f and R_f appropriately, one should therefore be able to force the last term in (111) to zero and thereby "compensate" iR_u.

In practice, adjustment of fR_f is a tricky business, since one generally does not know what value R_u has. Often, the potentiostat becomes unstable and breaks into oscillation near the point of full compensation. Thus, a common practice is to dial the potentiometer controlling f until oscillation occurs, then to back off to a figure about 10–20% smaller. Generally, this approach gives partial compensation, but it can lead even to overcompensation. Anyone thinking about using positive feedback really ought to examine the literature carefully before doing so. There are some nice reviews [36, 37, 39, 129]. Ahlberg and Parker have very recently advanced some rather unconventional ideas about achieving full compensation [163, 164]. Their practice could prove generally useful.

Acknowledgments

The author is grateful to Dr. Marcin Majda for giving the manuscript a careful reading and for offering many suggestions for improvement. Ms. Glenna Wilsky's painstaking preparation of the typescript is sincerely appreciated. Additional thanks go to Mr. Eldon Boatz for his work on the graphics. The National Science Foundation has generously supported some of this work under Grant CHE-81-06026.

References

1. J. O'M. Bockris and A. K. N. Reddy, "Modern Electrochemistry," 2 vols. Plenum, New York, 1970.
2. B. E. Conway, "Theory and Principles of Electrode Processes." Ronald Press, New York, 1965.
3. J. Koryta, J. Dvorak, and V. Bohackova, "Electrochemistry." Methuen, London, 1966.
4. D. A. MacInnes, "The Principles of Electrochemistry." Dover, New York, 1961. (Corrected version of 1947 ed.)
5. J. S. Newman, "Electrochemical Systems." Prentice-Hall, Englewood Cliffs, New Jersey, 1972.
6. K. J. Vetter, "Electrochemical Kinetics." Academic Press, New York, 1967.
7. R. N. Adams, "Electrochemistry at Solid Electrodes." Dekker, New York, 1969.
8. A. M. Bond, "Modern Polarographic Methods in Analytical Chemistry." Dekker, New York, 1980.
9. A. J. Bard and L. R. Faulkner, "Electrochemical Methods." Wiley, New York, 1980.
10. G. Charlot, J. Badoz-Lambling, and B. Tremillon, "Electrochemical Reactions." Elsevier, Amsterdam, 1962.
11. B. B. Damaskin, "The Principles of Current Methods for the Study of Electrochemical Reactions." McGraw-Hill, New York, 1967.

12. P. Delahay, "New Instrumental Methods in Electrochemistry." Wiley (Interscience), New York, 1954.
13. Z. Galus, "Fundamentals of Electrochemical Analysis." Horwood, Chichester, Sussex, England, 1976.
14. J. J. Lingane, "Electroanalytical Chemistry," 2nd ed. Wiley (Interscience), New York, 1958.
15. D. D. Macdonald, "Transient Techniques in Electrochemistry." Plenum, New York, 1977.
16. C. N. Reilley and R. W. Murray, "Electroanalytical Principles." Wiley (Interscience), New York, 1963.
17. E. Gileadi, E. Kirowa-Eisner, and J. Penciner, "Interfacial Electrochemistry—An Experimental Approach." Addison-Wesley, Reading, Massachusetts, 1975.
18. D. T. Sawyer and J. L. Roberts, Jr., "Experimental Electrochemistry for Chemists." Wiley (Interscience), New York, 1974.
19. M. E. Peover, *Electroanal. Chem.* **2,** 1 (1967).
20. M. M. Baizer, "Organic Electrochemistry." Dekker, New York, 1973.
21. C. K. Mann and K. K. Barnes, "Electrochemical Reactions in Nonaqueous Systems." Dekker, New York, 1970.
22. D. J. G. Ives and G. J. Janz, eds., "Reference Electrodes." Academic Press, New York, 1961.
23. J. N. Butler, *Adv. Electrochem. Electrochem. Eng.* **7,** 77 (1970).
24. R. G. Bates, "Determination of pH," 2nd ed. Wiley (Interscience), New York, 1973.
25. W. M. Latimer, "The Oxidation States of the Elements and Their Potentials in Aqueous Solutions." Prentice-Hall, Englewood Cliffs, New Jersey, 1952.
26. A. J. Bard and H. Lund, eds., "The Encyclopedia of the Electrochemistry of the Elements." Dekker, New York, 1973.
27. G. Milazzo and S. Caroli, "Tables of Standard Electrode Potentials." Wiley (Interscience), New York, 1977.
28. R. Parsons *et al., Pure Appl. Chem.* **37,** 503 (1974).
29. P. Delahay, "Double Layer and Electrode Kinetics." Wiley (Interscience), New York, 1965.
30. W. J. Albery, "Electrode Kinetics." Oxford Univ. Press (Clarendon), London and New York, 1975.
31. T. Erdey-Gruz, "Kinetics of Electrode Processes." Wiley (Interscience), New York, 1972.
32. J. T. Maloy and A. J. Bard, *J. Am. Chem. Soc.* **93,** 5968 (1971).
33. W. M. Schwarz and I. Shain, *Anal. Chem.* **35,** 1770 (1963).
34. D. E. Smith, *Electroanal. Chem.* **1,** 1 (1966).
35. R. R. Schroeder, *in* "Computers in Chemistry and Instrumentation. Vol. 2: Electrochemistry" (J. S. Mattson, H. B. Mark, Jr., and H. C. MacDonald, Jr., eds.), Chap. 10. Dekker, New York, 1972.
36. D. E. Smith, *Crit. Rev. Anal. Chem.* **2,** 247 (1971).
37. J. E. Harrar and C. L. Pomernacki, *Anal. Chem.* **45,** 47 (1973).
38. D. Garreau and J. M. Saveant, *J. Electroanal. Chem.* **86,** 63 (1978).
39. D. Britz, *J. Electroanal. Chem.* **88,** 309 (1978).
40. V. G. Levich, "Physicochemical Hydrodynamics." Prentice-Hall, Englewood Cliffs, New Jersey, 1962.
41. J. Newman, *Electroanal. Chem.* **6,** 187 (1973).
42. J. Newman, *Adv. Electrochem. Electrochem. Eng.* **5,** 87 (1967).
43. R. B. Bird, W. E. Stewart, and E. N. Lightfoot, "Transport Phenomena." Wiley, New York, 1960.

44. J. J. Lingane and I. M. Kolthoff, *J. Am. Chem. Soc.* **61**, 1045 (1939).
45. D. C. Grahame, *Chem. Rev.* **41**, 441 (1947).
46. R. Parsons, *Mod. Aspects Electrochem.* No. 1, 103 (1954).
47. D. M. Mohilner, *Electroanal. Chem.* **1**, 241 (1966).
48. R. Payne, *in* "Techniques of Electrochemistry" (E. Yeager and A. J. Salkind, eds.), Vol. 1, pp. 43–140. Wiley (Interscience), New York, 1972.
49. F. C. Anson, *Acc. Chem. Res.* **8**, 400 (1975).
50. H. Gerischer, *in* "Physical Chemistry" (H. Eyring, D. Henderson, and W. Jost, eds.), Vol. 9A, pp. 463 ff. Academic Press, New York, 1970.
51. V. A. Myamlin and Yu. V. Pleskov, "Electrochemistry of Semiconductors." Plenum, New York, 1967.
52. A. J. Nozik, *Annu. Rev. Phys. Chem.* **29**, 189 (1978).
53. R. Memming, *Electroanal. Chem.* **11**, 1 (1979).
54. G. Gouy, *J. Phys. Radium* **9**, 457 (1910); *C. R. Hebd. Seances Acad. Sci.* **149**, 654 (1910); D. L. Chapman, *Philos. Mag.* **25**, 475 (1913); O. Stern, *Z. Elektrochem.* **30**, 508 (1924).
55. R. F. Lane and A. T. Hubbard, *J. Chem. Phys.* **77**, 1401 (1973); B. F. Watkins, J. R. Behling, E. Kariv, and L. L. Miller, *J. Am. Chem. Soc.* **97**, 3549 (1975); A. P. Brown, C. Koval, and F. C. Anson, *J. Electroanal. Chem.* **72**, 379 (1976); R. J. Lenhard and R. W. Murray, *J. Electroanal. Chem.* **78**, 195 (1977); M. Fujihira, A. Tamira, and T. Osa, *Chem. Lett.* p. 367 (1977); A. Merz and A. J. Bard, *J. Am. Chem. Soc.* **100**, 3222 (1978); J. Zagal, R. Sen, and E. Yeager, *J. Electroanal. Chem.* **83**, 207 (1977); A. M. Yacynych and T. Kuwana, *Anal. Chem.* **50**, 640 (1978); J. P. Collman, M. Marrocco, P. Denisevich, C. Koval, and F. C. Anson, *J. Electroanal. Chem.* **101**, 117 (1979); H. O. Finklea, H. Abruna, and R. W. Murray, *Adv. Chem. Ser.* No. 184, 253 (1980); M. S. Wrighton, A. B. Bocarsly, J. M. Bolts, M. G. Bradley, A. B. Fischer, N. S. Lewis, M. C. Pallazzotto, and E. G. Walton, *Adv. Chem. Ser.* No. 184, 269 (1980); K. D. Snell and A. G. Keenan, *Chem. Soc. Rev.* **8**, 259 (1979); R. W. Murray, *Acc. Chem. Res.* **13**, 135 (1980).
56. A. C. Riddiford, *Adv. Electrochem. Electrochem. Eng.* **4**, 47 (1966).
57. W. J. Albery and M. L. Hitchman, "Ring-Disc Electrodes." Oxford Univ. Press (Clarendon), London and New York, 1971.
58. S. Bruckenstein and B. Miller, *Acc. Chem. Res.* **10**, 54 (1977).
59. D. Pletcher, *Chem. Soc. Rev.* **4**, 471 (1975).
60. D. H. Evans, *Acc. Chem. Res.* **10**, 313 (1977).
61. K. Vesely and R. Brdicka, *Collect. Czech. Chem. Commun.* **12**, 313 (1947).
62. I. M. Kolthoff and J. J. Lingane, "Polarography," 2nd ed. Wiley (Interscience), 1952.
63. L. W. Marple, L. E. I. Hummelstadt, and L. B. Rogers, *J. Electrochem. Soc.* **107**, 437 (1960).
64. D. L. Akins and R. L. Birke, *Chem. Phys. Lett.* **29**, 428 (1974).
65. T. Kuwana, M. Fujihira, K. Sunakawa, and T. Osa, *J. Electroanal. Chem.* **88**, 299 (1978).
66. J. P. Collman, M. Marrocco, P. Denisevich, C. Koval, and F. C. Anson, *J. Electroanal. Chem.* **101**, 117 (1979).
67. W. J. Albery, M. L. Hitchman, and J. Ulstrup, *Trans. Faraday Soc.* **64**, 2831 (1968).
68. F. G. Cottrell, *Z. Phys. Chem.* **42**, 385 (1902).
69. F. C. Anson, *Anal. Chem.* **38**, 54 (1966).
70. T. Kambara, *Bull. Chem. Soc. Jpn.* **37**, 523 (1974).
71. M. K. Hanafey, R. L. Scott, T. H. Ridgway, and C. N. Reilley, *Anal. Chem.* **50**, 116 (1978).
72. W. V. Childs, J. T. Maloy, C. P. Keszthelyi, and A. J. Bard, *J. Electrochem. Soc.* **118**, 874 (1971).

73. J. H. Christie, R. A. Osteryoung, and F. C. Anson, *J. Electroanal. Chem.* **13**, 236 (1967).

74. J. H. Christie, *J. Electroanal. Chem.* **13**, 79 (1967).

75. F. C. Anson, J. H. Christie, and R. A. Osteryoung, *J. Electroanal. Chem.* **13**, 343 (1967).

76. L. Meites, "Polarographic Techniques," 2nd ed. Wiley (Interscience), New York, 1958.

77. J. Heyrovsky, *Chem. Listy* **16**, 256 (1922).

78. D. Ilkovic, *J. Chim. Phys.* **35**, 129 (1938).

79. J. J. Lingane, *Ind. Eng. Chem., Anal. Ed.* **15**, 588 (1943).

80. R. S. Nicholson and I. Shain, *Anal. Chem.* **36**, 706 (1964).

81. J. M. Saveant and E. Vianello, *Electrochim. Acta* **12**, 1545 (1967).

82. R. S. Nicholson, *Anal. Chem.* **37**, 1351 (1965).

83. R. H. Wopschall and I. Shain, *Anal. Chem.* **39**, 1514 (1967).

84. J. Bacon and R. N. Adams, *J. Am. Chem. Soc.* **90**, 6596 (1968).

85. A. J. Bard and K. S. V. Santhanam, *Electroanal. Chem.* **4**, 215 (1970).

86. L. Meites, *in* "Techniques of Chemistry" (A. Weissberger and B. Rossiter, eds.), Part IIA, Vol. 1, Chap. 9. Wiley (Interscience), New York, 1971.

87. J. E. Harrar, *Electroanal. Chem.* **8**, 1 (1975).

88. C. N. Reilley, *Pure Appl. Chem.* **18**, 137 (1968).

89. A. T. Hubbard and F. C. Anson, *Electroanal. Chem.* **4**, 129 (1970).

90. A. T. Hubbard, *CRC Crit. Rev. Anal. Chem.* **2**, 201 (1973).

91. E. Laviron, *J. Electroanal. Chem.* **112**, 1 (1980).

92. P. J. Peerce and A. J. Bard, *J. Electroanal. Chem.* **114**, 89 (1980).

93. L. Meites and P. Zuman, eds., "Handbook of Organic Electrochemistry." CRC Press, Boca Raton, Florida, 1977 (continuing from Vol. 1).

94. L. Meites, P. Zuman, and A. Narayanan, "Handbook of Inorganic Electrochemistry," CRC Press, Boca Raton, Florida, 1980 (continuing from Vol. 1).

95. J. J. Lingane, *Anal. Chim. Acta* **44**, 411 (1969).

96. G. C. Barker and A. W. Gardner, *Z. Anal. Chem.* **173**, 79 (1960).

97. E. P. Parry and R. A. Osteryoung, *Anal. Chem.* **37**, 1634 (1964).

98. J. G. Osteryoung and R. A. Osteryoung, *Am. Lab.* **4**(7), 8 (1972).

99. J. B. Flato, *Anal. Chem.* **44**(11), 75A (1972).

100. W. M. Peterson, *Am. Lab.* **11**(12), 69 (1979).

101. A. M. Bond, *J. Electroanal. Chem.* **118**, 381 (1981).

102. W. Kemula, Z. Kublik, and S. Glodowski, *J. Electroanal. Chem.* **1**, 91 (1959).

103. I. Shain, *in* "Treatise on Analytical Chemistry" (I. M. Kolthoff and P. J. Elving, eds.), Part I, Vol. 4, Chap. 50. Wiley (Interscience), New York, 1963.

104. E. Barendrecht, *Electroanal. Chem.* **2**, 53 (1967).

105. T. R. Copeland and R. K. Skogerboe, *Anal. Chem.* **46**, 1257A (1974).

106. K. Z. Brainina, "Stripping Voltammetry in Chemical Analysis." Halsted, New York, 1974.

107. F. Bydra, K. Stulik, and E. Julakova, "Electrochemical Stripping Analysis." Halsted, New York, 1977.

108. K. Z. Brainina and M. B. Vydrevich, *J. Electroanal. Chem.* **121**, 1 (1981).

109. T. M. Florence, *J. Electroanal. Chem.* **27**, 273 (1970).

110. D. DeFord and J. W. Miller, *in* "Treatise on Analytical Chemistry" (I. M. Kolthoff and P. J. Elving, eds.), Part I, Vol. 4, Chap. 49. Wiley (Interscience), New York, 1963.

111. *Anal. Chem.* Biennial reviews on electroanalysis.

112. R. A. Durst, ed., "Ion Selective Electrodes," NBS Spec. Publ. No. 314. U.S. Gov. Print. Off., Washington, D.C., 1969.

113. J. Koryta, "Ion-Selective Electrodes." Cambridge Univ. Press, London and New York, 1975.

114. H. Freiser, ed., "Ion-Selective Electrodes in Analytical Chemistry," Vols. 1 and 2. Plenum, New York, 1978 and 1980.
115. A. K. Covington, ed., "Ion-Selective Electrode Methodology," Vols. 1 and 2. CRC Press, Boca Raton, Florida, 1979.
116. P. L. Bailey, "Analysis with Ion-Selective Electrodes," 2nd ed. Heyden, London, 1980.
117. R. P. Buck, in "Ion-Selective Electrodes in Analytical Chemistry" (H. Freiser, ed.), Vol. 1, Chap. 1. Plenum, New York, 1978.
118. A. K. Covington, in "Ion-Selected Electrode Methodology" (A. K. Covington, ed,), Vol. 1, Chap. 1. CRC Press, Boca Raton, Florida, 1979.
119. R. A. Durst, in "Ion-Selective Electrodes in Analytical Chemistry" (H. Freiser, ed.), Vol. 1, Chap. 5. Plenum, New York, 1978.
120. G. G. Guilbault, *Pure Appl. Chem.* **25,** 727 (1971).
121. G. A. Rechnitz, *Chem. Eng. News* **53**(4), 29 (1975).
122. R. K. Kobos, in "Ion-Selective Electrodes in Analytical Chemistry" (H. Freiser, ed.), Vol. 2, Chap. 1. Plenum, New York, 1980.
123. P. Vadgama, in "Ion-Selective Electrode Methodology" (A. K. Covington, ed.), Vol. 2, Chap. 2. CRC Press, Boca Raton, Florida, 1979.
124. C. R. Powley, R. F. Geiger, Jr., and T. A. Nieman, *Anal. Chem.* **52,** 705 (1980).
125. N. Winograd and T. Kuwana, *Electroanal. Chem.* **7,** 1 (1974).
126. W. N. Hansen, *Adv. Electrochem. Electrochem. Eng.* **9,** 1 (1973).
127. T. Kuwana and W. R. Heineman, *Acc. Chem. Res.* **9,** 241 (1976).
128. W. R. Heineman, *Anal. Chem.* **50,** 390A (1978).
129. P. T. Kissinger and W. R. Heineman, eds., "Laboratory Techniques in Electrochemistry." Dekker, New York, 1983. In press.
130. M. L. Meckstroth, B. J. Norris, and W. R. Heineman, *Bioelectrochem. Bioenerg.* **8,** 63 (1981).
131. I. B. Goldberg and A. J. Bard, in "Magnetic Resonance in Chemistry and Biology" (J. N. Herak and K. J. Adamic, eds.), Chap. 10. Dekker, New York, 1975.
132. T. M. McKinney, *Electroanal. Chem.* **10,** 97 (1977).
133. N. Winograd and T. Kuwana, *J. Am. Chem. Soc.* **93,** 4343 (1971).
134. R. H. Muller, *Adv. Electrochem. Electrochem. Eng.* **9,** 167 (1973).
135. J. Kruger, *Adv. Electrochem. Electrochem. Eng.* **9,** 227 (1973).
136. J. D. E. McIntyre, *Adv. Electrochem. Electrochem. Eng.* **9,** 61 (1973).
137. D. L. Jeanmaire and R. P. Van Duyne, *J. Electroanal. Chem.* **66,** 235 (1975).
138. R. P. Van Duyne, in "Chemical and Biological Applications of Lasers" (C. B. Moore, ed.), Vol. 4, pp. 101–185. Academic Press, New York, 1979.
139. T. Davidson, B. S. Pons, A. Bewick, and P. P. Schmidt, *J. Electroanal. Chem.* **125,** 237 (1981).
140. P. F. Kane and G. B. Larrabee, eds., "Characterization of Solid Surfaces." Plenum, New York, 1974.
141. A. W. Czanderna, ed., "Methods of Surface Analysis." Elsevier, Amsterdam, 1975.
142. C. A. Evans, Jr., *Anal. Chem.* **47,** 818A, 855A (1975).
143. B. G. Baker, *Mod. Aspects Electrochem.* **10,** 93 (1975).
144. M. Sluyters-Rehbach and J. H. Sluyters, *Electroanal. Chem.* **4,** 1 (1970).
145. L. Pospisil and R. de Levie, *J. Electroanal. Chem.* **22,** 227 (1969).
146. H. Moreira and R. de Levie, *J. Electroanal. Chem.* **29,** 353 (1971); **35,** 103 (1972).
147. R. D. Armstrong, M. F. Bell, and A. A. Metcalfe, *Electrochemistry* **6,** 98 (1978).
148. B. Breyer and H. H. Bauer, "Alternating Current Polarography and Tensammetry" (P. J. Elving and I. M. Kolthoff, eds.), Chemical Analysis, Vol. 13. Wiley (Interscience), New York, 1963.

149. D. E. Smith, *Anal. Chem.* **48,** 221A, 517A (1976).
150. D. G. Davis, *Electroanal. Chem.* **1,** 157 (1966).
151. R. P. Van Duyne and C. N. Reilly, *Anal. Chem.* **44,** 142, 153 (1972).
152. S. Gilman, *Electroanal. Chem.* **2,** 111 (1967).
153. R. E. Panzer and P. J. Elving, *Electrochim. Acta* **20,** 635 (1975).
154. M. A. Dayton, J. C. Brown, K. J. Stutts, and R. M. Wightman, *Anal. Chem.* **52,** 946 (1980).
155. J. E. Anderson, D. E. Tallman, D. J. Chesney, and J. L. Anderson, *Anal. Chem.* **50,** 1051 (1978).
156. F. C. Cowland and J. C. Lewis, *J. Mater. Sci.* **2,** 507 (1967).
157. A. J. Bard, *Science* **207** 139 (1980).
158. A. Heller, *Acc. Chem. Res.* **14,** 154 (1981).
159. C. K. Mann, *Electroanal. Chem.* **3,** 57 (1969).
160. L. R. Faulkner and A. J. Bard, *Electroanal. Chem.* **10,** 1 (1977).
161. D. J. Fisher, W. L. Belew, and M. T. Kelley, *in* "Polarography, 1964" (G. J. Hills, ed.), Vol. 2, p. 1043. Wiley (Interscience), New York, 1966.
162. J. G. Graeme, G. E. Tobey, and L. P. Huelsman, eds., "Operational Amplifiers—Design and Applications" (McGraw-Hill, New York, 1971.
163. E. Ahlberg and V. D. Parker, *J. Electroanal. Chem.* **107,** 197 (1980).
164. E. Ahlberg and V. D. Parker, *J. Electroanal. Chem.* **121,** 57 (1981).

Global Optimization Strategy for Gas-Chromatographic Separations

R. J. Laub

Department of Chemistry
San Diego State University
San Diego, California

I. Introduction

There can be little doubt today that no other analytical technique has had quite the impact throughout chemistry with which chromatography is credited. Indeed, testimony to the utility of (liquid–solid) chromatography as expressed more than forty years ago by the 1937 Nobel Laureate Karrer must moreover be said to apply equally to the present [1]: "No other discovery has exerted as great an influence and widened the field of investigation of the ... chemist as much as Tswett's chromatographic analysis."

Much of the conspicuous success of chromatography, despite utilization almost exclusively on an empirical basis, can be attributed to the inherent efficiency of the technique. Thus, for example, it is a simple matter to construct a GC column which exhibits 10,000 theoretical plates or an LC column of 5,000 theoretical plates. (Brief reflection, in fact, reveals that it is physically impossible to construct a chromatographic column of less than 36 plates.) Chromatography therefore offers the prospect of separations which, for example, are superior to the staging of separatory funnels or gas bottles.

Despite a number of seemingly recurrent periods of "rebirth," chromatography is a far older analytical method than is commonly supposed: Although Day [2] and Tswett [3] are generally given credit for the invention of modern-day chromatography [4], use of the word† can in fact be traced back at least 250 years [6]. Indeed, the technique has likely been employed in some form or other for the better part of 2000 years [7].

Although the label is not likely to be altered in the foreseeable future, "color-writing" is, of course, today a misnomer in most instances wherein

---

† An alternative suggestion as to the origin of application of the term chromatography has been offered by Purnell [5], who suggests that Tswett (Russian for color) may well have indulged his sense of humor in naming the technique after himself.

chromatography is employed. A modern operational definition is therefore required, such as that supplied recently by Laub and Pecsok [8, p. 3]:

> All chromatographic methods have the following features in common: two mutually immiscible phases are brought into contact (possess a common interface) wherein one (the mobile phase) is made to flow over the other (the stationary phase), which remains static. The surface area of the stationary phase which is exposed to the mobile phase is generally large. When a third component, called the solute, is introduced (injected) into the system, it is partitioned between the two phases. It is also carried (eluted) through the system by the mobile (carrier) phase. While being eluted it is partitioned between the mobile and stationary phases many times and, in most situations, equilibration is achieved.
>
> The mobile and stationary phases can, in principle, be gases, liquids, or solids; in practice, the mobile phase is either a gas or a liquid and the stationary phase is either a liquid or a solid. Thus there are four possible techniques: gas–liquid (GLC), gas–solid (GSC), liquid–liquid (LLC), and liquid–solid (LSC) chromatography.

Insofar as analytical separations are concerned, the choice of utilization of GC or LC is generally straightforward and, practical considerations aside, is predicated for the most part upon criteria such as the volatility of the solutes. Once, however, a decision has been reached regarding the particular chromatographic technique to be employed, the analyst is then faced with the task of optimizing the various parameters that affect the separation, e.g., temperature, pressure, flow rate, and composition of the stationary and mobile phases. It is in this area that chromatography has been utilized almost entirely on an empirical basis. Moreover, while various and at times seemingly sophisticated approaches have been advocated for selection of appropriate values for each system parameter, the analyst must inevitably rely upon his experience, as guided, via the literature, by the experience of others. Indeed, it appears at times that, in fact, little progress has been made since the turn of the century as regards replacement of this particular aspect of the "artistry" of chromatography with principles based upon fundamentally sound, yet simple and logical chemical science.

This chapter comprises an effort to bridge the gap between empirical and rational approaches to gas chromatography and to demonstrate in addition that the situation is not as problematical as is generally supposed.

A.   The Column: Heart of All Chromatographic Systems

By definition, separations are achieved in chromatography by virtue of differences in equilibration (partitioning) of solutes between the two phases present within the system, the latter collectively referred to here as the *column*.† Since, however, each solute is taken to be everywhere at equilibrium within the system, the chemical potential of any solute in the mobile phase, μ_M, must equal that of the solute in the stationary phase, μ_S, viz.,

$$\mu_M = \mu_S \tag{1}$$

where the chemical potential of solute in the ith phase is given by

$$\mu_i = \mu_i^\circ + RT \ln a_i \tag{2}$$

where a_i is the solute activity and μ_i° is the solute chemical potential at some standard state (however this is defined). Substitution of the quantities from Eq. (2) into Eq. (1), followed by replacement of activities by concentrations, produces

$$C_S/C_M = \exp(\Delta\mu^\circ/RT) = K_R \tag{3}$$

where the ratio of solute concentrations in the stationary (C_S) and mobile (C_M) phases is termed the partition coefficient K_R. Since, ideally, $\Delta\mu^\circ$ is a constant, K_R is predicted according to Eq. (3) to be independent of the amounts of solute and phases extant in the system. In reality, of course, these approximations hold only under conditions such that the solute sorption isotherm is linear, i.e., such that the solute is effectively at infinite dilution in each phase. One of the major advantages of the "normal elution" mode of gas chromatography is that, with few exceptions, these are just the conditions under which separations are most commonly carried out. Thus, it may be assumed here that for the purposes of optimization of separations Eq. (3) does indeed represent an adequate description of partitioning extant throughout the chromatographic system.

The most commonly employed method of representation of the separation of two solutes i and j is defined by the expression

$$\alpha_{i/j} = K_{R_i}/K_{R_j} = C_{S_i}C_{M_j}/C_{S_j}C_{M_i} \tag{4}$$

Effecting a chromatographic separation between two solutes thus amounts to maximization of the ratios C_{S_i}/C_{S_j} and C_{M_j}/C_{M_i}. [In point of fact, it can be demonstrated that Eq. (4) forms the basis of any separation technique wherein solutes are partitioned between phases.]

It is important to note that no terms appear in Eq. (4) that take into

---

† The *column* may thus consist of a stationary phase of filter paper (paper chromatography) or a layer of silica on a glass microscope slide (thin-layer chromatography), a mobile phase being drawn along the longitudinal axis of the stationary bed by virtue of capillary action. Alternatively, the column may actually be a tube which is packed with a porous sorbent stationary phase and through which is pumped (e.g., liquid chromatography) or blown (e.g., gas chromatography) the mobile phase.

account the manner in which the injection and detection processes are carried out. That is, injections are assumed to be performed in such a way that solutes are placed at the beginning of the column as an infinitely thin "plug" and that detection is such that an accurate (nondistortional) representation of solute bands is obtained as they emerge from the end of the column. Clearly, conditions other than these can only result in a decrease in the perceived separation (due largely to band broadening). Thus, some (at times not inconsiderable) importance must be attached to the extra-column devices with which solutes are injected and detected. The *column*, however, is responsible for whatever *separation* is achieved. Obviously, therefore, if the column does not effect a separation of the mixture at hand, one cannot expect, e.g., the detector to do so. It must, of course, be recognized that extra-column devices of a highly specialized type (e.g., a mass spectrometer) may in fact possess the capability of distinguishing between components that are coeluted from the chromatographic system. Indeed, provided that the mixture is sufficiently simple and the extra-column device sufficiently sophisticated, a physical separation may well prove to be unnecessary for the analysis at hand. However, one is then no longer within the purview of separations; furthermore and without exception, preseparation of mixtures into pure-component bands can only enhance the utility of such devices. Hence, the column must be regarded as the "heart" of any chromatographic system, and achievement (or improvement) of separations thereby becomes in principle solely a matter of selection of appropriate system parameters which from the standpoint of thermodynamics (and implicitly, kinetics) govern the ratios of Eq. (4).

B. Need for Quantitative Procedures Relating to Prediction of Retentions

Although the column is clearly of paramount importance in any chromatographic system, comparatively little has changed in the manner of, for example, the selection of GC mobile and stationary phases since the inception of the technique in 1952. Thus, a particular set of phases and conditions is chosen for a given separation today, as was true 30 years ago, because the analyst experienced some success with similar systems and conditions in the past, because the system and conditions were recommended in the literature, because the system and conditions are readily available and/or are the most convenient to put to use, or because of a combination of these criteria. When, however, the chosen phases and conditions fail to effect the desired separation, the analyst must then modify (perhaps a number of times) his initial decisions and repeat the analysis. If the results are still unsatisfactory from whatever point of view, the analyst has in the past had no other recourse but to be satisfied with the final version of the separation, or to consider other analytical methods.

Open-tubular GC capillary columns are, for example, often viewed as offering the prospect of separation "efficiency" (however this is defined) which is superior to that which can normally be achieved with packed columns. Thus, when a packed column containing a certain stationary phase proves to be inadequate, from whatever standpoint, for a given separation, recourse is commonly made to an open-tubular column containing the same (or equivalent) phase. If the column is 20 m in length and still proves to be unsuccessful, a column of double the length may next be tried at reduced temperature, and so on, *ad infinitum*.

This trial-and-error method of selection of columns and conditions appropriate for a given separation has consistently been cited as the greatest source of difficulty in chromatography and, in fact, weighs heavily in favor of the pejorative view that the method (whether GC or LC) is indeed much more an art than a science. There can moreover be no doubt that there is some substance to such claims insofar as solute solubility in liquid (or on adsorptive solid) systems is not well understood. Thus, formulation of even the most empirical of guidelines for the *a priori* selection of chromatographic columns and conditions represents a welcomed advance.

C.   Failure of Conventional Approaches

Broadly speaking, the approaches taken in chromatography to meet this need can be divided into those, on the one hand, which involve prediction (via Eq. (3)) of absolute partition coefficients, followed by calculation of relative retentions (α values) via Eq. (4); and those whereby relative retentions are forecast directly from data pertaining to the behavior of homologous series of solutes, coupled with tabulated properties of stationary phases. For example, the liquid–gas partition coefficient of a solute at infinite dilution in GLC is related to the properties of the solute and the phases [8, Chap. 2] via Henry's law and the ideal gas law: the solute vapor pressure p_i above the (liquid) stationary phase S is, for the region in which Henry's law applies,

$$p_i = \gamma_p^\infty x_i^S p_i^\circ \tag{5}$$

where γ_p^∞ is the pressure-based mole fraction activity coefficient of the solute of bulk vapor pressure p_i° which is present in the stationary phase at mole fraction x_1^S. Since, however, x_1^S is vanishingly small, $x_1^S \approx n_1^S/n_S$, where n_S is the number of moles of S present within the chromatographic system. Hence, dividing Eq. (5) by V_S, the total volume of stationary phase, and rearranging,

$$n_1^S/V_S = C_S = n_S p_1/\gamma_p^\infty p_1^\circ V_S \tag{6}$$

Since, further, from the ideal gas law,

$$n_1^M/V_M = C_M = p_1/RT \tag{7}$$

where M refers to the mobile phase, the partition coefficient is given by

$$K_R = C_S/C_M = RT/\gamma_p^\infty p_1^\circ \bar{V}_S \tag{8}$$

where $\bar{V}_S$ is the molar volume of the stationary phase. It can further be shown (e.g., [8, Chap. 2]) that

$$K_R = \underbrace{(t_R - t_A)}_{t_R'} \left[\underbrace{F_0 \left(\frac{T}{T_0}\right)\left(\frac{p_0 - p_w}{p_0}\right)}_{F_c}\right] \underbrace{\left\{\frac{3}{2}\left[\frac{(p_i/p_0)^2 - 1}{(p_i/p_0)^3 - 1}\right]\right\}}_{j} \underbrace{\left(\frac{\rho_S}{w_S}\right)}_{V_S^{-1}} \tag{9a}$$

$$= \quad t_R' \qquad\qquad F_c \qquad\qquad\qquad j \qquad\qquad V_S^{-1} \tag{9b}$$

where t_R and t_A are the retention times of retained and nonsorbed solutes (the latter referred to at times as the "air" peak), F_0 is the mobile-phase flow rate at the column outlet, T and T_0 are the column and outlet (usually ambient) temperatures, p_0 and p_w are the outlet and ambient water-vapor pressures (incorporated if a soap-bubble flow meter is used to measure F_0), p_i is the column inlet pressure, and ρ_S and w_S are the density and mass, respectively, of the stationary (liquid) phase in the column. Combination of Eqs. (8) and (9) produces

$$k' = K_R V_S/V_M = K_R/\beta \tag{10a}$$

$$V_R^\circ = jV_R = V_M + K_R V_S \tag{10b}$$

where k' is termed the capacity factor $[(t_R - t_A)/t_A]$, β the phase ratio (V_M/V_S), V_R° the "corrected" solute retention volume $(jF_c t_R)$, and V_M the "adjusted," "dead," or "air" volume $(jF_c t_A = jV_A)$. Equations (9) and (10) thus relate the solute partition coefficient to all experimental parameters of a GC system.

Combination of Eqs. (8) and (10) clearly shows that if γ_p^∞ could be predicted, then even raw solute retentions t_R could be calculated for a given flow rate, column temperature, phase ratio, etc. Moreover, since, from Eqs. (4) and (8),

$$\alpha_{i/j} = (\gamma_p^\infty p_1^\circ)_j/(\gamma_p^\infty p_1^\circ)_i \tag{11}$$

relative retentions (α values) may also be predicted, provided that γ_p^∞ and p_1° can somehow be estimated.

The seemingly insuperable obstacle to implementation of such an approach is prediction of γ_p^∞. Indeed, little can be said regarding *a priori* calculation of activity coefficients for other than *n*-alkane solvents, nor, in fact, has the situation improved noticeably over the past 50 years [9, 10]. The fully corrected (for fugacity and virial effects) activity coefficient is considered to be composed of two terms, viz.,

$$\ln \gamma_1^\infty = \ln \gamma_1^\infty(\text{config}) + \ln \gamma_1^\infty(\text{interact}) \tag{12}$$

where the solute configuration and interaction contributions to γ_1^∞ are considered to be separate and measurable quantities. The first term on the right-hand side of Eq. (12) is generally conceded to be represented adequately by the expression

$$\ln \gamma_1^\infty(\text{config}) = \ln\left(\frac{1}{r}\right) + \left(1 - \frac{1}{r}\right) \tag{13}$$

where r is the solvent/solute molar volume ratio. Insofar as the interactional contribution to γ_1^∞ is concerned, theoretical approaches closely allied to classical thermodynamics predict, as a first approximation and for those situations in which mixing volumes are negligible and in which random molecular distribution obtains,

$$\ln \gamma_1^\infty(\text{interact}) = \overline{V}_1 A_{12}/RT \tag{14}$$

where $\overline{V}_1$ is the molar volume of the solute and where A_{12} is defined by

$$A_{12} = (\Delta E_2^v/\overline{V}_2)^{1/2} - (\Delta E_1^v/\overline{V}_1)^{1/2} \tag{15}$$

where ΔE_i^v ($i = 1$ or 2, the latter pertaining to the solvent) is an internal energy of vaporization. (The approximation $\Delta E^v = \Delta H^v - RT$ is often used since excess heats of solution are frequently found to be positive and since solute activity coefficients pertaining to such mixtures appear at times to follow a semiempirical regular variation as a function of mole fraction.) Expressing the quantity $\overline{V}_1 A_{12}/RT$ as χ (an interaction parameter) and $(\Delta E_i^v/\overline{V}_i)^{1/2}$ as δ (a "solubility" parameter) produces the relation

$$\ln \gamma_1^\infty(\text{interact}) = \chi = \frac{\overline{V}_1}{RT}(\delta_2 - \delta_1)^2 \tag{16}$$

Equation (16) has been evaluated comprehensively by, for example, Cruickshank, Gainey, and Young [11]; the results are, however, disappointing, as shown in Table I, where calculated χ data for a number of solutes with n-octadecane solvent at 35°C are compared with those determined by GLC. χ^{expt} exceeds χ^{calc} on average by a factor of two or three for the n-alkane solutes but is somewhat smaller than χ^{calc} for the branched-chain hydrocarbons. The situation is reversed for the olefins and for cyclohexane and benzene. "Regular" solution theory therefore appears to be of limited utility for the prediction of gas–liquid chromatographic retentions in even the simplest case of n-alkane solutes with n-alkane stationary phases.

It should, however, be noted that there have been several recent and

TABLE I

Comparison of Calculated (Eq. (16)) with Experimental Solute Interaction
Parameters χ with n-Octadecane Solvent at 35°C[a]

Solute	$\bar{V}(cm^3/mol)$	α (cal/cm^3)$^{1/2}$	χ Calculated	χ Experimental
n-Hexane	131.598	7.2735	0.125	0.177
n-Heptane	147.456	7.4312	0.088	0.131
n-Octane	163.530	7.5504	0.040	0.101
2,2-Dimethylbutane	133.712	6.7309	0.371	0.242
2,3-Dimethylbutane	131.156	6.9811	0.238	0.131
2,4-Dimethylpentane	149.925	6.9626	0.282	0.196
1-Hexene	125.892	7.376	0.094	0.216
1-Heptene	141.744	7.503	0.070	0.150
1-Octene	157.850	7.609	0.032	0.129
Cyclohexane	108.744	8.1946	0.045	0.295
Benzene	89.399	9.1582	0.185	0.553
n-Octadecane	329.771	8.050	—	—

[a] From Cruickshank *et al.* [11].

notable advances [12, 13] wherein the concept of factorization of the activity
coefficient has successfully been extended and amplified. The most fruitful of
these is that provided by Janini and Martire [14], as discussed at length by
Laub, Martire, and Purnell [15, 16] for n-alkane systems. Moreover, while
their treatment has yet to be generalized to include other than n-alkanes,
it provides nevertheless the prospect of what appears to be a useful avenue
of approach to the study of solute/solvent solubility. In the meantime,
measurement and tabulation of interaction parameters (hence retentions) for
widely diverse systems should continue to be made in the hope that nonelectrolyte solutions theory will one day prove to be comprehensive enough to
rationalize what remains only a poorly understood area of thermodynamics
[17–20].

II. A Global Optimization Strategy for Gas-Chromatographic Separations

A. *Empirical Prediction of Retentions with Mixed Phases*

It is hardly surprising, in view of the limited success of conventional
approaches with pure phases, that prediction of retentions with mixed

phases would seem to be intractable or, at best, impractical. Nevertheless, a number of studies have been devoted to the potential advantages of methods involving combinations of sorbents. The earliest of such reports dates back to 1952 with the work of Martin and James [21] and James, Martin, and Smith [22] wherein mixed phases were employed in order to reduce gas–liquid interfacial adsorption. Mixed-phase systems were also fabricated by placing pure-phase columns in tandem by, e.g., Keulemans, Kwantes, and Zaal [23] in 1955 and Fredericks and Brooks [24] in 1956.

A variety of related studies of multiple-solvent systems were also reported at about this time (see, e.g., the review by Laub and Wellington [25]), but it was not until 1959 that the use of such columns was placed on a quantitative basis by Primavesi [26]. Solute-specific retention volumes V_g° with mixtures of triisobutylene (S) with silver nitrate/ethylene glycol (A) were found empirically to conform to the relation (since verified by many others)

$$V_g^\circ = w_A V_{g(A)}^\circ + w_S V_{g(S)}^\circ \tag{17}$$

(where w_i is the weight fraction of the ith component in the mixed phase, the subscripts indicating pure phase A or pure phase S) regardless of the manner of fabrication of the mixed-phase system with mechanically blended packings or with tandem pure-phase columns. The prediction of retentions with mixed phases from data pertaining to each of the pure solvents thus appears to be feasible on the basis of Eq. (17).

B.   *The Diachoric Solutions Hypothesis: Physicochemical Background*

In addition to empirical studies relating to solute/pure-solvent and solute/mixed-solvent systems, advances in the optimization of analytical GLC systems have also been brought about as a result of physicochemical investigations of weak intermolecular interactions relating to charge-transfer and to hydrogen-bonding phenomena. Many examples of the use of neat and blended phases which exhibit selectivity due to complexation have in fact been reported, but by far the most frequently cited have been those in which silver nitrate has been utilized in admixture with ethylene glycol for the analysis of olefins. However, it was not until the work of Gil-Av and Herling [27] and Muhs and Weiss [28] that the partition coefficients of interactive (e.g., olefin) solutes were shown to vary in a linear fashion with the concentration C_A of silver nitrate in the silver nitrate (additive, A)/ethylene glycol (solvent, S) solution in accordance with the relation

$$K_R = K_{R(S)}^\circ + K_{R(S)}^\circ K_1 C_A \tag{18}$$

where $K^\circ_{R_{(S)}}$ is the solute liquid–gas partition coefficient with notionally inert pure solvent, and K_1 is the putative stoichiometric equilibrium constant for the interaction of solute with silver nitrate.

According to Eq. (18), plots of K_R against C_A should produce straight lines from which K_1 may be deduced from the slope/intercept quotient. Muhs and Weiss [28], for example, measured K_1 data via Eq. (18) for well over a hundred solutes comprising diverse forms of unsaturation, and obtained stability constants that ranged from 0.1 to >1000 l/mol. Solutes which exhibit *negative* slopes with various additive–solvent combinations have also been noted, which, however, have generally been disregarded as due to "salting-out" effects.

The GLC method of investigation of intermolecular complexation has since been used by many others for the study of both ionic and nonionic systems and has been reviewed comprehensively by Wellington [29], Laub and Pecsok [8, 30], Vigdergauz and his colleagues [31, 32], deLigney [33] and, most recently, by Laub and Wellington [25]. However, as pointed out by Purnell [34] in 1966, there are a number of limitations implicit in Eq. (18), as well as practical difficulties associated with its implementation. Chief among the latter is the poor miscibility of most additives commonly associated by spectroscopists with complexation in stationary phases that are of sufficiently low vapor pressure (~ 0.01 Torr) to find application in gas chromatography. For example, 2,4,7-trinitrofluorenone (TNF), a Lewis acid commonly employed with aromatic hydrocarbons, is soluble only to the extent of $\sim 0.2\ M$ in dialkyl phthalate stationary phases. Furthermore, while other solvents more suited from the standpoint of solubility could conceivably be employed with additives such as TNF, the question then arises whether such solvents can, for the purposes of Eq. (18), be considered to be inert, i.e., noncomplexing.

Several alternatives [25] have been proposed in attempts at overcoming the problem of additive solubility; the most widely reported and evaluated is that initiated by Martire and Riedl [35] in 1968 wherein K_1 is said to be calculable from data derived solely from two GLC columns: one containing pure supposedly inert solvent and the other neat complexing phase. The surprising feature of these studies is tkat K_R is, in effect, said to vary linearly over the entire range $C_A = 0$ to $C_A = \overline{V}_A^{-1}$, the inverse molar volume of additive. Purnell and Vargas de Andrade [36] were the first to point out that, in fact, K_R is linear in C_A over extended ranges of concentration for a widely diverse set of solutes and solvent systems; Laub and Purnell [37] then showed that virtually every binary-solvent GLC system described in the literature also conformed to one or another form of Eq. (18). Liao, Martire, and Sheridan [38] demonstrated further that K_1 data derived from their relations were essentially identical to those calculated via Eq. (18), and Laub and Wellington [25] have recently proved that the Martire–Riedl and Gil-Av–Herling approaches are indeed formally equivalent.

Purnell and Vargas de Andrade [36] and Laub and Purnell [37] were the first to consider the physicochemical consequences of the linearity of plots of K_R against C_A: the stability constant of a solute with a mixed (binary) stationary phase is, according to Eq. (18), given by

$$K_1 = (K_R - K^\circ_{R_{(S)}})/K^\circ_{R_{(S)}}C_A \tag{19}$$

However, the slope/intercept quotient at the end-point (i.e., $K^\circ_{R_{(A)}}$) of a straight line must be equal to that at any other point on the line and so,

$$K_1 = (K^\circ_{R_{(A)}} - K^\circ_{R_{(S)}})\overline{V}_A/K^\circ_{R_{(S)}} \tag{20}$$

must also be true. Equating (19) and (20); solving for K_R; and recognizing that, in the absence of mixing volumes, $\overline{V}_A C_A = \phi_A$, a volume fraction, produces

$$K_R = \phi_A \Delta K^\circ_R + K^\circ_{R_{(S)}} = \phi_A K^\circ_{R_{(A)}} + \phi_S K^\circ_{R_{(S)}} \tag{21}$$

Equation (21) is a surprising result since it indicates that the variation of K_R with C_A is a consequence of nothing more than the volume dilution of A by S. Further, no terms appear in K_1, and in addition, negatively sloped plots of K_R versus C_A are explained on the basis that such solutes are simply less soluble in A than in S.

Since, as shown by Laub and Purnell [37], Laub and Pecsok [8], and Laub and Wellington [25], Eq. (21) applies equally to several *hundred* ternary systems (with one component at infinite dilution) to at worst $\pm 10\%$, Laub and Purnell [37] attempted to formulate models of solutions from which the relation could be derived. They concluded, as had Young [39] in an earlier study, that the equation could not be arrived at via conventional solution theories but that it could be explained on the basis of microscopic partitioning of A and of S. (Parenthetically, the notion of microscopic immiscibility, while at times provoking entertainingly emotive comment, is by no means new [cf. 8, 25].) Equation (21) was thus christened the diachoric ($\delta\iota\alpha\chi\omega\rho\iota\sigma\mu\omega\varsigma$, partitioned-volume) solutions hypothesis and the consequent implications the microscopic partition (MP) theory of solutions [37].

Equation (21) or its variants have since been discussed by a number of workers; recent studies by, e.g., Laub and co-workers [20] and Ashworth and Hooker [40] have cast some doubt upon the catholicity of the diachoric solutions hypothesis for intimately blended phases (see later). Nevertheless, it is clear at this point that whatever alternative models arise, each must, in light of the data tabulated thus far [25, 37], yield some form or other of Eq.

(21) as a limiting first approximation of solute retentions with mixed-solvent systems. For example, in the absence of excess volumes of mixing,

$$\phi_A \rho_A^\circ / \rho_{A,S} = w_A \qquad (22)$$

Further V_g° and K_R are related by

$$V_g^\circ = K_R(273/T\rho_{A,S}) \qquad (23)$$

Thus Eqs. (17), (18), and (21) are formally equivalent. Since, further, Eq. (17) arises from empirical observation while Eqs. (18) and (21) find foundation in complexation theory, the relations appear both to be internally consistent and to conform to experiment. (Whatever controversy attends the latter hence must also be attached to the former.)

In any event, the utility of Eqs. (17) and (21) from the standpoint of analysis has been established and indisputably justified and, at worst, the relations provide a significant improvement over less well-attended conventional methods. Furthermore, the approach advocated for the optimization of gas-chromatographic separations is clearly defined: rather than search, often over an extended period of time, for a pure solvent that will provide alpha values of a desired magnitude which are compatible with the chromatographic system at hand, relative retentions are optimized via Eq. (17) or (21) by adjustment of the composition of the (mixed) multicomponent stationary phase. In addition, no effort is made to determine on an *a priori* basis retentions with pure phases; rather, use is made of the variation of retentions with mixed phases which, according to Eqs. (17) and (21), may be calculated in advance from those arising with pure solvents. Exposition of this *particular* approach in relation to the optimization of gas-chromatographic separations forms the remainder of this chapter.

C. Calculation of Separations: Alpha Values

Equations (17) and (21) provide immediate access to information regarding relative retentions with blended GLC phases or with mixed GSC adsorbents. Recalling that the parameter of fundamental importance in separations is α,

$$\alpha_{i/j} = \frac{K_{R_i}}{K_{R_j}} = \frac{\left[\phi_A K_{R_{(A)}}^\circ + \phi_S K_{R_{(S)}}^\circ\right]_{\text{solute } i}}{\left[\phi_A K_{R_{(A)}}^\circ + \phi_S K_{R_{(S)}}^\circ\right]_{\text{solute } j}} \qquad (24a)$$

Or, in terms of specific retention volumes and sorbent component weight fractions,

$$\alpha_{i/j} = \frac{V_{g_i}^\circ}{V_{g_j}^\circ} = \frac{\left[w_A V_{g_{(A)}}^\circ + w_S V_{g_{(S)}}^\circ\right]_{\text{solute } i}}{\left[w_A V_{g_{(A)}}^\circ + w_S V_{g_{(S)}}^\circ\right]_{\text{solute } j}} \qquad (24b)$$

Equations (24a) and (24b) rely upon the linearity of plots of retention against composition. In the event of nonlinearity, a function such as $\log K_R = \Phi_A \log K_{R_{(A)}}^{\circ} + \phi_S \log K_{R_{(S)}}^{\circ} + (\text{constant}) \, \phi_A^2$ may be found to apply. Alternatively, an nth-order polynomial may be used. In any case, the relation

$$\alpha_{i/j} = \frac{K_{R_i}}{K_{R_j}} = \frac{f\left[\phi_A, \phi_S, K_{R_{(A)}}^{\circ}, K_{R_{(S)}}^{\circ}\right]_{\text{solute } i}}{f'\left[\phi_A, \phi_S, K_{R_{(A)}}^{\circ}, K_{R_{(S)}}^{\circ}\right]_{\text{solute } j}} \qquad (24c)$$

must always be valid and requires only the generation of functions f and f' in order to be utilized in what follows.

Considering for the moment only partition coefficients and volume fractions, and recognizing that $\phi_A + \phi_S = 1$, Eq. (24a) may be written as

$$\alpha_{i/j} = \frac{\left[\phi_A \Delta K_R^{\circ} + K_{R_{(S)}}^{\circ}\right]_{\text{solute } i}}{\left[\phi_A \Delta K_R^{\circ} + K_{R_{(S)}}^{\circ}\right]_{\text{solute } j}} \qquad (25)$$

where $\Delta K_R^{\circ} = K_{R_{(A)}}^{\circ} - K_{R_{(S)}}^{\circ}$. The separation of solutes i and j with mixtures of A with S may therefore be predicted from the retentions with pure phases A and S provided, of course, that K_R is linear in ϕ_A.

Figure 1a illustrates plots [41] of K_R against ϕ_A for four hypothetical solutes, the partition coefficients with pure S and with pure A being provided in Table II. The values are not atypical of those found in practice and, in addition, this seemingly simple mixture is in fact quite complicated. Note, first, that with pure solvent S, solutes W and Y overlap completely; that is, $\alpha = 1.00$. Thus, this solute pair could be separated with pure S only with a column of infinite efficiency. Stationary phase S would therefore normally be rejected as inappropriate for the mixture at hand, and a second phase, solvent A, tried as an alternative. However, X and Y overlap with pure A, so this phase too will not baseline-resolve all components of the solute mixture. Finally, a 50:50 mixture of A with S could be employed in an attempt to take advantage of the selectivity of both. Reference to Fig. 1a shows, however, that solutes Y and Z overlap at this composition. Thus, one is left according to convention with the prospect of repeating the analysis with different stationary phases with, however, little rationale provided as to their selection and, of course, no guarantee of success.

Alternatively, application of Eq. (24) provides a straightforward strategy for effecting the separation: alpha values are calculated at selected intervals of ϕ_A from the pure-phase data for each solute pair. From these values the optimum ϕ_A can then be ascertained. In the present instance, the number of solute pairs, given by the expression

$$\text{No. of pairs} = \frac{n!}{2(n-2)!} \qquad (26)$$

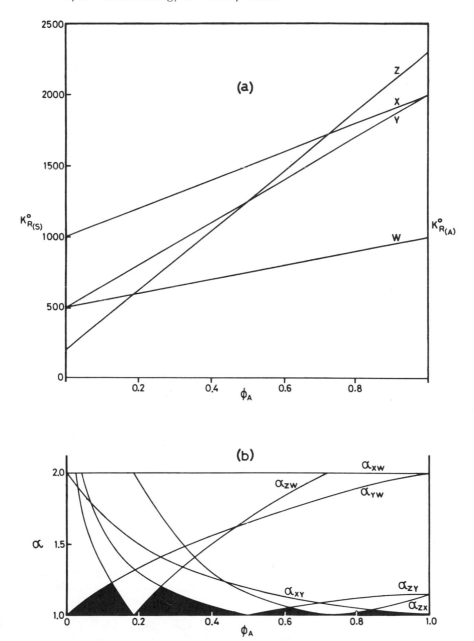

Fig. 1 (a) Plots of K_R against ϕ_A for four hypothetical solutes, where the linearity of Eq. (21) has been assumed. (b) Plots of alpha against ϕ_A for the data of (a). The best separation is predicted to occur at $\phi_A = 0.12$. [From Laub and Purnell [41].]

TABLE II

Partition Coefficients for Four Hypothetical Solutes
of Fig. 1 with Pure Phase S $(K^{\circ}_{R_{(S)}})$
and with Pure Phase A $(K^{\circ}_{R_{(A)}})^{a}$

Solute	$K^{\circ}_{R_{(S)}}$	$K^{\circ}_{R_{(A)}}$	ΔK°_{R}
W	500	1000	500
X	1000	2000	1000
Y	500	2000	1500
Z	200	2300	2100

[a] From Laub and Purnell [41].

is six, where n is the number of solutes. Table III lists the alpha values for all solute pairs at intervals of ϕ_A of 0.1 where, for clarity, the i/j assignment of solutes is inverted where required so as to maintain $\alpha \gtrless 1.00$.

The data of Table III show that for the first pair, W with X, separation is constant at all compositions of A with S at $\alpha = 2.00$. In contrast, separation of the second pair, W with Y, is impossible with pure S but improves steadily upon the addition of A to a maximum of $\alpha = 2.00$ with pure A. On the other hand, the pair W and Z is well separated with pure S, but resolution deteriorates when A is added to the stationary phase, with full overlap occurring at $\sim \phi_A = 0.2$. As the amount of A is increased further, separation again improves with, however, inversion of elution order, until at pure A, $\alpha = 2.300$.

Consideration of the remaining solute pairs provides further insight into the separation: X and Y are well separated with S, but overlap completely with A; X and Z are also well separated with S, overlap at $\phi_A \approx 0.8$, but are

TABLE III

Alpha Values for Solute Pairs of Fig. 1 and Table II at Increments of ϕ_A of 0.1[a]

Solute pair	α at $\phi_A =$										
	0.0	0.1	0.2	0.3	0.4	0.5	0.6	0.7	0.8	0.9	1.0
W/X	2.000	2.000	2.000	2.000	2.000	2.000	2.000	2.000	2.000	2.000	2.000
W/Y	1.000	1.182	1.333	1.462	1.571	1.667	1.750	1.824	1.889	1.947	2.000
W/Z	2.500	1.342	1.033	1.277	1.486	1.667	1.825	1.965	2.089	2.200	2.300
X/Y	2.000	1.692	1.500	1.368	1.273	1.200	1.143	1.097	1.059	1.027	1.000
X/Z	5.000	2.683	1.936	1.566	1.346	1.200	1.096	1.018	1.044	1.100	1.150
Y/Z	2.500	1.585	1.290	1.145	1.058	1.000	1.043	1.077	1.106	1.130	1.150

[a] Note that the i/j assignment of identity of solutes is, where necessary, inverted so as to maintain $\alpha \gtrless 1.00$.

TABLE IV

Alpha Values of Table III Ordered in Terms of the Poorest
Separation Obtained at Increments of ϕ_A of 0.1

ϕ_A	0.0	0.1	0.2	0.3	0.4	0.5	0.6	0.7	0.8	0.9	1.0
Alpha	1.000	1.182	1.033	1.145	1.058	1.000	1.043	1.018	1.044	1.027	1.000
values	2.000	1.342	1.290	1.277	1.273	1.200	1.096	1.077	1.059	1.100	1.150
	2.000	1.585	1.333	1.368	1.346	1.200	1.143	1.097	1.106	1.130	1.150
	2.500	1.692	1.500	1.462	1.486	1.667	1.750	1.824	1.889	1.947	2.000
	2.500	2.000	1.936	1.566	1.571	1.667	1.825	1.965	2.000	2.000	2.000
	5.000	2.683	2.000	2.000	2.000	2.000	2.000	2.000	2.089	2.200	2.300
Most difficult pair	W/Y	W/Y	W/Z	Y/Z	Y/Z	Y/Z	Y/Z	X/Z	X/Z	X/Y	X/Y

resolved with pure A ($\alpha = 1.150$). Finally, Y and Z are fully resolved both with S and with A, but cannot be separated at $\phi_A = 0.5$.

In order to determine the overall *optimum* value of ϕ_A, the data of Table III must be rearranged in order of the *worst* value of alpha at each column composition, as shown in Table IV. Note that the *worst* alpha is considered since, if the separation of the corresponding pair can be achieved, then *all other* pairs must also be resolved. Thus, the set of most difficult alpha at each volume fraction provides an envelope of local minima from which that region which presents the least difficulty can be selected, i.e., the mixture of A with S that will yield the best separation. In the present instance, the ϕ region of 0.1 to 0.4 appears to offer the greatest resolution for all solutes.

D. Graphical Representation of Alpha Data: The Window Diagram

A major difficulty arises with the separation scheme as presented thus far in that only a *region* of ϕ is specified within which the best separation *appears* to be attainable. Self-evidently, reduction of the ϕ interval to, say, 0.001 will narrow the ϕ region to the limit of accuracy with which a mixed phase can be fabricated. However, the number of calculations required increases accordingly and, while the entire procedure can be computerized, it becomes somewhat tedious for more than a few solutes.

Graphical representation of the alpha data is clearly called for, as demonstrated first by Laub and Purnell [41] in 1975. Plots of α against ϕ_A for the solutes of Fig. 1a are shown in 1b, which provide illustration of the variation of the separation of each solute pair as a continuous function of the column composition. The advantage of maintaining $\alpha \gtrsim 1$ now also becomes apparent: the V-shaped plots appear to be a set of inverted and partially overlapped triangles where points of solute overlap form apices which touch the abscissa. Furthermore, those areas in which no overlaps occur are clearly delineated (black areas in Fig. 1b), and provide regions or "windows" within which the separation becomes feasible.

The first window is formed by the lines of alpha representing the solute pairs W/Y and W/Z. The line for the former begins, for $\phi_A = 0$, at $\alpha = 1.00$ since the solutes overlap fully with pure stationary phase S. As, however, the volume fraction of A is increased, the separation improves as reflected by the rise in the alpha line. In contrast, the pair W/Z is well separated with pure S, but the separation decreases upon addition of more A to the stationary phase until, at $\phi_A = 0.18$, full overlap is encountered. Proceeding beyond that point (adding more A) effects an increase in the separation, and the alpha line rises again. Subsequent windows are formed in the same manner by the intersection of alpha lines of the remaining solute pairs.

The window diagram thus is clearly more informative than Table IV insofar as the former provides a detailed and continuous picture of the separation at hand, while the latter, depending upon the increment of ϕ_A chosen, can offer only a discontinuous (hence, less informative) tabulation of resolution of the solutes as a function of column composition.

E. Information Available from the Window Diagram

Inspection of Fig. 1b indicates four windows of separation, namely, those centered at $\phi_A = 0.12, 0.25, 0.62,$ and 0.85, the first of these clearly being the tallest. (Note that the breadth of a window is unimportant since maximization of the dependent variable α is sought, as opposed to the independent variable ϕ_A.) The optimum column composition thus is indicated to be $\phi_A^{opt} = 0.12$. $\alpha = 1.23$ at this value, which corresponds to the most difficult separation at ϕ_A^{opt}, all other values of alpha being larger. Therefore, if the chromatographic system efficiency can be adjusted to resolve the pairs W/Y and W/Z which form the window, all other pairs of solutes will also be separated. The first and most important result to be deduced from the window diagram hence is specification of the exact optimum stationary-phase composition with which the separation at hand may be effected. Location of the best window is done simply by inspection and the optimum column composition is found by dropping a perpendicular from its maximum to the abscissa.

Extension of a horizontal line from the top of the window to the left-hand ordinate provides the precise value of the most difficult alpha at ϕ_A^{opt}; here $\alpha = 1.23$. It is now apparent why the tallest window affords the best separation: the most difficult alphas corresponding to the maxima of all other windows are less than 1.23, so separation of the solute mixture with those corresponding compositions of ϕ_A would be more difficult than that at ϕ_A^{opt}. Thus, the window diagram also specifies the minimum efficiency that must be achieved by the chromatographic system.

Provision of the most difficult alpha in addition allows deduction of the number N_{req} of theoretical plates required in order to effect the separation. The relation first derived by Purnell [42] in 1960,

$$N_{\text{req}} = 36 \left(\frac{\alpha}{\alpha - 1} \right)^2 \left(\frac{k' + 1}{k'} \right)^2 \tag{27}$$

indicates that, for 6σ resolution at $k' > 10$ and $\alpha = 1.23$, the separation requires 900 plates. Further, since the analyst will in the course of experiment have determined the column efficiency that can be expected for the system at hand, the column length required can also be calculated. In the present example and with commonly employed packed-column procedures, which yield 1000 plates per foot, a column 0.9 ft long is required for baseline resolution of all solutes.

Finally, reference to the straight-line diagram (Fig. 1a) yields the order of elution at ϕ_A^{opt}, here, Z, W, Y, and X at $\phi_A^{\text{opt}} = 0.12$.

In summary, the window diagram procedure provides upon inspection: (a) the optimum column composition, (b) the most difficult alpha, (c) the most difficult solute pairs, (d) the number of plates required to effect baseline resolution, (e) the column length required, and (f) the order of elution.

F. Simplification of the Window Diagram Procedure

Self-evidently, generation of a sufficient number of data points for construction of Fig. 1b is not difficult for a few solutes but becomes problematic for more than five or six. A number of simplifications can, however, be invoked which facilitate its implementation. First, in gas chromatography, alpha values in excess of 1.2 represent trivial separations. The upper limit of ordinate alpha can therefore be limited to 1.2, which in essence assumes that a column of 1300 plates can be constructed. Secondly, close inspection of window diagrams indicates that over the range $\alpha = 1.0$ to 1.2, the sides of the windows approximate straight lines. Thus, the V-shaped plots can be drawn as triangles. Thirdly, and making use of the above approximations, it is easier to calculate values of ϕ_A at fixed points of alpha rather than vice versa. Equation (25) is therefore rearranged to

$$\phi_A = (K_{R_{(S)_i}}^\circ - \alpha K_{R_{(S)_j}}^\circ)/(\alpha \Delta K_{R_j}^\circ - \Delta K_{R_i}^\circ) \tag{28}$$

Consider, for example, the first eleven solutes of Table V (squalane and dinonyl phthalate solvents at 100°C), where the straight-line diagram for all solutes is shown in Fig. 2. According to Eq. (26) there are 55 pairs of solutes which must be considered but most of these can be eliminated upon inspection; for example, solute 8 elutes well after solute 1 at all compositions of S + A and can hence be eliminated for the purposes of construction of the window diagram. In contrast, the lines for solutes 8 and 11 cross and so must be considered as a relevant pair. Similar examination of each solute pair in turn leads to elimination of all but seven pairs, the data for which are tabulated in Table VI.

Within the approximation that the sides of the windows are straight, only three points need now be calculated for each triangle, namely, ϕ_A at $\alpha = 1.1$, at 1.0, and, where inversion occurs, at 1.1 opposite the point of inversion. The data calculated from Eq. (28) are presented in this fashion in Table VI, with which construction of the resultant window diagram, Fig. 3, is now straightforward. Note that the sides of the windows have been extended in the figure to $\alpha = 1.20$ so that each alpha line intersects either an ordinate or another (adjacent) window.

TABLE V

Partition Coefficients for Listed Solutes with Squalane (S)
and Dinonyl Phthalate (A) Phases at 100°C[a]

	Solute	$K^{\circ}_{R_{(S)}}$	$K^{\circ}_{R_{(A)}}$	ΔK°_{R}
1	n-Hexane	33.7	24.5	−9.2
2	Benzene	41.2	58.6	17.4
3	n-Heptane	67.2	49.0	−18.2
4	n-Octane	140	97.1	−42.9
5	Ethylbenzene	200	254	54
6	n-Nonane	301	202	−99
7	Ethoxybenzene	416	745	329
8	Acetophenone	618	1906	1288
9	n-Decane	630	414	−216
10	n-Butylbenzene	830	980	150
11	n-Undecane	1315	847	−468
12	Toluene	108	138	30
13	p-Xylene	221	276	55
14	Benzaldehyde	281	884	603
15	n-Propylbenzene	372	462	90

[a] From Laub and Purnell [41].

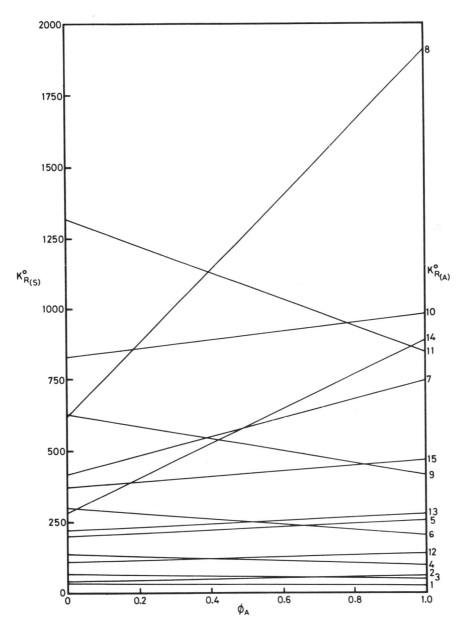

Fig. 2 Plots of K_R against ϕ_A constructed from the end-point $(K^{\circ}_{R_{(S)}}, K^{\circ}_{R_{(A)}})$ data for the solutes of Table V. [From Laub and Purnell [41].]

TABLE VI

Values of ϕ_A at α = 1.1, 1.0, and 1.1 for Relevant Pairs of
First Eleven Solutes of Table V

Solute pair	ϕ_A at α =		
	1.1	1.0	1.1
2/3	0.5860	0.7303	0.8744
5/6	0.5114	0.6601	0.8048
7/9	0.2983	0.3927	0.4889
8/9	(α = 1.0194 at ϕ_A = 0.0)	0.0080	0.0492
8/10	0.1186	0.1863	0.2627
8/11	0.3370	0.3969	0.4596
10/11	0.6351	0.7848	0.9274

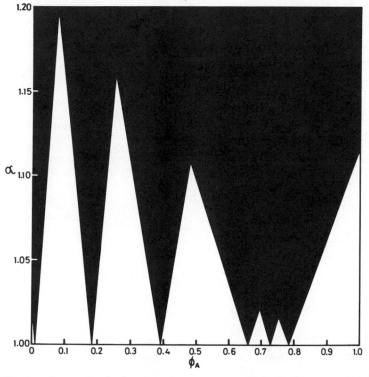

Fig. 3 Window diagram for the first eleven solutes of Table V. [From Laub and Purnell [41].]

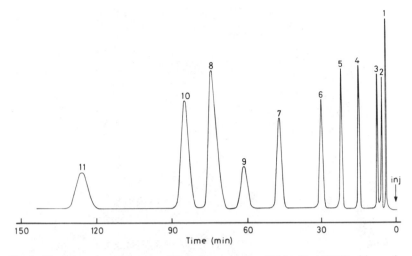

Fig. 4 First-time chromatogram of first eleven solutes of Table V at 100°C with squalane + dinonyl phthalate mixed phases such that $\phi_A = 0.076$. [From Laub and Purnell [41].]

The optimum column composition for separation of the solute mixture is predicted to be $\phi_A^{opt} = 0.075$, and the *first-time* chromatogram obtained with the requisite number of theoretical plates at ϕ_A^{opt} is given in Fig. 4.

G. Accuracy of the Window Diagram Procedure

The window diagram [41] for all fifteen solutes of Table V is shown in Fig. 5, where the optimum column composition is predicted to occur at $\phi_A = 0.075$ and where $N_{req} = 4000$ plates. Figure 6 illustrates the chromatograms obtained: that with pure squalane is shown in (a), while that with pure dinonyl phthalate is given in (b). Chromatogram (c) illustrates the separation found with an intimate blend of squalane and dinonyl phthalate solvents such that $\phi_A = 0.0755$, that is, (squalane + dinonyl phthalate) plus support. Chromatogram (d), on the other hand, is that obtained with mechanically mixed pure-phase packings such that $\phi_A = 0.0755$, that is, (support + squalane) plus (support + dinonyl phthalate). In addition to virtually complete resolution of all solutes in (c) and (d), the figure shows that the elution times and order of the solutes in the latter chromatograms are indistinguishable. Thus, the intimately blended mixture appears to offer the same solute retentions as those arising from mechanically constituted packings for these solvent systems.

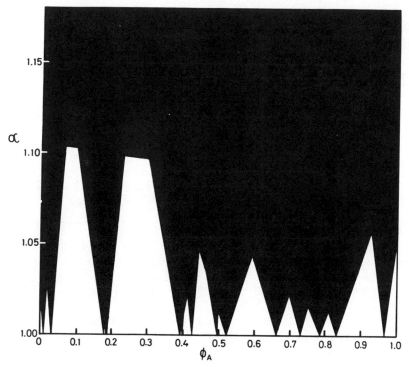

Fig. 5 Window diagram for all solutes of Table V. [From Laub and Purnell [41].]

A more comprehensive test [43] of this important feature of the optimization strategy is given in Fig. 7, where the chromatograms of thirty solutes of various types are presented with columns containing (a) pure squalane and (b) pure dinonyl phthalate. The solute mixture (see Table I of [43]) is such that separation of all solutes with any commonly employed single phase is highly improbable: solutes 13 and 14 (2,3,4-trimethylpentane and toluene) overlap fully, for example, with the boiling-point phase squalane, as do solutes 20 and 21 (1-nonene and o-xylene). With the more selective phase dinonyl phthalate, on the other hand, the pairs 1,3-pentadiene/4-methyl-1-pentene (numbers 3 and 4), 2-methyl-1,3-pentadiene/cyclohexane (numbers 7 and 9), benzene/2,4,4-trimethyl-2-pentene (numbers 8 and 11), 1-nonene/n-nonane (numbers 20 and 22), and 1-decene/n-decane (numbers 27 and 28) are inseparable. All components are, however, completely resolved with columns (c), squalane + dinonyl phthalate blended intimately, and (d), squalane + dinonyl phthalate packings mixed mechanically, ϕ_A in both (c) and (d) corresponding to $\phi_A^{opt} = 0.0807$. In addition, the mechanically mixed packing

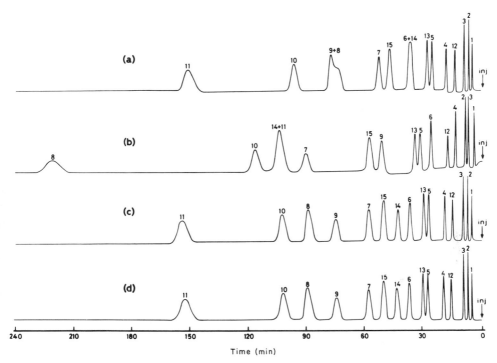

Fig. 6 Chromatograms of the solutes of Table V with the phases (a) pure squalane, (b) pure dinonyl phthalate, (c) an intimate blend of squalane + dinonyl phthalate such that $\phi_A = 0.0755$, and (d) a mechanical mixture of pure-phase packings such that $\phi_A = 0.0755$. [From Laub and Purnell [41].]

again appears to provide retentions which are identical to those obtained with the blended-phase column.

The controversy surrounding Eqs. (17) and (21) has nevertheless cast some doubt upon the equivalence of Figs. 6c and d and Figs. 7c and d. Figures 8 and 9 show, by way of illustration, plots [20] of K_R against ϕ_A for eight solutes with intimate blends of squalane with dinonyl phthalate at 30.0°C. The data are of high accuracy insofar as can be ascertained [44] by comparison with those arising from static (i.e., nonchromatographic) techniques, and clearly indicate deviations from linearity of up to 10%. How, then, is it possible that blended phases give rise to chromatograms which are to all intents and purposes identical with those obtained with mechanically mixed pure-phase packings?

Table VII lists [25] the absolute and relative partition coefficients at 30.0°C for *n*-hexane, benzene, methylcyclohexane, toluene, and *n*-octane

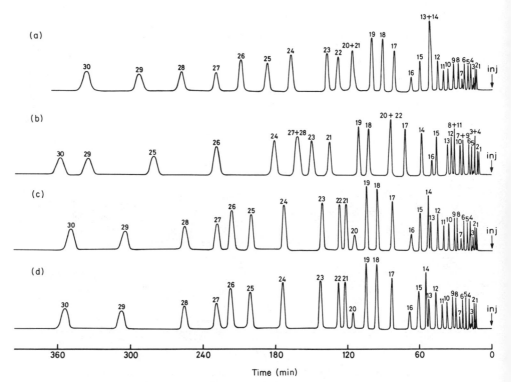

Fig. 7 Chromatograms of thirty solutes of various type with (a) pure squalane, (b) pure dinonyl phthalate, (c) squalane + dinonyl phthalate blended intimately ($\phi_A = 0.0807$), and (d) squalane + dinonyl phthalate mixed packings (ϕ_A identical to (c)). [Reprinted with permission from R. J. Laub and J. H. Purnell, *Anal. Chem.* **48**, 799. Copyright 1976 American Chemical Society.]

solutes with a mechanical mixture of squalane and dinonyl phthalate packings, where the optimum blend for separation of these compounds occurs at $\phi_A^{opt} = 0.270$. The actual and predicted alpha data agree quite closely (as do the partition coefficients) even though the latter were determined at a flow rate (~ 400 ml/min) that is an order of magnitude higher than that which is normally employed in GC studies with high-precision apparatus. Nevertheless, mechanical mixtures of the pure-phase packings give in this instance precisely those retentions which are expected on the basis of Eq. (21).

The packing used in the above test was carefully removed from the column, slurried ("intimized") with methylene chloride, dried, and repacked. The resultant absolute and relative retentions are given in Table VIII: the actual and predicted partition coefficients disagree by upwards of 10%.

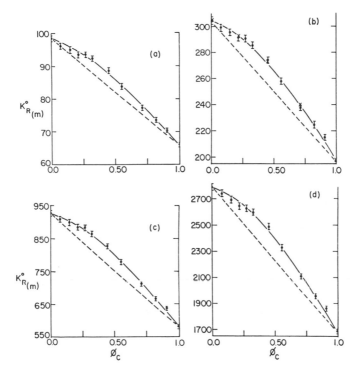

Fig. 8 Plots of K_R against ϕ_A for (a) n-pentane, (b) n-hexane, (c) n-heptane, and (d) n-octane solutes with intimate blend of squalane and dinonyl phthalate solvents at 30.0°C. Error bars reflect $\pm 0.9\%$. [Reprinted with permission from M. W. P. Harbison *et al.*, *J. Phys. Chem.* **83**, 1262. Copyright 1979 American Chemical Society.]

Further, and as shown in Figs. 8 and 9, the experimental data are higher in all cases than those predicted from the diachoric solutions hypothesis. The actual and predicted alpha values, on the other hand, show agreement which matches that found in Table VII. Thus, whatever curvature is exhibited by the K_R data appears in effect to be divided out upon calculation of *relative* retentions.

The chromatograms [25] found in this test of the window diagram procedure, Fig. 10, offer further illustration of the equivalence of the relative retentions obtained with each type of packing.† Solutes 1 and 2 are partially overlapped with pure squalane (chromatogram (a)), while solutes 3 and 4

---

† The flow rate was adjusted to ~ 400 ml/min so as to produce 325 theoretical plates for each column. This is just sufficient to provide baseline resolution of the solutes according to the worst alpha (1.497) at the optimum column composition predicted from the window diagram, the latter generated as usual solely from the pure-phase partition coefficients [40].

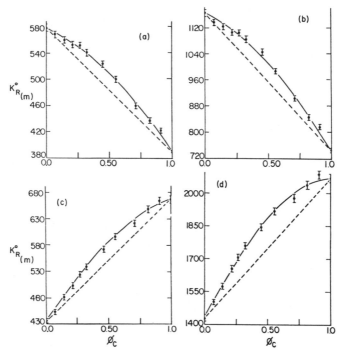

Fig. 9 Plots of K_R against ϕ_A for (a) cyclohexane, (b) methylcyclohexane, (c), benzene, and (d) toluene solutes with intimate blends of squalane and dinonyl phthalate solvents at 30.0°C. Error bars reflect $\pm 0.9\%$. [Reprinted with permission from M. W. P. Harbison *et al., J. Phys. Chem.* **83**, 1262. Copyright 1979 American Chemical Society.]

TABLE VII

Comparison of Experimental with Predicted (Eq. (21)) K_R and α Data for Listed Solutes with Mechanical Mixture of Packings of Pure Squalane and Pure Dinonyl Phthalate Comprising $\phi_A^{opt} = 0.270$ at 30.0°C[a]

	Solute	K_R		α	
		Actual	Predicted	Actual	Predicted
1	*n*-Hexane	270.4	275.7	0.17	0.17
2	Benzene	490.1	496.3	0.30	0.31
3	Methylcyclohexane	1043	1051	0.65	0.66
4	Toluene	1611	1598	(1.00)	(1.00)
5	*n*-Octane	2450	2493	1.58	1.56

[a] With permission from R. J. Laub and C. A. Wellington, "Molecular Association" (R. Foster, ed.), Vol. 2, Chap. 3. Copyright 1979 by Academic Press Inc. (London) Ltd.

TABLE VIII

Comparison of Experimental with Predicted (Eq. (21)) K_R and α Data for
Listed Solutes with Intimized Packing of Table VII ($\phi_A = 0.270$) at 30.0°C[a]

	Solute	K_R		α	
		Actual	Predicted	Actual	Predicted
1	n-Hexane	287.0	275.7	0.17	0.17
2	Benzene	515.5	496.3	0.30	0.31
3	Methylcyclohexane	1084	1051	0.64	0.66
4	Toluene	1706	1598	(1.00)	(1.00)
5	n-Octane	2676	2493	1.57	1.56

[a] With permission from R. J. Laub and C. A. Wellington, "Molecular Association" (R. Foster, ed.), Vol. 2, Chap. 3. Copyright 1979 by Academic Press Inc. (London) Ltd.

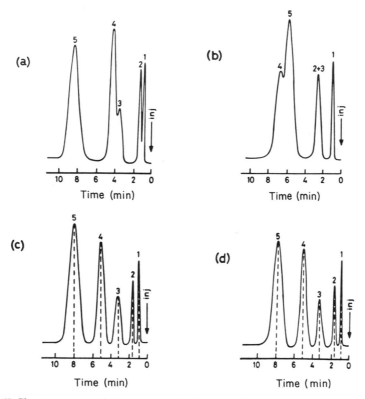

Fig. 10 Chromatograms of (1) n-hexane, (2) benzene, (3) methylcyclohexane, (4) toluene, and (5) n-octane with: (a) pure squalane, (b) pure dinonyl phthalate, (c) mechanical mixture of pure-phase packings such that $\phi_A = 0.270$, and (d) packing of (c) after intimized with methylene chloride, all at 30.0°C. Vertical dashed lines indicate peak positions predicted from Eq. (24a). [With permission from R. J. Laub and C. A. Wellington, "Molecular Association" (R. Foster, ed.), Vol. 2, Chap. 3. Copyright 1979 by Academic Press Inc. (London) Ltd.]

are virtually inseparable with these experimental conditions. Chromatogram (b), that obtained with pure dinonyl phthalate, shows that solutes 1 and 2 are now resolved but that solute 2 overlaps with 3. In addition, the elution order of solutes 4 and 5 has been inverted but separation of this pair is poor. (Presentation of the chromatograms of simple mixtures of solutes in this fashion provides a convenient illustration of the changes in retentions from one pure phase to another. For example, solute 2 is seen to move slightly to the left (increased retention) from (a) to (b), while solute 3 moves to the right. Solute 4 almost doubles in retention in (b), while solute 5 decreases in elution time to such an extent that it is eluted before solute 4 with pure dinonyl phthalate.) Chromatogram (c), obtained with mechanically mixed packings constituted such that $\phi_A = 0.270$, shows baseline resolution of the solutes, where the vertical dashed lines show that the predicted

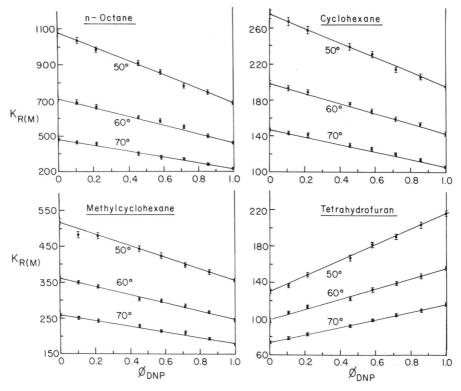

Fig. 11 Plots of K_R against ϕ_A for indicated solutes and temperature with mechanical mixtures of pure squalane and pure dinonyl phthalate packings. Error bars reflect $\pm 2\%$. [Reprinted with permission from C.-F. Chien *et al.*, *Anal. Chem.* **52**, 1402. Copyright 1980 American Chemical Society.]

peak positions correspond with those found to within 1%. Chromatogram (d) illustrates the separation given by the intimized packing, where not only are the predicted and experimental solute elutions coincident, but, in addition, they are exactly equivalent to those given by the mechanically fabricated phase, chromatogram (c).

In any event, and as argued first by Laub and Wellington [25] and later by Tiley [45], retentions with mechanically mixed packings *must* conform to those calculated on the basis of Eq. (21).† Comprehensive proof of the validity of this proposition has in fact recently been supplied by Chien, Kopecni, and Laub [46], who determined partition coefficients of a variety of types of solutes with mechanical mixtures of packings of squalane and dinonyl phthalate over the temperature range 50 to 70°C. Representative plots of their data are shown in Fig. 11 for *n*-octane, cyclohexane, methylcyclohexane, and tetrahydrofuran solutes, where the error bars correspond to ±2% and where the average of the linear least-squares regression correlation coefficients is 0.999.

Thus, Eq. (21) (or its equivalents) may be applied in GLC separations with confidence and irrespective of the use of mechanically or intimately blended solvents, since in any event only *relative retentions* are considered in the window strategy.

H. Criteria for Selection of Windows

Practical considerations (such as the relative costs of stationary phases) aside, the primary criterion for selection of a window applicable to a given analysis is that which offers the best alpha, since the corresponding requisite column efficiency is thereby minimized within the approximation that $k' > 10$. However, while the latter condition is generally the case in conventional packed-column GLC, it may well not be true when open-tubular (capillary) columns are employed. Suppose, for example, that the phase ratio for two open-tubular columns containing pure squalane and pure dinonyl phthalate is in each case 100. The capacity factors of the solutes of Table V would therefore be 1/100th of the listed partition coefficients and the number of plates required to effect the separation hence would be altered.

Figure 12 illustrates plots of $10^5/N_{req}$ against ϕ_A for the first eleven solutes of Table V as calculated from Eq. (27) with an assumed phase ratio of 100; the optimum ϕ_A is now 0.28, for which N_{req} is 4350 plates. Furthermore, the most difficult solute pairs are 8/10, 1/2, and 8/11. Thus, because of the capacity factors of the first-eluting compounds (1 and 2), the optimum

† Equation (21) can, in fact, be derived on the basis that A and S are completely immiscible liquids.

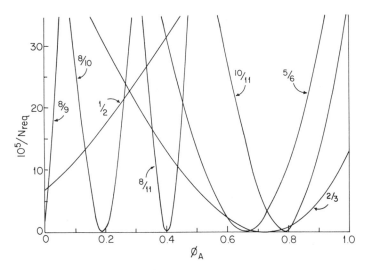

Fig. 12 Plots of $10^5/N_{req}$ against ϕ_A for first eleven solutes of Table V with an assumed phase ratio of 100.

window is no longer that at $\phi_A = 0.12$ as indicated in Fig. 3 for which k' was assumed to be negligible with respect to the right-hand term of Eq. (27).

Figure 12 in fact suggests that other parameters which are a function of alpha may also be used for window diagrams with equal success. For example, the resolution between two solutes can be expressed as

$$R_S = \frac{1}{4}\left(\frac{\alpha - 1}{\alpha}\right)\left(\frac{k'}{k' + 1}\right)N^{1/2} \tag{29}$$

where the time of analysis is taken implicitly into account by inclusion of the capacity-factor term. Plots of R_S against ϕ_A hence offer, as does Fig. 12, what amounts to the simultaneous optimization of separation at minimum time for a given column phase ratio [47].

I. Computerization of the Window Diagram Procedure

Figure 13 illustrates plots [48] of K_R versus ϕ_A for forty solutes with squalane and dinonyl phthalate solvents at 100°C. Self-evidently, manual construction of a window diagram, even with the approximations cited above, would be difficult for this mixture. Laub, Purnell, and Williams [48] therefore devised a computer program for such complex samples, the logic of which was published in 1977. First, each pair of data points (i.e., $K_{R_{(S)}}^\circ$ and $K_{R_{(A)}}^\circ$) for each solute is stored. The alpha values of all pairs are then calculated in

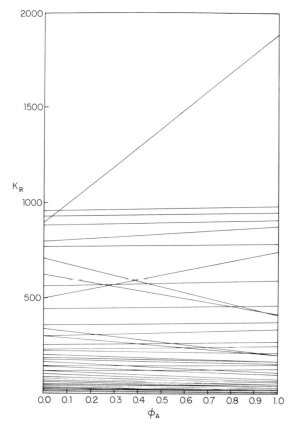

Fig. 13 Plots of K_R against ϕ_A for forty solutes with squalane and dinonyl phthalate at 100°C. [From Laub *et al.* [48].]

turn with pure solvent S and the worst of these retained in a continuously updated alpha memory. For example, the first calculation for the data of Table III yields an alpha of 2.000, which is placed in memory. The second calculation for the next pair gives an alpha of 1.000; since this is smaller, it becomes the current worst alpha and displaces the stored value. Completion of all calculations with regard to pure S specifies a worst alpha at $\phi_A = 0$; this forms the first point on the window diagram boundary. ϕ_A is then increased to, say, $\phi_A = 0.01$, and the process repeated. However, since the phase now contains *some* A, Eq. (21) must be employed in order first to calculate solute K_R data at the new column composition.

Implementation of this procedure can easily be done with a programmable pocket calculator and requires no more than a few hundred keystrokes. The approach is somewhat by "brute force," however, and, since for complex mixtures the number of pairs of solutes may be large, the overall time for determination of a window diagram boundary may be excessive. For example, calculations for ten solutes at a ϕ interval of 0.01 require approximately ten hours for completion. Clearly, the entire procedure could be expedited if the number of *relevant* pairs of solutes could be identified prior to determination of the alpha data. Examination of the data of Table V showed, for example, that of the 55 pairs of solutes only seven were in fact relevant to the window boundary. Thus, program analysis time can be diminished considerably by inclusion of a sort-and-reject subroutine at the start of a program in order to eliminate pairs of solutes whose alpha values exceed some preset limit (e.g., 1.2) at all column compositions.

The logic formulated by Laub *et al.* [48] makes use of such a sort procedure, which, in addition, takes into account the possibility of plateau-shaped window tops. The computer-drawn window diagram for the solutes of Fig. 13 is shown, for example, in Fig. 14, where a region of alpha of 1.032 is specified at from $\phi_A = 0.10$ to 0.17. Figure 15 shows the *first-time* chro-

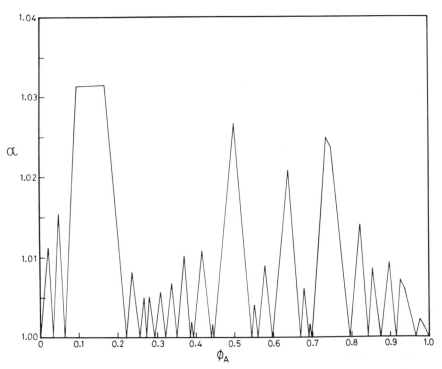

Fig. 14 Computer-drawn window diagram for the solutes of Fig. 13. [From Laub *et al.* [48].]

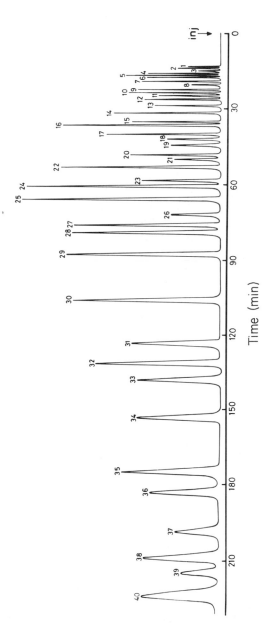

Fig. 15 First-time chromatogram of the solutes of Fig. 13 with a predicted optimum stationary phase comprising ϕ_A (DNP) of 0.168. [From Laub *et al.* [48].]

matogram [48] of the solute mixture with a column containing mechanically mixed packings such that $\phi_A = 0.168$ which exhibits the appropriate number of theoretical plates as calculated from Eq. 27 (here, $N_{req} = 40,000$), where all components have been resolved.

III. Practical Application of the Optimization Strategy

A. Selection of Phases

The fact that retentions with mechanically mixed packings correspond precisely to those predicted from Eq. (21) greatly facilitates practical implementation of the window diagram procedure: a few pure-phase packings (or open-tubular columns) for use with the range of samples likely to be encountered (could be) kept at hand. A few runs of a new sample with two or three of the pure-solvent systems would then be sufficient to establish the proper temperature, column length, and composition of mixed phases required to effect a separation.

Consider by way of illustration Fig. 16, where the chromatograms [43] of five solutes at 100°C are presented with (a) pure squalane, (b) pure dinonyl phthalate, and (c) pure dipropyl tetrachlorophthalate phases. Solutes 20 and 21 (1-nonene and o-xylene) overlap completely with squalane. In contrast, solutes 27 (1-decene) and 28 (n-decane) cannot be separated with pure dinonyl phthalate. Furthermore, the order of elution has changed from that observed in (a). Chromatogram (c) shows overlap of solutes 19 (p-xylene) and 27, but separation of solutes 20, 21, and 28 has been achieved. The elution order is again different from that found in (a) and (b).

Figure 16 underscores the difficulty associated with selection of a pure phase for a given separation: since the solute mixture contains an alkane, two olefins, and two aromatic hydrocarbons, one would normally attempt this separation first with a phase that provides resolution on the basis of vapor pressure. Full overlap of solutes 20 and 21 with pure squalane shows, however, that such phases are unlikely to resolve the mixture fully, even with columns of very high efficiency. At this point, one might suppose that a solvent which can interact selectively with the olefins and aromatic hydrocarbons would be successful. Such a phase is dinonyl phthalate, but, as shown in Fig. 16b, solute 27 overlaps with 28 with this solvent. Finally, a phase which exhibits charge-transfer complexation with aromatic hydrocarbons, such as dipropyl tetrachlorophthalate, could be expected to resolve solutes 27 and 28, which, as shown in chromatogram (c), turns out to be the case. However, solute 27 now overlaps with 19. Thus, the search for a pure solvent

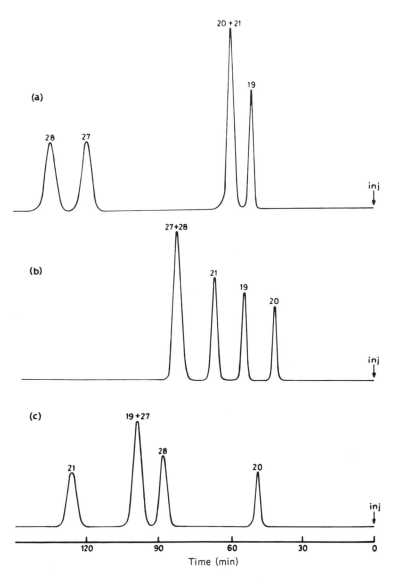

Fig. 16 Chromatograms of (19) *p*-xylene, (20) 1-nonene, (21) *o*-xylene, (27) 1-decene, and (28) *n*-decane with columns containing (a) pure squalane, (b) pure dinonyl phthalate, and (c) pure dipropyl tetrachlorophthalate at 100°C. [Reprinted with permission from R. J. Laub and J. H. Purnell, *Anal. Chem.* **48,** 799. Copyright 1979 American Chemical Society.]

for the separation of the solutes at hand could be protracted. In addition, of course, solute mixtures normally encountered in gas chromatography are more complex than the example presented here; that is, the mixture may contain heterofunctional solutes, for each class of which a particular phase may prove useful but which, for the total sample, is inappropriate. Parenthetically, this difficulty alone restricts severely the usefulness of approaches advocated, e.g., by Snyder [49], wherein attempts are made to classify solvents according to one or another selectivity scheme: such tabulations do indeed help to clarify and to simplify the choice of a phase for a particular mixture containing only one or two types of solutes, but fail badly when samples containing a number of compounds of a variety of chemical types are encountered, the latter, in the author's experience, comprising the vast majority of those dealt with in gas chromatography. In short, there are at present no universal sets of pure phases available which, when employed individually as neat solvents, will separate all members of all classes of chemical compounds. Thus, phases can be ordered [49] on the basis of selectivity for at most a few chemical types for which, even then and in all cases examined to date, notable exceptions occur.

Alternatively, the window diagram procedure provides an immediate and *quantitative* solution to separation of the mixture at hand. Figure 17 gives the straight-line diagrams [43] for the three pairs (cf. Eq. (26)) of phases of Fig. 16, where a number of reversals (crossing lines) of elution order occur with each solvent combination. The corresponding window diagrams [43] are presented in Fig. 18: that in (a), for squalane with dinonyl phthalate, provides two large windows, namely, those at $\phi_A = 0.17$ and 0.48, with the former clearly being superior (in terms of alpha) to the latter. Squalane with dipropyl tetrachlorophthalate, Fig. 18b, also gives several windows of favorable alpha, with that at $\phi_A = 0.05$ being the tallest. Figure 18c shows, however, that the best alpha attainable with dinonyl phthalate/dipropyl tetrachlorophthalate is only 1.07, i.e., considerably less than those found both in (a) and in (b).

Figure 19 gives the first-time chromatograms obtained with (a) an intimate blend of squalane with dinonyl phthalate of $\phi_A = 0.1648$, (b) mechanically mixed packings of the same volume fraction, (c) an intimate blend of squalane with dipropyl tetrachlorophthalate of $\phi_A = 0.0522$, and (d) a mechanical mixture of packings corresponding to the composition in (c). Not only are all solutes baseline-resolved but, once again, retentions with intimately blended solvents prove to be identical to those arising with mechanically fabricated packings. Further, of course, a lengthy search for appropriate pure phases has been avoided (only three pure-phase columns were employed) and separation has been achieved precisely as predicted from inspection of Fig. 18. Note in addition that the time of analysis with squalane/dipropyl tetrachlorophthalate is somewhat longer than that for squalane/dinonyl phthalate, which could have been determined by cursory estimation of the partition coefficients of solute 28 (the last-eluted solute in the chromatograms) at the values of ϕ_A^{opt} in Figs. 16a and b. Thus, when two or more windows arise with one or more pairs of solvents that offer approximately equivalent alpha (recalling that maximization of this parameter is of primary interest in separations), secondary criteria such as analysis time may then be invoked, that is, form the basis upon which a choice between one or another column

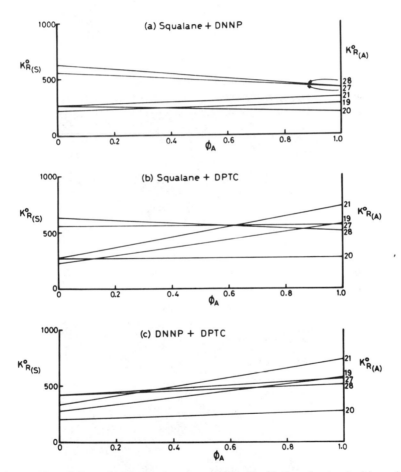

Fig. 17 Plots of K_R against ϕ_A for the solutes of Fig. 16 with the solvent pairs (a) squalane with dinonyl phthalate, (b) squalane with dipropyl tetrachlorophthalate, and (c) dinonyl phthalate with dipropyl tetrachlorophthalate. [Reprinted with permission from R. J. Laub and J. H. Purnell, *Anal. Chem.* **48,** 799. Copyright 1976 American Chemical Society.]

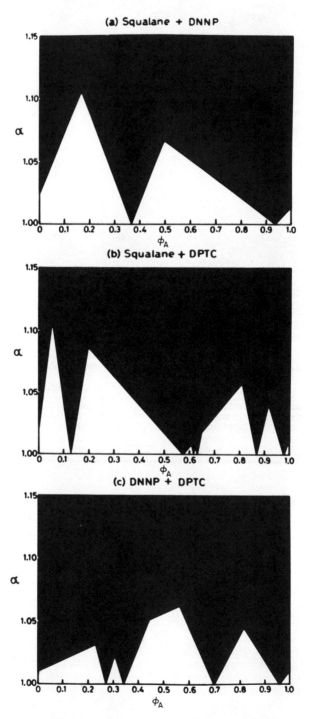

Fig. 18 Window diagrams for the solutes of Fig. 16 with the indicated pairs of solvents. [Reprinted with permission from R. J. Laub and J. H. Purnell, *Anal. Chem.* **48,** 799. Copyright 1976 American Chemical Society.]

composition is made. A number of other criteria can, of course, be cited in addition to time, these including availability and cost of particular phases, column efficiency inherent with each, and volatility.

The simple example presented above provides illustration of how and why particular phases should be chosen for use in admixture for a given analysis by gas chromatography. The success of the window diagram procedure is predicated upon location of pairs of phases that provide markedly different

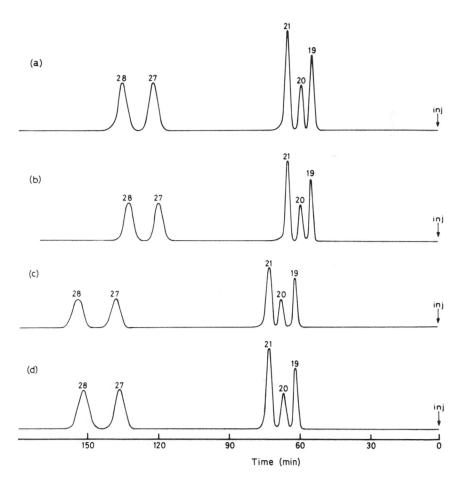

Time (min)

Fig. 19 Chromatograms of the solutes of Fig. 16 at 100°C with (a) an intimate blend of squalane and dinonyl phthalate phases ($\phi_A = 0.1648$), (b) mechanically mixed pure-phase packings of squalane and dinonyl phthalate ($\phi_A = 0.1648$), (c) an intimate blend of squalane and dipropyl tetrachlorophthalate phases ($\phi_A = 0.0522$), and (d) mechanically mixed pure-phase packings of squalane and dipropyl tetrachlorophthalate ($\phi_A = 0.0522$). [Reprinted with permission from R. J. Laub and J. H. Purnell, *Anal. Chem.* **48,** 799. Copyright 1976 American Chemical Society.]

orders of elution for this or that set of solutes. Figure 16 showed, for example, that the relative retentions of all five solutes could be altered considerably by changing the pure phase. In contrast, one would expect similar retention behavior and elution order with phases that retain solutes to the same extent because of similar solute–solvent interactions; that is, two "boiling-point" phases would be expected to give relative retentions that are similar, irrespective of the solute type and even though the *raw* retentions may differ considerably from one such phase to the other.

For this reason, one can expect that a boiling-point solvent will suffice as one of the phases in most instances, since rarely does a real mixture consist of solutes whose vapor pressures are identical. Thus, for a given mixture, *some* separation can be anticipated for *some* of the solutes with solvents of this type. The second phase is then chosen from those available which are selective for the solutes at hand on *other* than the basis of vapor pressure. For example, in the above instance dinonyl phthalate and dipropyl tetrachlorophthalate were used because the solutes contained sites of unsaturation, and each of the straight-line plots with squalane indeed indicated a number of inversions of retention order. In contrast, if a particular polyester phase had been employed with squalane, and if the straight-line and window plots showed that all column compositions were unsatisfactory because of low alpha, it would likely prove to be of little use to try another polyester or another hydrocarbon phase. That is to say, if a phase representative of a particular class of solvents is found not to be useful, the same can in all probability be said for all phases of that class. The window diagram procedure thus is predicated upon a few phases chosen to represent spectra of solute–solvent interactions.

Two additional and minor considerations require clarification at this stage with regard to mixed packings. First, if the volatility of one of the phases differs appreciably from that of the other at the system temperature employed, the volume fraction of the mixture will change with time. The rate of change will, of course, depend upon the bleed rate of each of the solvents but deterioration of the separation (hence, column composition) can easily be monitored by inspection of the resultant chromatograms. Since, however, one of the criteria for initial selection of phases is minimal vapor pressure, the problem is generally not likely to be of major concern and can in any event be remedied by lowering the column temperature or by substitution of phases that exhibit equivalent selectivity yet lower volatility. Secondly, temperature programming may be required for a given set of solutes in order to elute the last component in a reasonable amount of time, which may be thought to alter the column composition because of changes in density of the phases. Fortunately, however, the change in volume fraction will be affected only minimally with mixtures of most phases commonly employed in gas chromatography because there is little difference in their expansion coefficients over extended ranges of temperature. For example, a mixture of squalane and dinonyl phthalate of $\phi_A = 0.500$ at $0°C$ will correspond to $\phi_A = 0.498$ at $100°C$, a change in volume fraction of only 0.002.

B. Replication of Pure Phases with Mixed Packings

An immediate benefit of the window approach is realization in practicable and quantitative terms of the long-sought [50–52] goal in gas chromatography of elimination of the erroneously perceived need in many laboratories to stock all 400 or 500 stationary phases that are currently available commercially. By way of illustration [53], consider the OV series of methylphenylsilicone stationary phases (eight in all), which are available in composition from 0% phenyl to 75% phenyl substition [54]. It appears, for example, reasonable to expect that for these phases, a 1:2 mixture of OV-101 (100% methylsilicone) + OV-25 (75% phenyl- 25% methylsilicone) would be equivalent to one of pure OV-17 (50% phenyl- 50% methylsilicone). [On a weight (rather than number or mole) basis, such a mixed phase would be expected to correspond more closely to OV-11 (46 wt % phenyl).] However, and in contrast to studies by Mann and Preston [55] and by Lynch, Palocsay,

TABLE IX

Specific Retention Volumes for Listed Solutes with OV-101, OV-11, and OV-25 Stationary Phases at $60°C^a$

	Solute	V_g° (cm^3/g)		
		OV-101	OV-11	OV-25
1	n-Hexane	61.95	44.31	21.95
2	n-Heptane	142.8	105.2	49.35
3	n-Octane	323.7	243.8	110.4
4	n-Nonane	727.5	563.3	244.7
5	Methylcyclohexane	172.9	153.1	86.74
6	Benzene	95.25	130.0	113.9
7	Toluene	222.2	308.1	254.6
8	o-Xylene	596.6	895.0	723.1
9	m-Xylene	510.7	709.7	555.3
10	p-Xylene	510.9	706.5	556.2
11	Chlorobenzene	406.1	655.0	569.2

^{a} Reprinted with permission from C.-F. Chien, M. M. Kopecni, and R. J. Laub, *Anal. Chem.* **52**, 1407. Copyright 1980 American Chemical Society.

and Leary [56], Parcher, Hansbrough, and Koury [54] showed that plots of specific retention volumes of ten solutes (including those which exhibit marked gas–liquid interfacial adsorption) did not vary linearly with percent phenyl content of the OV phases and in point of fact exhibited maxima in the neighborhood of OV-11 and OV-17.

Table IX lists [53] the specific retention volumes of ten solutes with OV-101, OV-11, and OV-25 at 60°C, for which maxima can be seen for the aromatic hydrocarbons with OV-11. The data are presented graphically and in accordance with Eq. (17) in Fig. 20 as plots of V_g° against w_A for the three possible pairs of these phases, namely OV-101 with OV-11 (a), OV-11 with

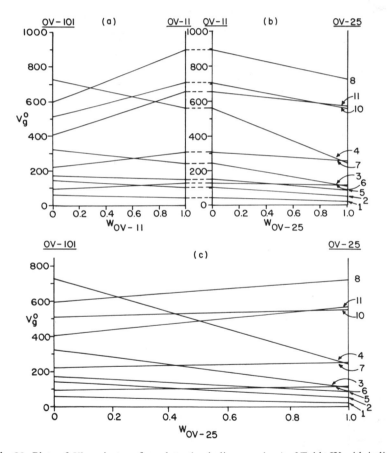

Fig. 20 Plots of V_g° against w_A for solutes (excluding *m*-xylene) of Table IX with indicated pairs of stationary phases at 60°C, constructed solely from the end-point data (i.e., under the assumption that Eq. (17) applies exactly). [Reprinted with permission from C.-F. Chien, M. M. Kopecni, and R. J. Laub, *Anal. Chem.* **52,** 1407. Copyright 1980 American Chemical Society.]

OV-25 (b), and OV-101 with OV-25 (c). The data in (c) clearly look nothing like those in (a) and (b); the raw retentions accordingly cannot be said to correspond at any composition to those found with pure OV-11.

In striking contrast, Fig. 21 shows the three corresponding window diagrams constructed from the data of Table IX: that in (c) appears to be almost an exact composite of those in (a) and (b), a result, again, of the use of relative (not absolute) retentions in the fabrication of window diagrams. Thus, the two largest windows A and B in (a) are duplicated by those in (c) (A' and B'), while the smaller windows in (a) and in (b) can also be identified in (c).

The chromatograms obtained with windows A and A' are shown in Fig. 22, while those with B and B' are given in Fig. 23. The only noticeable difference between chromatograms (a) and (b) of Fig. 22 is that solute 4 is eluted about 30 sec later in (a) than in (b). Slightly greater differences can be seen between the chromatograms in (a) and (b) of Fig. 23: the solute pair 6/2 is marginally better resolved in (a) than in (b), but the overall analysis time is about 6 min shorter in (b), reflecting differences in absolute retentions between the two

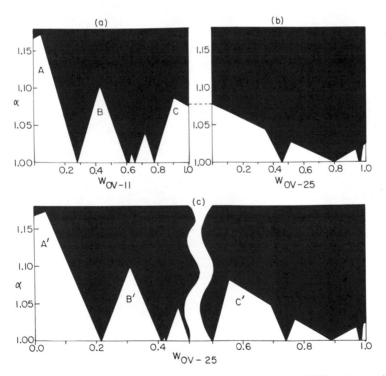

Fig. 21 Window diagrams for solutes of Fig. 20 with indicated pairs of OV stationary phases at 60°C. [Reprinted with permission from C.-F. Chien, M. M. Kopecni, and R. J. Laub, *Anal. Chem.* **52,** 1407. Copyright 1980 American Chemical Society.]

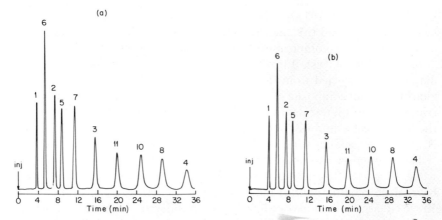

Fig. 22 Chromatograms at 60°C of the solutes of Fig. 20 with (a) mechanical mixture of pure-phase packings of OV-101 and OV-11 such that $w_A = 0.044$ (window A of Fig. 21), and (b) mechanical mixture of pure-phase packings of OV-101 and OV-25 such that $w_A = 0.040$ (window A′ of Fig. 21). [Reprinted with permission from C.-F. Chien, M. M. Kopecni, and R. J. Laub, *Anal. Chem.* **52**, 1407. Copyright 1980 American Chemical Society.]

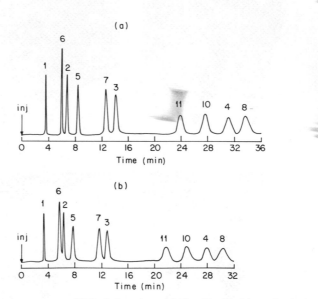

Fig. 23 Chromatograms at 60°C of the solutes of Fig. 20 with (a) mechanical mixture of pure-phase packings of OV-101 and OV-11 such that $w_A = 0.430$ (window B of Fig. 21), and (b) mechanical mixture of pure-phase packings of OV-101 and OV-25 such that $w_A = 0.310$ (window B′ of Fig. 21). [Reprinted with permission from C.-F. Chien, M. M. Kopecni, and R. J. Laub, *Anal. Chem.* **52**, 1407. Copyright 1980 American Chemical Society.]

combinations of phases. Nevertheless, Figs. 21, 22, and in particular 23 demonstrate with little doubt that while *absolute* retentions with OV-11 cannot be duplicated with mixtures of OV-101 with OV-25, *relative* retentions obtained with either provide identical separations. Furthermore, while the generality of this proposition has yet to be tested exhaustively with all OV phases of this type with a wide range of classes of solutes, the viability of such an approach must be regarded as highly likely.

C. Multicomponent Phases

On the assumption that blended stationary phases are immiscible or can be fabricated as such from mechanically mixed packings, Eq. (21) can be generalized to

$$K_R = \sum_{i=1}^{n} \phi_i K_{R_{(i)}}^{\circ} \tag{30}$$

that is, retentions with multicomponent phases or multiphase packings can be predicted from those pertaining to each of the pure-phase systems. Multidimensional window diagrams may be employed in such cases in order to optimize the amount of each sorbent used to effect a separation. Laub, Purnell, and Williams [57] pointed out, for example, that for the utilization of three stationary phases, volume compositions could be represented on equilateral triangular plots with each apex corresponding to $\phi_i = 1$. Retentions would then be plotted perpendicular to the surface of the composition triangle, with the highest three-dimensional point of intersection of alpha versus ϕ_i giving the optimum column composition and the corresponding most favorable alpha. Pyramidal windows would thus be formed by the intersection of three or more tie-planes of alpha/ϕ_i and the boundaries of the ϕ base.

In principle, the approach can be expanded to include n stationary phases, where representation of retention data is facilitated by considering each to be a lattice point on an n-dimensional surface that is bounded by the condition that $\sum \phi_i = 1$. Considerable difficulties arise, however, even in the case of three solvents, in determining which point of intersection corresponds to the best window; this is exacerbated by the several complications (such as plateaus) mentioned above in connection with binary solvents. As a result, Laub *et al.* [57] resorted to the simpler method (utilized as well for hand-held calculators) in which, after identification of all relevant pairs of solutes, alpha data are calculated at discrete intervals of ϕ. As before, the alpha file is

TABLE X

Number of Possible Mixtures p Available from Combinations of n
Solvents as a Function of Volume Fraction Interval $\delta\phi_i{}^a$

$\delta\phi_i$			n	
	2	3	4	5
1.00	2	3	4	5
0.50	3	6	10	15
0.33	4	10	20	35
0.25	5	15	35	70
0.20	6	21	56	126
0.04	26	351	3,276	23,751
0.02	51	1378	24,804	341,055
0.01	101	5151	176,851	4,598,126

^{a} From Laub, *et al.* [57].

continually updated so that, upon conclusion of all calculations, a best
alpha at the optimum column composition is returned.†

Utility of computer logic and programs (that by Laub *et al.* [57] being
dubbed DIACHOR) can be evaluated with the hypothetical solute retentions
with the five stationary phases lised in Table XI. (These data are also suitable
for the evaluation of alternative optimization strategies since separation of
the solutes with any one of the pure phases is impossible.) For example, pure
solvent A would give two peaks for this mixture even with columns of
moderately high efficiency: the first peak would be solute 2 while the second
peak would contain solutes 1 and 3–6. Pure phase B would exhibit partial
separation of solutes 2 and 3, but 1, 5, 4, and 6 would not be resolved. Solvent
C would also give three peaks corresponding to solutes 2, 3, and 5, solutes
1 and 4, and solute 6. Solvent D would, however, show only two peaks,
comprising solutes 2, 3, 5, and 6, and solutes 1 and 4. Solvent E would yield
no improvement, since solutes 1–3 would coelute with solutes 5 and 6,

---

† While choice of the ϕ interval is defined solely by the computer time available to the user,
it should, however, be noted that the number of calculations increases significantly with the
number of phases considered. The number of mixtures p obtainable from n solvents divided
into m intervals is, in fact,

$$p = (m + n - 1)!/m!(n - 1)!.$$

Table X gives, for example, the number of possible mixtures available from up to five phases
at various intervals of ϕ. Recall, in addition, that each value of p must be multiplied by the
number of relevant pairs in order to determine the total number of calculations required.

TABLE XI

Partition Coefficients of Hypothetical Solutes Employed for
Test of DIACHOR Multisolvent Optimization Program[a]

Solute	$K^{\circ}_{R_{(A)}}$	$K^{\circ}_{R_{(B)}}$	$K^{\circ}_{R_{(C)}}$	$K^{\circ}_{R_{(D)}}$	$K^{\circ}_{R_{(E)}}$
1	162.1	219.4	126.6	201.9	91.93
2	102.1	138.0	78.37	127.7	93.12
3	158.3	143.9	79.93	124.9	94.22
4	159.7	228.1	125.4	198.8	150.1
5	159.2	221.6	82.96	122.2	90.25
6	164.0	229.2	130.2	120.9	91.04

[a] From Laub *et al.* [57].

followed by solute 4. Thus, none of the pure phases will provide, even remotely, separation of all solutes.

Table XII gives the computer-optimized best column compositions for the solutes of Table XI as well as the best alphas; also presented are the solute partition coefficients, order of elution (in parenthesis), and most difficult solute pair (asterisked) with each [57]. As indicated, the overall best column composition (i) consists of all five solvents and gives a most-difficult alpha of 1.0949, where the order of elution of the solutes is 2, 3, 5, 6, 1, and 4 (2 and 3 being the most difficult pair). In contrast, the best combination of any four of the phases (ii) provides an alpha of only 1.038. Assuming that 500 plates per foot is obtained for each phase, the column lengths required for $k' > 10$ for solvent mixtures (i) and (ii) are therefore 10 and 54 ft, respectively.

Further inspection of Table XII also shows that in (ii) only a trace of solvent A is required, while in (iii), (iv), and (v) only three solvents are used; mixture (vi) uses only about 4% of solvent D. In fact, in the author's experience, no more than three solvents need be utilized in single-column single-run analysis of samples of moderate complexity, which simplifies considerably application of the use of multicomponent phases and packings as well as reducing the computational time and memory capacity required of whatever computer device is employed (a consequence of the fact that the range of selectivities which can be achieved even with combinations of as few as three solvents is enormous).

The choice of solvents appropriate for a given set of solutes is, as with mixtures of binary stationary phases, predicated solely on the basis that each provide different orders of elution. Suppose, for example, that no combination of two solvents will resolve satisfactorily a particular solute mixture, as determined by chromatographing the solutes with each of the two (say, A and

TABLE XII

Calculated Best Solvent Combinations and Partition Coefficients for
Solutes of Table XI with Phases A–E (Solvent Mixture (i)) and with
the Five Possible Combinations of Four Phases (Solvent Mixtures (ii)–(v))[a]

Solvent mixture	$\delta\phi$	Most difficult alpha	Solvent composition				
			ϕ_A	ϕ_B	ϕ_C	ϕ_D	ϕ_E
(i)	0.02	1.0949	0.16	0.16	0.22	0.20	0.26
(ii)	0.01	1.0380	0.01	0.65	0.19	—	0.15
(iii)	0.01	1.0365	—	0.67	0.18	0.00	0.15
(iv)	0.01	1.0301	0.00	0.81	0.01	0.18	—
(v)	0.01	1.0294	0.00	0.82	—	0.18	0.00
(vi)	0.01	1.0287	0.09	—	0.10	0.04	0.77

Solvent mixture	K_R for solute no.[b]					
	1	2	3	4	5	6
(i)	153.2(5)	105.4(1)*	115.4(2)*	168.4(6)	127.1(3)	139.4(4)
(ii)	182.1(4)	119.6(1)	124.4(2)	196.2(6)	174.9(3)	189.0(5)*
(iii)	183.6(4)	120.5(1)*	124.9(2)*	197.9(6)	176.9(3)	190.7(5)
(iv)	215.3(5)*	135.5(1)	139.8(2)	221.8(6)*	202.3(3)	208.7(4)
(v)	216.3(5)	136.1(1)	140.5(2)	222.8(6)	203.7(3)*	209.7(4)*
(vi)	106.1(5)	93.84(1)	99.79(3)*	150.4(6)	97.00(2)*	102.7(4)

[a] From Laub et al. [57].
[b] Order of elution is given in parentheses; asterisks indicate most difficult solute pairs.

B) pure phases (which requires, of course, only two experiments). A third
run with another phase (say, C) that is thought to give selectivity which
differs from either of the first two would then provide three new combinations
of solvents in addition to A with B, namely, A with C, B with C, and A with B
with C. Thus, only a single additional experiment triples the number of
possible combinations. A fourth run with another phase produces seven new
possibilities, and so on.

D. Employment of Relative Retentions

From the standpoint of practical implementation of Eq. (17) or (21), use of
partition coefficients or specific retention volumes presents a serious impedi-
ment: Eq. (9b), for example, necessitates accurate measurement of the inlet
and outlet pressures, flow rate, stationary-phase density, and so forth, in
order to determine K_R data. The latter are particularly difficult to ascertain e.g., for, silicone gum phases at elevated temperatures. Furthermore,

even though the use of specific retention volumes with Eq. (17) in part obviates this need, the required time and equipment may, from a practical viewpoint, still preclude measurement of V_g° data. In addition, the vast majority of retentions which are tabulated in the GC literature are reported as relative to some standard or other. As pointed out by Hildebrand and Reilley [58], however, alpha values alone cannot be used to predict optimum column compositions for the separation of a given mixture unless the reference solute itself exhibits absolute retentions that are identical with each of the pure phases. Thus, account must be taken of the fact that the reference solute may be retained to different degrees from one solvent to another, which will in any event likely be the case if the phases have been chosen (as advocated above) on the basis that differences in solute elution order are extant with each.

Laub, Purnell, Summers, and Williams [59] devised a simple solution to this difficulty in 1978: recall Eq. (17) for binary stationary phases A and S in terms of specific retention volumes for solute i:

$$V^\circ_{g(A,S)_i} = w_A V^\circ_{g(A)_i} + w_S V^\circ_{g(S)_i} \tag{31}$$

Dividing through Eq. (31) by the specific retention volume of the chosen standard solute z with pure phase S ($V^\circ_{g(S)_z}$),

$$\frac{V^\circ_{g(A,S)_i}}{V^\circ_{g(S)_z}} = w_A \left(\frac{V^\circ_{g(A)_i}}{V^\circ_{g(S)_z}} \right) + w_S \left(\frac{V^\circ_{g(S)_i}}{V^\circ_{g(S)_z}} \right) \tag{32}$$

The right-hand term contains $\alpha^\circ_{(S)_{i/z}}$, that is, the retention of solute i relative to solute z with pure phase S, but the other terms require still further modification. Multiplying the left-hand quantity by unity, chosen in this case to be $(V^\circ_{g(A,S)_z} | V^\circ_{g(A,S)_z})$, and the middle term by the unit ratio $(V^\circ_{g(A)_z} | V^\circ_{g(A)_z})$ yields

$$\left(\frac{V^\circ_{g(A,S)_i}}{V^\circ_{g(A,S)_z}} \right) \left(\frac{V^\circ_{g(A,S)_z}}{V^\circ_{g(S)_z}} \right) = w_S \left(\frac{V^\circ_{g(S)_i}}{V^\circ_{g(S)_z}} \right) + w_A \left(\frac{V^\circ_{g(A)_i}}{V^\circ_{g(A)_z}} \right) \left(\frac{V^\circ_{g(A)_z}}{V^\circ_{g(S)_z}} \right) \tag{33}$$

The first bracketed ratio on the left-hand side of Eq. (33) is just $\alpha_{(A,S)_{i/z}}$, that is, the retention of i with respect to z with the mixed phase. The second term is the ratio of retentions of z with the mixed phase and with pure S. For columns† of equal stationary-phase weight and dead volume per unit length, this is just the capacity-factor ratio for solute z:

$$V^\circ_{g(A,S)_z} / V^\circ_{g(S)_z} = k'_{(A,S)_z} / k'_{(S)_z} \tag{34}$$

which Laub and co-workers chose to call β. Similarly, the ratio on the far right of Eq. (33) is the capacity-factor quotient of solute z with pure phases A

---

† The columns need not, however, be of the same length nor be used with identical flow rates.

and S and so is termed $\beta^{\circ}_{(A)}$. Finally, $(V^{\circ}_{g(A)_i} | V^{\circ}_{g(A)_z})$ is just $\alpha^{\circ}_{(A)_{i/z}}$. Equation (33) can now be written as

$$\alpha_{(A,S)_{i/z}} \cdot \beta = w_S \cdot \alpha^{\circ}_{(S)_{i/z}} + w_A \cdot \alpha^{\circ}_{(A)_{i/z}} \cdot \beta^{\circ}_{(A)} \tag{35}$$

and Eq. (17) has been expressed in terms of quantities measureable solely from raw and dead retention times, i.e., those available directly from chromatograms.

Table XIII presents, by way of illustration of Eq. (35), the specific and relative retentions for five hypothetical solutes with phases A and S. Solutes 1 (the standard) and 2 cannot be resolved with pure S, while solutes 1 and 5 cannot be separated with pure A; furthermore, a number of inversions of retention can be seen at intermediate compositions of S + A. Note in addition that α and $\alpha \cdot \beta$ are permitted to fall below unity in the table, i.e., are not inverted in order to reflect accurately the variation of *absolute* retentions of each of the solutes.

According to Eq. (35), the appropriate ordinate parameters to be used for construction of plots of retention against column composition are $\alpha^{\circ}_{(S)_{i/z}}$ (for pure S) and $\alpha^{\circ}_{(A)_{i/z}} \cdot \beta^{\circ}_{(A)}$ (for pure A), which are equivalent to $V^{\circ}_{g(S)_i}$ and $V^{\circ}_{g(A)_i}$. The plots are shown in Fig. 24 with both sets of ordinates; also shown is a horizontal dashed line which has been drawn from the left-hand intercept of solute 1, the reference solute, across the plot. The reason that $\alpha^{\circ}_{(S)}$ and $\alpha^{\circ}_{(A)}$ data alone do not correctly reflect retentions can now be discerned graphically: solute 1 is retained to a greater degree with pure A than with pure S, so its retention line slopes positively. Now, if all other retentions were plotted relative to solute 1, that of the latter would be a horizontal line and the slopes of the remaining solutes would be in error by an amount equal to the ratios of their true slopes with that of solute 1.

TABLE XIII

Specific and Relative (No. 1 = 1.000) Retentions for Five
Hypothetical Solutes with Stationary Phases S and A[a]

Solute	$V^{\circ}_{g(S)}$ (cm^3/g)	$\alpha^{\circ}_{(S)_{i/1}}$	$V^{\circ}_{g(A)}$ (cm^3/g)	$\alpha^{\circ}_{(A)_{i/1}}$	$\alpha^{\circ}_{(A)_{i/1}} \cdot \beta^{\circ}_{(A)}$
1	75	1.000	100	1.000	1.333
2	75	1.000	150	1.500	2.000
3	100	1.333	60	0.600	0.800
4	125	1.667	160	1.600	2.133
5	200	2.667	100	1.000	1.333

[a] From Laub *et al.* [59].

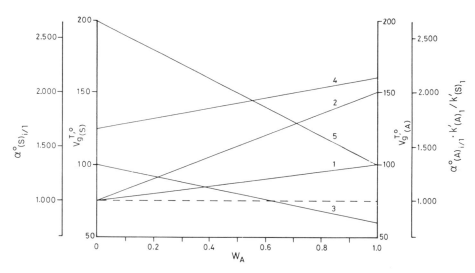

Fig. 24 Plots of (inner ordinates) V_g° and (outer ordinates) $\alpha_{(S)_{i/1}}^\circ$ (left) and $\alpha_{(A)_{i/1}}^\circ \cdot \beta_{(A)}^\circ$ (right) against w_A for solutes of Table XIII. Dashed line originating at left intercept of solute 1 drawn parallel to the abscissa. [From Laub *et al.* [59].]

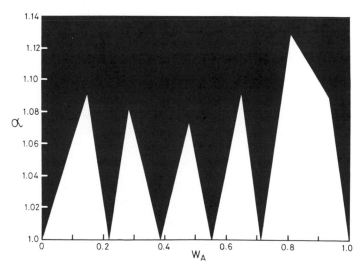

Fig. 25 Window diagram constructed from data of Table XIII (columns 3 and 6) with solute 1 as the reference compound. [From Laub *et al.* [59].]

In contrast, multiplication of $\alpha^\circ_{(A)}$ by $\beta^\circ_{(A)}$ takes into account the change in retentions of the reference solute in essence by definition of a *hypothetical* horizontal line to which all retentions (even the reference compound) are related. Window diagrams fabricated from $\alpha \cdot \beta$ data will therefore provide the correct column composition for the separation of all solutes *including* the standard. In the present instance, that for the data of Table XIII is shown in Fig. 25, which indicates that $w_A^{\text{opt}} = 0.810$, at which column composition $\alpha = 1.128$ ($N_{\text{req}} = 2800$ plates for $k' > 10$).

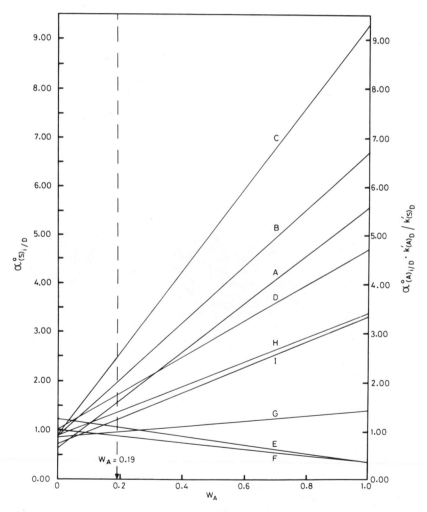

Fig. 26 Plots of $\alpha \cdot \beta$ against w_A for data of Table XIV. Vertical dashed line at $w_A = 0.19$ is the optimum column composition for separation of all solutes. [From Laub *et al.* [59].]

TABLE XIV

Relative Retention Data for Listed Solutes Relative to
N,N-Dimethylaniline (D) with SE-30 and 2,4,7-Trinitrofluorenone
(TNF) Stationary Phases at 180°C[a]

Solute	Name	$\alpha^{\circ}_{(SE\text{-}30)_{i/D}}$	$\alpha^{\circ}_{(TNF)_{i/D}} \cdot \beta^{\circ}_{(TNF)}$ [b]
A	Aniline	0.609	5.622
B	N-Methylaniline	0.840	6.687
C	o-Toluidine	0.903	4.716
D	N,N-Dimethylaniline	1.000	4.716
E	cis-Decalin	1.221	0.354
F	$trans$-Decalin	1.007	0.354
G	Indane	0.856	1.424
H	Indene	0.876	3.410
I	2,3-Benzofuran	0.722	3.330

[a] From Laub *et al.* [59] and Cooper *et al.* [60].
[b] $\beta^{0}_{(TNF)} = 4.716$.

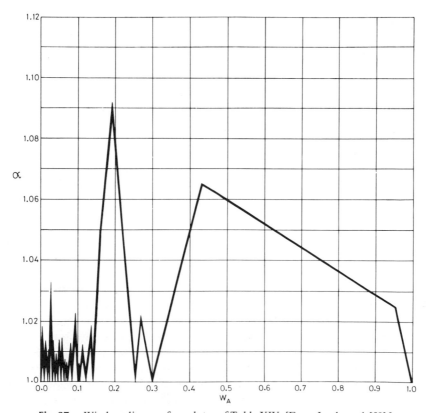

Fig. 27   Window diagram for solutes of Table XIV. [From Laub *et al.* [59].]

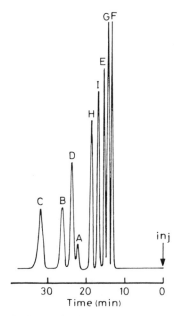

Fig. 28 Chromatogram of solutes of Table XIV with mechanical mixture of pure-phase packings of SE-30 and TNF such that $w_{TNF} = 0.20$. [From Laub *et al.* [59].]

An example [59] of relative retention data typical of those found in the literature [60] is given in Table XIV for aromatic amine and hydrocarbon solutes with SE-30 and 2,4,7-trinitrofluorenone (TNF) stationary phases. The straight-line plot is shown in Fig. 26 and the window diagram in Fig. 27, where the optimum column composition is predicted to occur at $w_{TNF}^{opt} = 0.19$ with an alpha of 1.092. Figure 28 shows the first-time chromatogram of the solutes with mixed packings such that $w_{TNF} = 0.19$, where virtually complete resolution has been achieved as predicted.

E. Mixed GSC and GLSC Sorbents

While there are a number of difficulties [61] associated with catalytic decomposition of solutes with gas–solid or gas–liquid–solid adsorbent phases, it is nevertheless clear from what has been said thus far that if retentions vary in a predictable manner [62] from one phase to the next, then mixtures of sorbents can also be used to provide enhanced separations. For example, Al-Thamir, Laub, and Purnell [61] employed mixtures of squalane on alumina with dinonyl phthalate on alumina for the separation of all commercially available C_1 to C_4 hydrocarbons. Mixed sorbents have, however, been little used in the past, primarily because of the batch-to-batch and column-to-column irreproducibility of retentions with pure-phase materials. Broadly speaking, on the other hand, a few types of samples (e.g., permanent gases) can be separated at present *only* with adsorbent stationary phases; employment of mixtures of the latter should therefore not be overlooked as a means of providing enhanced resolution of the former.

F. *Blended GC Mobile Phases*

The window optimization strategy has thus far been employed in only one instance with binary GC mobile phases: Pretorius [63] found in 1978 that mixtures of nitrogen with steam improved markedly the separation of sterols and that solute retentions could in fact be predicted from those obtained with the pure mobile phases. It has been recognized for a number of years that the use of different carriers in GC will alter solute elution times. It has also been noted that in rare cases these (virial) effects can cause inversions in retention order [64]. However, given the logarithmic variation of K_R as a function of solute–carrier virial interactions [8], it is somewhat surprising that Pretorius found K_R to vary linearly with mixtures of nitrogen with steam. In any event, situations in which mixed carrier phases might well prove to be of some utility can easily be visualized. Suppose, for example, that a particular separation is so complex that even after optimization of the stationary phase the most difficult solute pair still corresponds to an alpha of 1.01. A mere 2% increase of alpha to 1.03 will reduce the column length required for resolution by a factor of ten. Blended mobile phases can therefore be resorted to as a kind of "fine tuning" for difficult and/or complex samples. [Parenthetically, virial effects can be increased (as can column efficiency [65]) at constant flow rate and pressure drop by adding a choke to the column outlet and increasing the inlet pressure.]

G. *Optimization of Temperature*

Temperature is not ordinarily considered as a parameter that can be used to enhance separations, although it is of course widely recognized that time of analysis can be shortened by temperature programming. However, when multiple sorption mechanisms account for retention, one or the other may predominate over a given range of temperature. In such cases, retention inversions may be expected to occur, at which point the window optimization methodology becomes useful.

As first recognized by Laub and Purnell [66], temperature alone can in fact be employed *quantitatively* to effect a particular separation; furthermore, this parameter is particularly simple to use with window diagrams since plots of log(retention) against $1/T$ (van't Hoff plots) are invariably linear. Figure 29 shows [66], for example, the van't Hoff plots [67] for n-alkane and benzene solutes with N,N-bis(2-cyanoethyl)formamide stationary phase, where both solution and gas–liquid interfacial adsorption contribute to retention. Benzene cannot be separated at either temperature extreme (40–180°C), although resolution appears to be feasible at intermediate values. Further-

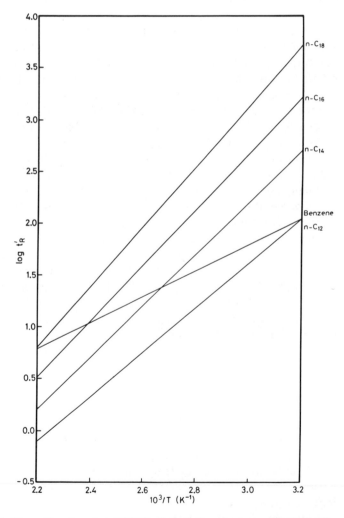

Fig. 29 Plots of log t'_R against $10^3/T$ (K^{-1}) for indicated solutes with N,N-bis(2-cyanoethyl)-formamide solvent. Data of Rogozinski and Kaufman [67]. [From Laub and Purnell [66].]

more, temperature-programmed elution of the solutes may or may not result in separation, depending upon the initial, gradient, and final values, since several retention inversions occur. In contrast, the window diagram [66] illustrated in Fig. 30 shows three useful temperatures, corresponding to $\alpha = 1.406$ ($T = 165°C$), $\alpha = 1.552$ ($T = 124°C$), and $\alpha = 1.900$ ($T = 78°C$). Thus, this is an instance in which *lowering* the column temperature from, say, 90 to 70°C or *raising* the temperature from 90 to 125°C will *in each case* bring about improved separation.

As it happens, Fig. 30 shows that all three windows provide very large alpha values, so there is seemingly little from which to choose on the basis of column length. Reference back to Fig. 29 indicates, however, that the time of analysis (at constant solvent volume) will be shortened by a factor of 60 with the temperature of window A. Thus, the secondary criterion of time of analysis predominates in this case and the optimum temperature of window A, although providing the lowest alpha, is overwhelmingly favored.

A second example [66] of the use of temperature alone to enhance separations is given in Fig. 31, where plots of log k' arc shown against $10^3/T$ for a number of solutes with a salt-coated alumina adsorbent phase. The

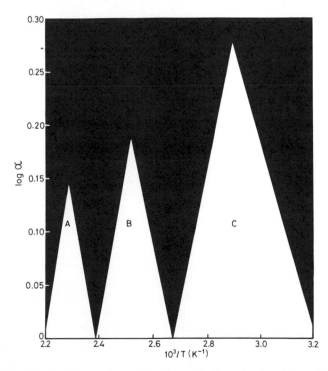

Fig. 30   Window diagram for solutes of Fig. 29. [From Laub and Purnell [66].]

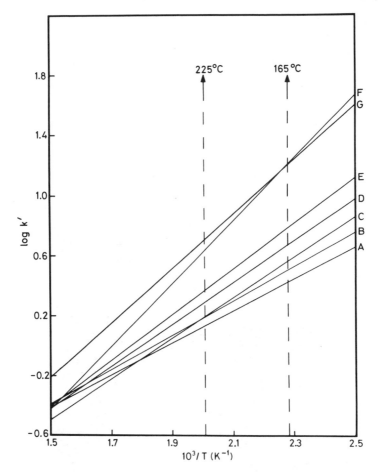

Fig. 31 Plots of log k' against $10^3/T$ (K^{-1}) for the solutes (A) *trans*-2-hexene, (B) 1-hexene, (C) *cis*-2-hexene, (D) *trans*-1,4-hexadiene, (E) *cis*-1,4-hexadiene, (F) *trans,trans*-2,4-hexadiene, and (G) benzene with the gas–solid adsorbent 10% (w/w) sodium sulfate on acid-washed F-1 alumina. Data of Sawyer and Brookman [68]. [From Laub and Purnell [66].]

vertical dashed lines indicate the temperature range over which the initial investigation was carried out by Sawyer and Brookman [68] but, since the plots are linear, the lines can be extended with confidence to the limits shown. However, it would be difficult to determine an optimum temperature solely by inspection of the van't Hoff plots. The window diagram [66] given in Fig. 32 shows immediately, however, that several optimal temperatures will provide resolution outside and in addition to the range considered originally. The lower temperature window (125°C) offers an alpha of 1.154

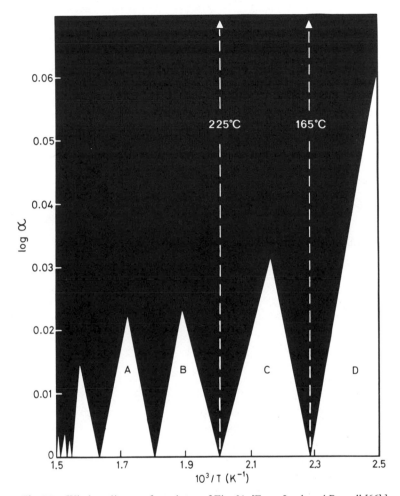

Fig. 32 Window diagram for solutes of Fig. 31. [From Laub and Purnell [66].]

(D), while that at C provides an alpha of 1.075, which, for a reported [68] column efficiency of ~250 plates per foot, require, respectively, 2000 and 8000 theoretical plates and 8- and 32-ft column lengths. The overall time of analysis in each case will, however, be about the same, so, apart from practical difficulties associated with construction of the 32-ft column and in contrast to the previous example, the lower temperature is seen here to be overall the more favorable.

The use of window diagrams for temperature optimization illustrates the considerable predictive power of the methodology and, in particular, means whereby subjective judgments may be placed on an objective basis. In addition, the generality of the technique should now be readily apparent: stated simply, any chromatographic system parameter may be optimized with respect to this or that separation so long as the latter can be described somehow or other as a function of the former. Furthermore, the answer(s) thereby obtained *must* be correct since, as by now must also be readily apparent, the window diagram itself is nothing more than a new graphical method of representation of retention data. Thus, if the data put into the procedure are accurate, the resultant optimum value of the system parameter obtained will, within the same limits of accuracy, also be correct. The situation could hardly be more advantageous, especially in view of the ease with which the entire procedure can be computerized.

IV. Use of the Optimization Strategy for the Analysis of Mixtures of Initially Unknown Content and Complexity

Rarely is an analyst faced in gas chromatography with a separation for which he has prior knowledge of the number, kind, and amount of each of the solutes present. Often, he may not even have access to the source, history, or prior treatments utilized to produce and/or modify (intentionally or otherwise) the physical and chemical properties of the sample at hand. Furthermore, while preliminary chromatographic runs with a few columns containing various phases and utilized at several temperatures are generally sufficient to establish the level of complexity of the mixture, few insights can thereby be gained as to appropriate phases for resolution of all components. As a result, other methods of instrumental analysis (e.g., mass spectrometry) are at this point often employed in attempts to assess at least the chemical nature of the mixture (i.e., mainly alcohols, amines, etc.), which, when coupled with data resulting from the trial chromatograms, offers something akin to a starting point. Thereafter, however, the analyst is faced not only with all the difficulties cited above for the selection of a pure phase and conditions appropriate for samples of known content, but is at the added disadvantage of still not knowing the precise degree of complexity (i.e., number and amount of constituents) of the mixture at hand.

The window diagram procedure provides a simple and direct approach for the analysis of mixtures of initially unknown content and complexity which is based upon the linear variation of retentions with column composition. Furthermore, the analyst need not know at the outset the number and kind of solutes present (or their chemical type), nor are any assumptions or other data (real or artificially deduced) necessary. Because this feature forms what in many respects may well be regarded as the most powerful approach yet devised for chromatographic separations, it is presented below in some detail and is illustrated with actual samples of industrial and environmental significance.

A. Employment of Partition Coefficient or Specific Retention Volume Data

Figure 33 shows a chromatogram [69] of an industrial still residue with a packed column containing squalane and dinonyl phthalate phases at 75°C. At least 43 peaks are visible, many of which, however, are minor in area compared to those which are off-scale at the attenuation used. A number of steps could be taken at this stage to enhance the separation: for example, the column temperature could be programmed from, say, ambient to 150°C; this presumably would provide enhanced resolution of the first-eluting solutes while producing narrower bands for the last compounds. As shown earlier, however, it cannot be assumed that the separation will thereby be improved; indeed, it may well become worse. Alternatively, one could employ a longer column or, as is more common, use an open-tubular column of high efficiency with this or that stationary phase; as with samples of known content, however, the search for such a phase may prove to be lengthy and, of course, offers no guarantee of success in locating a useful solvent.

The sample as shown is clearly of a level of complexity such that the usual approaches advocated in dealing with such mixtures are at best inapplicable. And yet Fig. 33 represents, in the view of the author, a case typical of those which are encountered routinely in gas chromatography. How, then, and on what basis, or with what criteria can one attack such separations?

The first step in dealing with mixtures of unknown content and complexity in accordance with the window diagram procedure is selection of a pair of stationary phases which offer different selectivity for most of the solutes. Just as with mixtures of known content, a "boiling-point" phase together with any other which does not provide resolution on the basis of vapor pressure will in all likelihood serve as a useful starting point. If the resultant (pure-phase) chromatograms are not *distinctly* different, a third phase would next be tried that differs chemically from the first two. For example, if a polyethylene glycol phase were the second solvent, a polyester might be the third phase utilized. After only three runs (each with a separate phase), therefore, those two phases that produce the greatest difference in retention order would be selected for further evaluation.

Note that because each phase, if chosen properly, differs considerably from the others and because real mixtures invariably contain peaks of widely

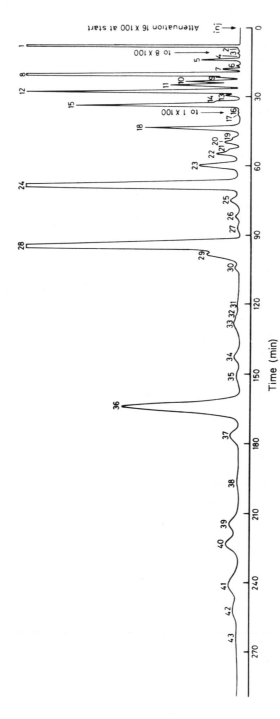

Fig. 33 Chromatogram of an industrial-still residue at 75°C with a column containing squalane and dinonyl phthalate mixed packings such that $\phi_A = 0.294$ (see text). [Reprinted with permission from R. J. Laub and J. H. Purnell, *Anal. Chem.* **48**, 1720. Copyright 1976 American Chemical Society.]

different areas, assessment of an appropriate pair of solvents via monitoring of the alteration of retentions will prove to be relatively easy. Furthermore, solute mixtures, while often containing a large number of compounds, may in fact show only a few major constituents; that is, 95% or so of the mixture may comprise only ten to twenty compounds. This not only facilitates recognition of the movement of solutes from one phase to the next, but also simplifies the overall separations task; thus, resolution only of the major constituents of the sample need initially be considered.

Figure 34 shows [69] chromatograms of the major constituents of the industrial sample with columns containing (a) dinonyl phthalate, (b) mechanically mixed pure-phase packings of squalane and dinonyl phthalate of $\phi_A = 0.628$, (c) as in (b) except that $\phi_A = 0.332$, and (d) pure squalane, all at 100°C. These phases were chosen precisely because of the criteria cited above and, in particular, because noticeable shifts in retention were observed from pure squalane to pure dinonyl phthalate.

Chromatogram (a) shows nine peaks, which are all resolved except for solutes 4 and 5. Chromatogram (b) again gives 9 peaks, where solute 5 has been separated from 4. Indeed, one might be tempted to stop at this point and claim that all of the major components have been separated. However, chromatogram (c) shows ten solutes, where solute 5 of (b) has now split into two peaks. Chromatogram (d), on the other hand, gives only eight peaks.

Chromatograms (a)–(d) of Fig. 34 establish a *floor* to the number of major constituents present in the sample: since ten peaks were seen (c) at one point, there must be *at least* ten major components in the mixture. If, in addition, for any reason it is thought that two constituents have virtually identical partition coefficients with the two phases (hence, form one peak at all intermediate compositions), an alternative set of phases may be employed that will provide resolution of that pair. Furthermore, one or more major constituents may be strongly retained, that is, overlooked at the chosen column temperature, this possibility being eliminated simply by chromatographing the mixture at elevated temperatures with one of the pure phases. In the present instance neither of the above difficulties appeared to have arisen with squalane/dinonyl phthalate and, since overlaps and retention inversions were clearly evident, this pair of phases was used for the described analysis. At this point, therefore, a column temperature and appropriate phases have been identified, as has a limiting lower number of major constituents in the sample.

The next step in the analysis is the determination of a *ceiling* for the number of major constituents, that is, the maximum number of solutes actually present in Fig. 34. In the present instance, the partition coefficients of all visible peaks were measured and plotted as shown [69] in Fig. 35a against ϕ_A, each numbered point corresponding to each peak in Fig. 34.

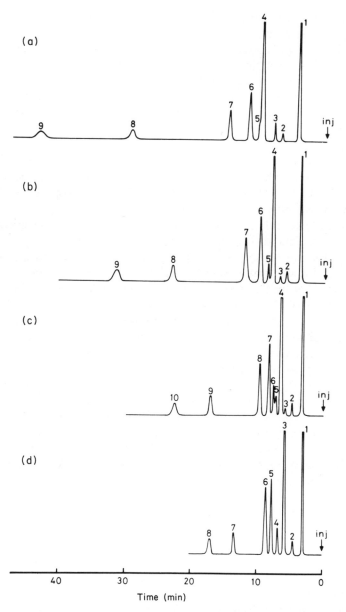

Fig. 34 Chromatograms of the sample of Fig. 33 with (a) pure dinonyl phthalate, (b) mechanically mixed pure-phase packings of squalane and dinonyl phthalate where $\phi_A = 0.628$, (c) as in (b) except that $\phi_A = 0.332$, and (d) pure squalane. Attenuation adjusted so that only the major constituents of the sample are visible. [Reprinted with permission from R. J. Laub and J. H. Purnell, *Anal. Chem.* **48**, 1720. Copyright 1976 American Chemical Society.]

The data shown in Fig. 35a appear to be a bewildering array. However, since mechanically mixed packings were used, each solute must be represented by a straight line which in turn must have four (but *only* four) points on it, corresponding to the four columns employed. Thus, identification of which points belong with which solute is in many cases straightforward. Consider, for example, the last-eluting solute in each of the chromatograms of Fig. 34: the areas appear by inspection to be nearly identical and, indeed, a

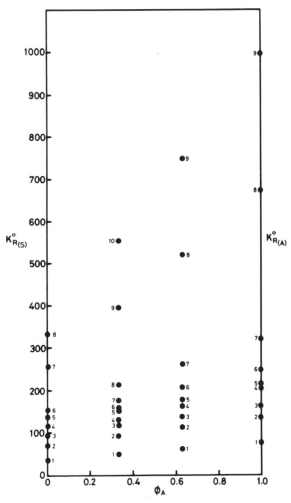

Fig. 35a Plots of K_R against ϕ_A for peaks of Fig. 34 without connecting lines. [Reprinted with permission from R. J. Laub and J. H. Purnell, *Anal. Chem.* **48**, 1720. Copyright 1976 American Chemical Society.]

straight line can be drawn through the last set of points in Fig. 35a, as shown in Fig. 35b. Thus, the peaks (and points) all appear to be due to the same solute. Similar considerations show that the penultimate peak in each chromatogram also exhibits approximately the same area; since the four points corresponding to these peaks can be connected by a straight line as shown in Fig. 35b, a second solute has thereby tentatively been identified. Consideration of the remainder of the peaks and data points yields the lines

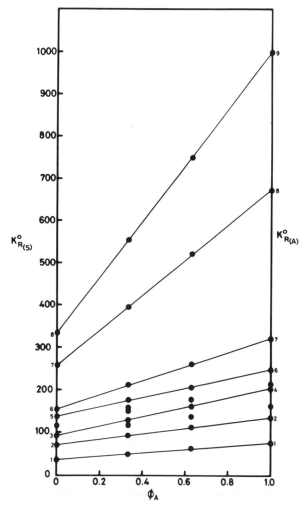

Fig. 35b Plots of Fig. 35a with lines drawn in for components identified by inspection. [Reprinted with permission from R. J. Laub and J. H. Purnell, *Anal. Chem.* **48,** 1720. Copyright 1976 American Chemical Society.]

drawn in the K_R/ϕ_A plots. Altogether, seven solutes have been identified by the procedure, which amounts only to visual inspection of the chromatograms and use of a rule in constructing lines.

However, several points remain unused in Fig. 35b; these data must correspond to solutes since they were measured from peaks seen in Fig. 34. Since further consideration of peak areas is of no help in deciding which points constitute what lines, *all possible* lines are shown utilizing *all points* in Fig. 35c, namely, straight lines have been drawn in without regard for peak

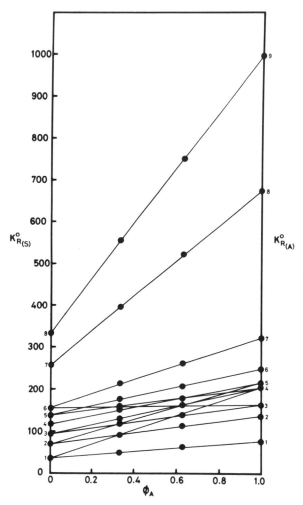

Fig. 35c Plots of Fig. 35a with all possible lines drawn in. [Reprinted with permission from R. J. Laub and J. H. Purnell, *Anal. Chem.* **48**, 1720. Copyright 1976 American Chemical Society.]

areas but within the constraints that each line must be straight (within the error limits of the chromatographic technique) and each line must have four but *only* four points on it. This does not, however, preclude a point from being used more than once. Several points are in fact used twice or even three times; i.e., they indicate the possibility that two or three solutes coelute at one or another column composition. Nevertheless, and within the boundary conditions noted, only thirteen lines can be constructed; that is, the *maximum* number of major constitutents in the sample is thirteen.

Visual inspection of Figs. 34 and 35 has thus established at this point that there may be as few as ten or as many as thirteen solutes present as major components. Furthermore, no effort has been made to identify their chemical type or structure, nor, for that matter, an optimum column composition for their separation. Instead, the number of compounds present has first been dealt with. The result is somewhat ambiguous, however, especially in view of the fact that several of the lines in Fig. 35c may be fictitious; that is, it may be an artifact of the data points that thirteen lines could be constructed. The next step in the analysis, therefore, is elimination of false lines, i.e., determination of the actual number of solutes present.

Brief reflection indicates that a window diagram will provide a column composition at which all thirteen solutes, if indeed there are thirteen, will be resolved. (If, of course, there are only ten, these would also be separated.) However, a simpler approach is provided by inspection of Fig. 35c: a column of moderate efficiency with ϕ_A of ~ 0.08 will not baseline-resolve all thirteen solutes, but all will, if present, be visible, which will be sufficient information to eliminate the fortuitous lines of Fig. 35c. The chromatogram is shown [69] in Fig. 36 where, in fact, only ten peaks can be seen. There must therefore be only ten major constituents in the mixture. The partition coefficients of the peaks are plotted in Fig. 37, where the fictitious lines have been removed. The actual number of major constituents in the sample (and the variation of K_R with ϕ_A for each) has thus been established, again without reference to the identify of any of the solutes.

Separation of the mixture is now straightforward. Figure 38 shows [69] the window diagram for the ten solutes, where two column compositions offer significant alpha values, namely, $\alpha = 1.081$ at $\phi_A = 0.294$ and $\alpha = 1.084$ at $\phi_A = 0.734$. Use of the former window will, however, provide significantly shorter elution times, and, since the two alphas are nearly equal, $\phi_A = 0.294$ was chosen.

Figure 39 shows the first-time chromatogram [69] of the solutes where all ten have successfully been resolved in 52 min.

Finally, recall that Fig. 33 shows the sample with the same column composition but at higher sensitivity and at 75°C. Most of the major constituents are off-scale and several overlaps with minor components are visible. To resolve the full mixture, however, would require only a reiteration of the above optimization procedure but at the higher sensitivity setting.

To summarize the optimization strategy for application with samples of initially unknown content and complexity;

(1) Two stationary phases are first identified which provide relative retentions that are different for the major constitutents of the sample. This presents no difficulty since only representative phases of classes of solvents need be considered, and it is facilitated by maintaining stock pure-phase columns in convenient lengths. The column temperature is adjusted to provide suitable elution times.

(2) The sample is chromatographed with two or three columns of intermediate composition (fabricated from mechanically mixed packings) and the partition coefficients of each peak are measured. This can be done most easily by measuring retentions relative to that of a standard whose

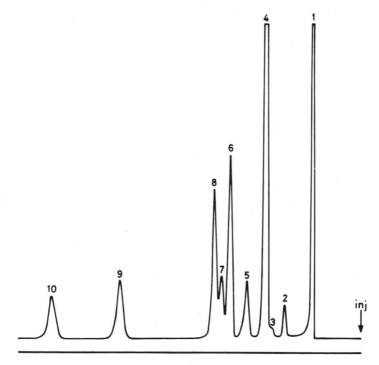

Fig. 36 Chromatogram of the major constituents of the sample of Fig. 33 at $\phi_A = 0.0749$. [Reprinted with permission from R. J. Laub and J. H. Purnell, *Anal. Chem.* **48,** 1720. Copyright 1976 American Chemical Society.]

partition coefficients with the pure phases are known accurately. (Alternatively, specific retention volumes and weight fractions may be employed if more convenient, since Eqs. (17) and (21) are formally identical.) The minimum number of major constituents in the sample is thereby identified.

(3) After plotting the retention data against column composition, all possible lines are drawn through all points, with the conditions that all lines must be straight and that each must have only that number of points on it

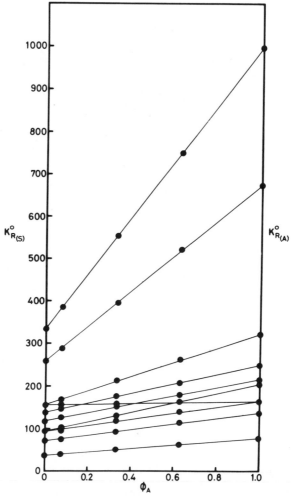

Fig. 37   Correct plots of K_R against ϕ_A for the 10 unambiguously identified major constituents of the sample of Fig. 33. [Reprinted with permission from R. J. Laub and J. H. Purnell, *Anal. Chem.* **48**, 1720. Copyright 1976 American Chemical Society.]

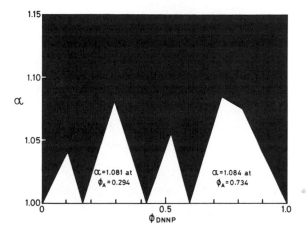

Fig. 38 Window diagram for the solutes of Fig. 37. [Reprinted with permission from R. J. Laub and J. H. Purnell, *Anal. Chem.* **48,** 1720. Copyright 1976 American Chemical Society.]

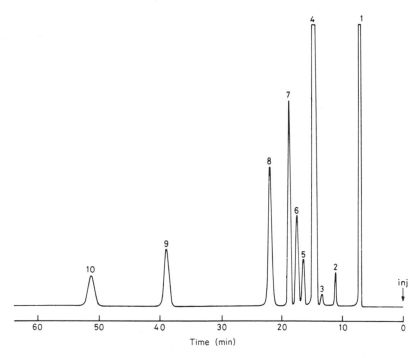

Fig. 39 First-time chromatogram showing complete separation of the major constituents of the sample of Fig. 33 with a column of ϕ_A of 0.2940. [Reprinted with permission from R. J. Laub and J. H. Purnell, *Anal. Chem.* **48,** 1720. Copyright 1976 American Chemical Society.]

corresponding to the number of columns employed. The number of lines is the maximum number of major constituents present in the sample.

(4) Since one or more of the lines constructed for the above plot may be fictitious, a decision must be reached regarding the actual number of solutes. A column composition must therefore be deduced (either by inspection or from a window diagram, depending upon the complexity of the mixture) at which all peaks will appear, and the sample must be chromatographed at that value of ϕ_A. The actual number of solutes is thereby found directly from the resultant chromatogram; if the column is of sufficient efficiency, all compounds will also be resolved.

(5) The fictitious lines are identified and removed from the straight-line diagram, and the actual window diagram is constructed for the real solutes. (The resultant optimum column composition may or may not correspond to that determined above.)

(6) If desired and/or necessary, the mixture is chromatographed again at the true optimum column composition, where all peaks will be baseline-resolved.

(7) The entire procedure is repeated at a higher sensitivity setting for minor (trace-level) constituents in the sample.

B. Employment of Relative Retention Data

Application of the foregoing strategy requires that partition coefficients or specific retention volumes be measured as a function of ϕ_A or of w_A. However, this presents certain practical difficulties when, for example, gum phases are employed at elevated temperatures. The alternative is use of relative retentions and capacity factors, that is, parameters that can be measured directly from chromatograms. Extension [70] of the methodology to the analysis of mixtures of initially unknown content and complexity in fact requires only that retentions be represented at intermediate weight-fraction column compositions, that is, ordering of relative retentions by multiplication by β, where

$$\beta = k'(\text{at } w_A)/k'_{(S)} \qquad (36)$$

for the standard solute. (If, in addition, the liquid loadings per unit column length are not equal, β must be multiplied by the ratio of solvent/additive loading weight percents.) The analysis of mixtures of unknown content thus reduces to the sequence noted above, except that relative retentions, capacity factors, and weight fractions (rather than partition coefficients and volume fractions) are employed.

Figure 40 presents the chromatograms [70] of a mixture of an initially unknown number of chlorinated phenols and cresols at 175°C with columns containing mechanically mixed packings of equal liquid loading of (a) pure OV-17 ($w_A = 0$), (b) $w_A = 0.333$, (c) $w_A = 0.667$, and (d) pure Carbowax 20M ($w_A = 1.0$). The phases were chosen on the basis that they have frequently been employed in the past for samples of this type and, more telling, because each gave chromatograms of the sample that were significantly different. Chromatogram (a) shows, for example, eight solutes and comparatively short retentions. Further, near-baseline resolution is found for all but solutes 5 and 6 and it appears (deceptively so) that mere fabrication of a

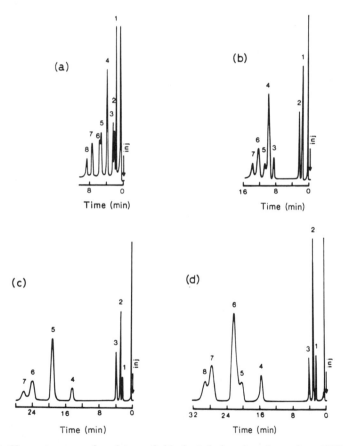

Fig. 40 Chromatograms of a mixture of chlorinated phenols and cresols at 175°C with (a) pure OV-17 ($w_A = 0$), (b) $w_A = 0.333$, (c) $w_A = 0.667$, and (d) pure Carbowax 20M ($w_A = 1.0$). [From Laub and Purnell [70].]

column of higher efficiency with pure OV-17 will enable the separation to be carried out with this solvent. Chromatogram (b) shows only seven peaks, as does chromatogram (c), and little appears to be gained by the addition of Carbowax to the stationary phase. However, chromatogram (d) again reveals eight peaks, but, judging from the height of each, a number of reversals in elution order have occurred on passing from (a) to (d). There must, therefore, be at least eight solutes in the mixture.

Choice of the standard (reference) solute is straightforward with this

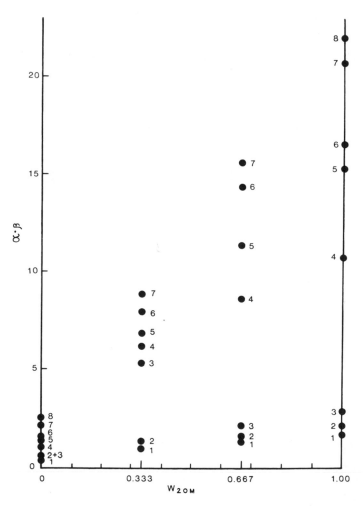

Fig. 41 Plots of retention data relative to peaks (a) 4, (b) 4, (c) 5, and (d) 6 of Fig. 40 against w_A. Points are numbered according to order of elution in Fig. 40. [From Laub and Purnell [70].]

sample, since one of the peaks is readily recognized in all chromatograms, namely, 4 (a), 4 (b), 5 (c), and 6 (d). If that had not been the case, an external standard would have been chromatographed separately with each column or an internal standard added to the mixture.

The $\alpha \cdot \beta$ data measured directly from each chromatogram are plotted against w_A in Fig. 41. As with the earlier example of the industrial still residue, little can be discerned from these plots alone. Furthermore, this sample represents a level of complexity such that it is difficult to draw in lines even

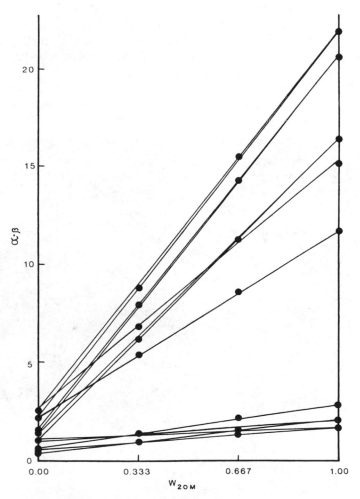

Fig. 42 All possible lines drawn through all possible combinations of sets of points of Fig. 41. [From Laub and Purnell [70].]

after comparison with the chromatograms. As a result, Fig. 42 shows all possible lines drawn through all possible sets of points; altogether there may be as many as thirteen solutes present, which is the ceiling number of compounds to be resolved.

Figure 43 shows the window diagram [70] constructed from the data for all possible solutes of Fig. 42, where the optimum weight fraction of Carbowax is predicted to be 0.02, that is, just 2%. Furthermore, the most difficult alpha at this column composition is 1.095, which is surprisingly good considering the complexity of Fig. 42.

Figure 44 shows [70] the chromatogram of the solute mixture at $w_A = 0.02$ and, for the first time, nine solutes are visible. Furthermore, all are baseline-resolved. Note in addition the spacings between peaks: if thirteen solutes had in fact been present, the extra four would have fallen between 3 and 4, 4 and 5, and 6 and 7.

Finally, having identified the actual number of solutes present via a chromatogram with the optimum solvent composition prescribed by the

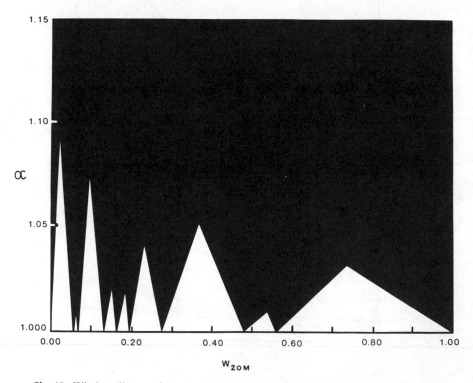

Fig. 43 Window diagram for solutes of Fig. 42. w_A^{opt} is predicted to be 0.02 of Carbowax 20M. [From Laub and Purnell [70].]

window diagram for all possible solutes, the only task left is increasing the speed of analysis. The column length may be shortened, the phase ratio reduced, and so forth, but the simplest method is simply raising the temperature and flow rate; the chromatogram obtained at 200°C at elevated inlet pressure is illustrated [70] in Fig. 45, where all solutes are still separated but where the time of analysis has been reduced from 45 to 17 min.

This analysis of mixtures of initially unknown content and complexity must be regarded as a significant advance insofar as use was made solely of

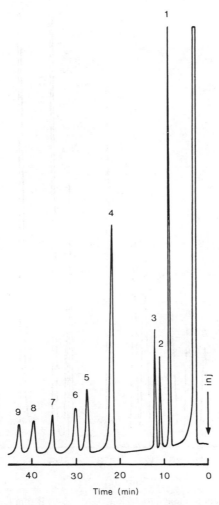

Fig. 44 First-time chromatogram of the chlorinated phenols and cresols mixture with $w_A = 0.02$ at 175°C. [From Laub and Purnell [70].]

relative retention and capacity factor data with polymeric stationary phases. In addition, the solutes employed in the example are considered generally to be difficult to chromatograph with mixed phases because of effects such as gas–liquid interfacial adsorption, yet they were easily resolved via the window diagram procedure.

Indeed, Fig. 44 might well be regarded as the ultimate example of the power of the optimization strategy for gas-chromatographic separations.

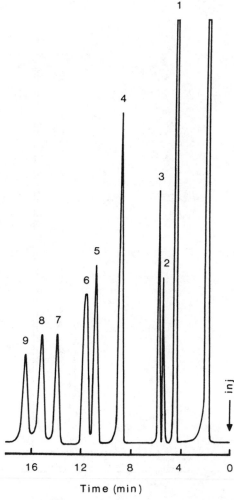

Fig. 45 Chromatogram of the solutes of Fig. 44 at 200°C. [From Laub and Purnell [70].]

V. Contemporary Developments

In the time between original submission and publication of this chapter, a number of important developments have arisen. These include, in particular, advances made in the understanding of mixtures of nonelectrolytes with reference to the diachoric solutions hypothesis, a novel interpretation of solutions thermodynamics (including the Gibbs–Duhem relation) as regards mixed solvents, reformulation of conventional theories to account for solvent self-association with concomitant solute–solvent complexation, and a variety of new insights and applications pertinent to the analytical utility of the plenary optimization strategy. Since each of these areas bears directly upon the subject matter of this chapter, they are briefly recounted here for the sake of completeness as well as for the convenience of the reader.

A. Diachoric Solutions

The bold hypothesis of the microscopic partition model of nonelectrolyte solutions is that mixtures of diachoric solvents, say A and S, yield partition coefficients for third components (solute probes at infinite dilution) which are described by Eq. (21) or its variants, Eq. (31) or (35). Since these relations can be derived only on the assumption that A and S are immiscible, and since, in all experimental studies carried out to date, the mixed phases have in fact exhibited macroscopic miscibility, local aggregation of A and S is said to persist, so that they are microscopically immiscible. The only alternative to this view is that A and S form mixtures which are completely ideal, which, given the range of solvent types investigated thus far, appears highly unlikely.

1. Macroscopically- Immiscible Solvents

While Eq. (21) is said to apply to solvent systems which are immiscible, no distinction is drawn between microscopic or macroscopic aggregation; that is, the relation should hold equally for solvents which exhibit, in intimate admixture, conjugate phases. Application of conventional thermodynamics to this situation by Tiley [45] in 1979 yielded the conclusion that plots of K_R against ϕ_A might or might not be curved in the two-phase region, depending upon the solvent–solvent interaction parameters, the molar volumes of solutes and solvents, and the solute partition coefficient ratio with each of the two pure phases composing the system. Laub, Purnell, and Summers [71] are the only workers thus far to have tested this proposition. Two solvent systems were utilized: tributyl phosphate + ethylene glycol (upper consolute temperature of 53°C at w_{TBP} of 35%) and ethyl benzoate + propylene glycol (upper consolute temperature of 47°C at w_{EB} of 75%; broad) with the solutes furan, methyl formate, and diethyl ether. Plots, first, of density against weight percent for the two systems were linear to well within experimental error (pycnometry), and gave no indication of conjugate solution. Secondly, plots of (GC) solute partition coefficients against volume fraction (21°C for TBP with EG; 30°C for EB with PG) for all systems studied were without question linear. Further, the lines within the two-phase regions (which of course must be straight) were beyond doubt contiguous with those of the single-phase regions.

It may in this instance be argued that the physical properties of the solvents were conceivably modified in the course of distribution on the surface of the GC support material to such an extent that the resultant solute partition coefficients were no longer representative of bulk solution. Two possibilities thereby arise: the two solvent components in each system studied were completely demixed or, alternatively, the two-phase region of each was extended very nearly over the complete volume-fraction range, $\phi_A = 0$ to 1. However, neither situation appears to be supportable in view of the partition coefficients of each solute with each pure phase, these falling almost precisely at the end-points of the (straight) lines.

Alternatively, it has on occasion been argued [72] that Eq. (21) would be expected to apply only in those instances wherein the solutes and solvents are of considerably disparate molecular weight and/or molar volume. It is therfore worth noting that the solute–solvent systems represented in this study are such that there is little difference in molecular size. Thus, in addition to conforming to the criterion of immiscibility, the results of this work belie dependence of the relation on solute or solvent bulk properties. Indeed, Klein and Widdecke [73] have shown that at the other end of the solvent spectrum, i.e., with polymeric stationary phases, the diachoric solutions equation also applies irrespective of formulation of the solvents as statistical homo- or block copolymers.

2. Curvature in Equation (21)

Since publication by Harbison, Laub, Martire, Purnell, and Williams [20] of plots of K_R of indisputable accuracy which exhibited noticeable ($\pm 10\%$) curvature against ϕ_A, a number of explanations have been proffered which purport to account for deviations from Eq. (21). On the one hand, conventional relations (e.g., Eq. (12)) have been extended to include empirically fitted variables, these being cast primarily as binary-solvent interaction parameters. Acree [74], for example, included empirical and judiciously selected parameters in a reformulation of the relations derived first by Martire and his colleagues [14, 35, 75, 76] in order to force conformity of some (but not all) of the data reported by Chien, Kopecni, Laub, Petkovic, and Smith [77–79]. In contrast, the latter group of workers has shown that expansion of Eq. (21) to include solute–solvent interaction and solvent self-association [25] accounts for the modest curvature of solute retention data observed with the self-interactive phase N,N-dibutyl-2-ethylhexanamide in admixture with n-hexadecane and n-octadecane diluents.

Finally, the peculiar notion that Eq. (21) violates the Gibbs–Duhem relation and that it cannot therefore be considered to be rigorously valid under any circumstances has been put forth recently by Snyder and Poppe [80]. Self-evidently, this argument is entirely specious on several grounds. First, of course, since the Gibbs–Duhem relation is formulated in terms of mixed *solvents*, it offers no information regarding the variation of partition coefficients of probe *solutes* as a function of blended-phase composition. Secondly, since in the diachoric solutions formulation the solvent components are assumed to be immiscible, the activity coefficients of each must be taken to be unity over all compositions. The Gibbs–Duhem relation is thereby obeyed exactly, although, since $d \ln \gamma_i = 0$, little further insight is gained. This latter point emphasises once again that for the purposes of Eq. (21), there is in fact no difference between completely ideal mixtures and those whose (diachoric) components are completely immiscible.

The foregoing broadly represents the situation at the present time. Little has arisen which has obviated the generality of Eq. (21) since it and its original data compilation were published by Laub and Purnell [37] in 1976, nor have there yet been formulated relations which are as nearly catholic for non-electrolyte solutions. On the other hand, several systems comprising intimately blended solvents have been shown lately to yield solute partition coefficients which are at some variance with the diachoric solutions equation as well as with more conventional approaches, irrespective of considerable and at times somewhat subjective modification of the latter. The matter thus remains clearly one which invites further vigorous and comprehensive study.

B. Analytical Applications of the Global Optimization Strategy

1. Reviews

Three reviews which in part comprise discussion of various aspects of the window-diagram optimization strategy have been published by Laub [81–83]. The second of these provides a referenced tabulation of the analytical applications of the strategy in liquid and in gas chromatography through 1980.

2. Column Fabrication

One of the most important features of the plenary optimization strategy is that the column efficiency N_{req} required to effect a particular separation is specified. This in turn dictates the column length L_{req} which must be fabricated. Further, it may be found that the separation at hand requires fewer than 30,000 plates; i.e., it can be achieved with packed (as opposed to capillary) columns. Of some importance, therefore, are procedures by which packed columns of ~800–1500 plates per foot can be assembled. Laub and Purnell [84] presented details of such a packing apparatus in 1980: the device consists essentially of a high-pressure, stainless-steel reservoir to which is attached a precoiled length (15-ft maximum) of $\frac{1}{8}$-in. column tube. Packing is added to the reservoir and then displaced by pressure (500 psig nitrogen) into the column. The key feature of the reservoir is a silicone O-ring seal at its top which is recessed to a depth of about two-thirds thickness so that the barrel/cap seal is made at it and not at the cap threads. Thus, only slight compression (finger-tightness) is required to seal the unit to better than 1000 psig. The packing apparatus consistently yielded columns of better than 800 plates per foot and, in one instance, a column in excess of 1400 plates per foot. A typical chromatogram obtained with a column of 50-ft by $\frac{1}{8}$-in. stainless steel (1.5% w/w mechanical mixture of squalane and dinonyl phthalate phases on 120/140-mesh Chromosorb G, AW-DMCS-treated) which was packed as described in 12.5-ft sections and which exhibited 40,000 effective theoretical plates is shown in Fig. 15.

There is, in contrast, no question that capillary (open-tubular) columns yield efficiencies (however these are defined) which exceed by several factors those available from packed columns. Moreover, the excitement generated by chromatograms resulting from true high-resolution open-tubular systems seems destined to be reproduced in every laboratory where packed columns have previously held sway (see, for example, Figs. 1 and 2 of [83]). Often overlooked, however, is the gain in information content which typically results. That is, data from open-tubular columns provide, in addition to enhanced analysis, correlation between chemical makeup and bulk physicochemical properties of the mixture of a kind and degree which are rarely possible with packed columns. For example, Fig. 46 shows a capillary chromatogram of neat unleaded gasoline [85], in which many hundreds of peaks are resolved in a little over an hour (the expanded region to the right of the figure shows 50 peaks and shoulders). Fingerprint patterns such as these are now employed routinely in the petroleum industry for the determination of octane number [86]. Prior to the advent of the capillary GC method, this task was performed by tuning of a standard engine with the fuel of interest with, of course, somewhat subjective results. Virtually the same can also be said of medical diagnostic information available from chromatograms of physiological fluids, of trace constituents in mixtures of importance in the flavor and fragrance industries, in pharmaceutics, in environmental studies, and so on. There can therefore be little doubt that the advent of open-tubular column GC is rightly regarded as an important development in chromatographic analysis and, of equal certainty, that the corresponding revelation of complexities not previously appreciated results in enhanced understanding of the chemical nature of a wide variety of materials.

Details regarding laboratory fabrication of glass open-tubular columns were provided as long ago as 1959 by Desty, Goldup, and Whyman [87] and Desty and Goldup [88], these efforts culminating in the description of a glass-drawing machine by Desty, Haresnape, and Whyman [89] in 1960 with which virtually any length of tube of whatever desired dimensions can easily be made. (Indeed, little improvement in this original apparatus has been effected in the versions which are commercially available today.) Once the glass tube has been drawn, it must be leached with HCl and silylated with

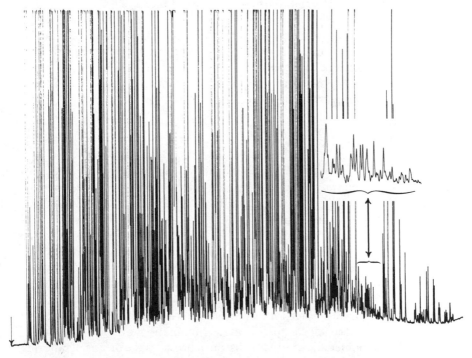

Fig. 46 Chromatogram of neat unleaded gasoline with a glass open-tubular column fabricated in the laboratory of the author [85].

this or that disilazane, the recipe given by Laub, Roberts, and Smith [90] being as effective as any other. Finally, the column is coated by filling with a solution of stationary phase in pentane or in methylene chloride, plugging one end (wood glues work well), and applying a vacuum at the other end until the volatile solvent has been completely removed. It is then ready for conditioning in the usual way.

An important advance in capillary-column technology is the recent introduction by Dandeneau, Bente, Rooney, and Hiskes [91] of pipes comprising fused amorphous silica. The tubes are rendered flexibile by an outer coating of polyimide and are inherently straight and nonfriable; few difficulties hence are encountered in their manipulation. Moreover, the leaching and silylating steps mentioned above for glass columns are obviated since this form of silica is apparently inert to all but the most labile of solutes. The only drawbacks to tubes of this type are that the temperature limit of the outer coating is at present only $\sim 280°C$, they cannot be fabricated in most research laboratories since costs associated with the drawing equipment are prohibitive, and the tubes are available only in stock sizes (generally 0.2- mm or larger) from commercial sources. Another difficulty is that stationary phases of moderate selectivity wet the walls of such tubes only poorly, which results in bead (as opposed to even-film) formation of the solvent, which in turn decreases dramatically the overall column efficiency. Nevertheless, all of these drawbacks can be expected to be overcome in the very near future, and it can be predicted with some confidence that the use of glass columns will at that time be supplanted almost entirely.

3.   Gas–Solid Chromatography

Mixed adsorbents have been used for many years as stationary phases in gas chromatography, where retentions conform strictly to Eq. (31) (cf. Section III.E). For example, Cadogan and Sawyer [92] utilized lanthanum chloride/(diluent) sodium chloride-coated silica for the study of charge-transfer interactions (Eq. (18)) of aromatic hydrocarbon solutes with the former salt. Inspection of Fig. 1 of [92] shows that columns of varying $LaCl_3/NaCl$ ratio provide, in addition, selectivity which differs from one aromatic solute to the next, which, of course, portends application of window diagrams for optimization of the sorbent composition. Kopecni, Milonjic, and Laub [93] have in fact recently described a range of highly selective alkali-metal-modified silica adsorbents; combination of one or more of these with organometallic or other substrates provides the means of tailoring stationary phases for virtually any separations task.

4. Gas–Liquid Chromatography

Several noteworthy developments have occurred since the initial demonstration by Chien, Kopecni, and Laub [53] of the utility of window diagrams for replacement of all GC solvents by a handful of standard stationary phases (cf. Section III.B). Foremost among these is the progress made toward development of such a set of phases, with which virtually any separation can be achieved. Chien, Kopecni, and Laub [94] chose first to evaluate phases which provide separations based almost exclusively on solute volatility; that is, those for which the activity-coefficient ratio of Eq. (11) is near unity. Their extensive tabulation of specific retention volumes of high accuracy demonstrates beyond doubt that the most useful of these, namely the polydimethylsiloxanes SE-30, OV-1, SP-2100, and OV-101, are indeed equivalent. Selection of one over the others can therefore be based upon secondary criteria such as formation and thermal stability of thin films on capillary walls, from which standpoint both OV-1 and SE-30 are satisfactory. The second phase, chosen by Laub and his co-workers to be as different as possible from "boiling-point" solvents, was the liquid crystal N,N'-bis(p-butoxybenzylidene)-α,α'-bi-p-toluidine (BBBT), which has a melting point of 159°C and a nematic–isotropic transition temperature of 303°C. However, BBBT exhibits considerable volatility (column "bleed") at temperatures in excess of 220°C; so, even though offering extraordinary selectivity, the material is of limited utility in routine analysis [95]. A simple solution to this difficulty was shown by Laub, Roberts, and Smith [90, 96] to consist of blending BBBT with a polydimethylphenylsiloxane or a polymethylphenyl-siloxane (e.g., SE-52). Several advantages thereby accrue: first, and perhaps somewhat surprisingly, the selectivity of the mixed phase varies linearly with composition. That is, the properties of this mesomorph are not destroyed upon admixture with silicone gums. Columns of virtually any selectivity, ranging from pure SE-30 to pure BBBT, can thereby be fabricated as if the liquid crystal were any other GC solvent. (This result of course implies that BBBT is dispersed (as opposed to dissolved) in the diluent solvent [97].) Secondly, the vapor pressure of BBBT appears to follow Raoult's law upon dilution in silicones, which has the practical consequence of decreasing accordingly its bleed from the GC column. Thus, the upper temperature limit of such mixed-phase columns is increased approximately in accord with the degree of dilution. Thirdly, while the efficiency of columns of pure BBBT is very poor, the overall efficiency of blends of, say, 20% (w/w) BBBT/80% SE-52 very nearly approaches that of pure SE-52.

These results can be expected generally to hold true in other instances where the mixed-solvents approach is employed. That is, the bleed of the more volatile phase will be reduced as its initial concentration in the blend is diminished, while the overall column efficiency will be an approximately linear but inverse function of the concentration of the poorer of the phases. The selectivity of the mixed solvent is predicted in the usual way from the diachoric solutions relations.

Ultimately, however, it must be recognized that liquid crystals such as BBBT represent a poor compromise between stationary-phase efficiency and selectivity. Finkelmann, Laub, Roberts, and Smith [98] therefore undertook the synthesis and evaluation of mesomorphic polysiloxane (MEPSIL) phases, the properties of which were designed to yield high column efficiency with concomitant high selectivity. Representative examples of this unique class of compounds were first detailed by these workers in 1982, when it was shown that MEPSIL solvents do indeed provide an unprecedented spectrum of adjustable parameters for quantitative control of solute retention and retention order, that is, means of optimization of separations. This work also demonstrates the practical feasibility of definition of a handful of standard stationary phases with which all gas-chromatographic separations can be achieved.

5. Reduction of Analysis Time

An interesting ancillary consequence of the use of blended phases is reduction of analysis time by dilution of sorbent packings with inert support. Al-Thamir, Laub, and Purnell [99, 100] presented several examples of the merits of this technique with regard to GLC, GSC, and GLSC in 1979–1980: they found that the kinetics of solute–stationary-phase mass transfer (the predominant detriment to column efficiency) could thereby be improved dramatically, and that solute capacity factors (hence the overall analysis time) could be adjusted to virtually any desired value. While the inert-dilution technique does not appear to be of consequence in gas–liquid chromatography, these findings are of particular importance in packed-column gas–solid and gas–liquid–solid chromatography where mass-transfer effects are in fact very troublesome.

6. Contemporary Modes of Data Presentation

Three types of retention index schemes for use in prediction of optimum stationary-phase mixtures were published recently by Laub [101]. The simplest of these will reduce in most cases to the approximation described by Pecsok and Apffel [102], wherein Kovats retention indices [103] are substituted for partition coefficients in Eq. (21). This form of data presentation is very attractive for use with window diagrams, since the pertinent values can be taken directly from strip-chart recordings. Furthermore, since retention indices correspond to log(retention ratios), any deviations from the diachoric relations will at the very least be diminished substantially if not masked altogether. The separation of 40 polychlorinated biphenyls provides a recent example where this approach has proved to be of value [104].

As discussed in Section II.H, different forms of ordinate alpha values can be used in plots (window diagrams) of relative retention against column composition. Figure 12 showed, for example, the use of $10^5/N_{req}$ plotted as a function of ϕ_A for several solutes of Table V, where the phase ratio had been increased to a hypothetical value of 100 (thus reducing the capacity factors to well below 10). Since formulation of this example, Jones and Wellington [105] have published a close analogy wherein a separation factor S is defined by the relation

$$S = 2R_S/N^{1/2} \tag{37}$$

where R_S is the resolution between two solutes as defined by Eq. (29). It is a simple matter to show that $1/N_{req} = \frac{1}{4}S^2$ or $S = 2/(N_{req})^{1/2}$ when $R_S = 1.5$ (6σ separation). Plots utilizing alpha, S, N_{req}^{-1}, or $N_{req}^{-1/2}$ ordinates will therefore appear to be quite similar and will in any event yield the same set of optimum abscissa values. The real advantage of use of the separation factor, however, arises upon realization that

$$S = \frac{t_{R(2)} - t_{R(1)}}{t_{R(2)} + t_{R(1)}} \tag{38}$$

that is, S is given by the ratio of the difference to the sum of raw retentions, where the dead time t_A need not be determined. This aspect of generation of window-diagram data is not overly important in GC (where ambiguity in measurement of the elution time of a nonretained solute rarely arises) but is of considerable significance in liquid chromatography. There is in the latter technique in fact some controversy regarding the best method of determination of void time (or volume), for which the procedure described by Al-Thamir, Laub, Purnell, and Wellington [106] has provided a solution. For purposes of generation of window diagrams, however, their method is obviated when the data are reduced in terms of the Jones–Wellington separation factor, Eq. (38).

Mention is required at this point also of recently introduced variants on the geometry of window diagrams. When blends comprising more than two solvents are to be optimized, two-dimensional Cartesian-coordinate plots no longer suffice for representation of relative retentions as a function of composition of the phase. Furthermore, it may be necessary to optimize simultaneously several interdependent variables for which relations such as

Eq. (21) are no longer applicable. Deming and his co-workers [107, 108] have considered situations of these types, and have demonstrated the utility of three-dimensional Cartesian window diagrams with semiempirical methods of data generation. Alternatively, and commensurate with these publications, Kirkland and his colleagues [109] utilized triangular coordinates for optimization of ternary solvents in liquid chromatography. In this method each corner of the triangle represents a pure mobile-phase component, and lines of solute relative retentions are then superimposed on the surface much like a contour map (this and other forms of topological analysis have, of course, been used for many years to describe ternary phase diagrams; cf. Figs. 5.26c and 5.27a of [110]). However, since the results from one coordinate system self-evidently must be identical to those derived from another, there appears to be little from which to choose other than on the basis of convenience of presentation. Issaq and his co-workers [111] have in any event recently provided a detailed analysis of these procedures, where, *inter alia*, their (listed) computer-data reduction and representation scheme appears to be very much superior to those described elsewhere (the system advocated by Lehrer [112], for example, fails when inversions occur in elution order).

Finally, the use of window diagrams has spread recently to other areas of analytical chemistry. The most notable examples are the optimization of solvent pH for the separation of waves in electrochemistry as described by Anderson and Laub [113], and application of the method in resolution of lanthanide-induced spectral shifts in NMR spectroscopy by Laub, Pelter, and Purnell [114]. Variants of the window-diagram technique should moreover prove useful in spectral search routines, where a correlation coefficient would be employed as the ordinate and selected parameters representing the spectra graphed according to a relevant abscissa. By extension, solutions to purely mathematical functions could be found in the same way. The advantage in doing so is avoidance of falling into local minima, which is a serious problem in many search routines (such as SIMPLEX). Further, of course, the strategy returns a *global set* of optima rather than a particular local minimum, where the latter may not in any event correspond to the overall best answer to the problem at hand. There can be little doubt, therefore, that the window-diagram methodology does in fact represent a universal optimization strategy, the utility of which appears at the present time to be unbounded.

Acknowledgments

Support of this work by the National Science Foundation and by the Department of Energy is gratefully acknowledged.

References

1. L. S. Ettre and C. Horvath, *Anal. Chem.* **47**, 422A (1975).
2. D. T. Day, *Proc. Am. Philos. Soc.* 36, 112 (1897); *Science* **17**, 1007 (1903).
3. M. Tswett, *Ber. Dtsch. Bot. Ges.* **24**, 316, 384 (1906).
4. L. S. Ettre, *in* "75 Years of Chromatography—A Historical Dialogue" (L. S. Ettre and A. Zlatkis, eds.), pp. 483–490. Elsevier, Amsterdam, 1979.
5. J. H. Purnell, "Gas Chromatography," p. 1. Wiley, New York, 1962.
6. T. I. Williams and H. Weil, *Nature (London)* **170**, 503 (1952).
7. K. C. Bailey, "The Elder Pliny's Chapters on Chemical Subjects." Arnold, London, 1929.
8. R. J. Laub and R. L. Pecsok, "Physicochemical Applications of Gas Chromatography," p. 3. Wiley (Interscience), New York, 1978.
9. G. Scatchard, *Chem. Rev.* **8**, 321 (1931).
10. J. H. Hildebrand and S. E. Wood, *J. Chem. Phys.* **1**, 817 (1933).
11. A. J. B. Cruickshank, B. W. Gainey, and C. L. Young, *Trans. Faraday Soc.* **64**, 337 (1968).
12. L. B. Kier and L. H. Hall, "Molecular Connectivity in Chemistry and Drug Research." Academic Press, New York, 1976.
13. A. Fredenslund, J. Gmehling, and P. Rasmussen, "Vapor–Liquid Equilibria Using UNIFAC." Elsevier, Amsterdam, 1977.
14. G. M. Janini and D. E. Martire, *J. C. S. Faraday Trans. 2* **70**, 837 (1974).
15. R. J. Laub, D. E. Martire, and J. H. Purnell, *J. C. S. Faraday Trans. 1* **73**, 1686 (1977).
16. R. J. Laub, D. E. Martire, and J. H. Purnell, *J. C. S. Faraday Trans. 2* **74**, 213 (1978).
17. E. A. Moelwyn-Hughes, "Physical Chemistry," p. 682. Pergamon, Oxford, 1957.
18. E. F. G. Herington, *in* "Vapour-Phase Chromatography" (D. H. Desty, ed.), p. 5. Butterworths, London, 1957.
19. A. J. Ashworth and D. H. Everett, *Trans. Faraday Soc.* **56**, 1609 (1960).
20. M. W. P. Harbison, R. J. Laub, D. E. Martire, J. H. Purnell, and P. S. Williams, *J. Phys. Chem.* **83**, 1262 (1979).
21. A. J. P. Martin and A. T. James, *Biochem. J.* **50**, 679 (1952).
22. A. T. James, A. J. P. Martin, and G. H. Smith, *Biochem. J.* **52**, 238 (1952).
23. A. I. M. Keulemans, A. Kwantes, and P. Zaal, *Anal. Chim. Acta* **13**, 357 (1955).
24. E. M. Fredericks and F. R. Brooks, *Anal. Chem.* **28**, 297 (1956).
25. R. J. Laub and C. A. Wellington, *in* "Molecular Association" (R. Foster, ed.), Vol. 2, Chap. 3. Academic Press, New York, 1979.
26. G. R. Primavesi, *Nature (London)* **184**, 2010 (1959).
27. E. Gil-Av and J. Herling, *J. Phys. Chem.* **66**, 1208 (1962).
28. M. A. Muhs and F. T. Weiss, *J. Am. Chem. Soc.* **84**, 4697 (1962).
29. C. A. Wellington, *Adv. Anal. Chem. Instrum.* **11**, 237 (1973).
30. R. J. Laub and R. L. Pecsok, *J. Chromatogr.* **113**, 47 (1975).
31. F. K. Nasyrova, R. S. Giniyatvillin, and M. S. Vigdergauz, *in* "Advances in Gas Chromatography" (M. S. Vigdergauz, ed.), Vol. IV, Part 1, p. 147. Akad. Nauk SSSR, Kazan, 1975.
32. F. K. Nasyrova and M. S. Vigdergauz, *in* "Advances in Gas Chromatography" (M. S. Vigdergauz, ed.), Vol. IV, Part 1, p. 157. Akad. Nauk SSSR, Kazan, 1975.
33. C. L. deLigney, *Adv. Chromatogr.* **14**, 265 (1976).
34. J. H. Purnell, *in* "Gas Chromatography 1966" (A. B. Littlewood, ed.), p. 3. Inst. Pet., London, 1967.
35. D. E. Martire and P. Riedl, *J. Phys. Chem.* **72**, 3478 (1968).
36. J. H. Purnell and J. M. Vargas de Andrade, *J. Am. Chem. Soc.* **97**, 3585, 3590 (1975).
37. R. J. Laub and J. H. Purnell, *J. Am. Chem. Soc.* **98**, 30, 35 (1976).

38. H.-L. Liao, D. E. Martire, and J. P. Sheridan, *Anal. Chem.* **45,** 2087 (1973).
39. C. L. Young, *J. Chromatogr. Sci.* **8,** 103 (1970).
40. A. J. Ashworth and D. M. Hooker, *J. Chromatogr.* **131,** 399 (1977).
41. R. J. Laub and J. H. Purnell, *J. Chromatogr.* **112,** 71 (1975).
42. J. H. Purnell, *J. Chem. Soc.* p. 1268 (1960).
43. R. J. Laub and J. H. Purnell, *Anal. Chem.* **48,** 799 (1976).
44. R. J. Laub, J. H. Purnell, P. S. Williams, M. W. P. Harbison, and D. E. Martire, *J. Chromatogr.* **155,** 233 (1978).
45. P. F. Tiley, *J. Chromatogr.* **179,** 247 (1979).
46. C.-F. Chien, M. M. Kopecni, and R. J. Laub, *Anal. Chem.* **52,** 1402 (1980).
47. J. R. Gant, J. W. Dolan, and L. R. Snyder, *J. Chromatogr.* **185,** 153 (1979).
48. R. J. Laub, J. H. Purnell, and P. S. Williams, *J. Chromatogr.* **134,** 249 (1977).
49. L. R. Snyder, *J. Chromatogr.* **92,** 223 (1974); *J. Chromatogr. Sci.* **16,** 223 (1978).
50. S. T. Preston, Jr., *J. Chromatogr. Sci.* **11,** 201 (1973).
51. J. J. Leary, J. B. Justice, S. Tsuge, S. R. Lowry, and T. L. Isenhour, *J. Chromatogr. Sci.* **11,** 201 (1973).
52. J. K. Haken, *J. Chromatogr. Sci.* **13,** 430 (1975).
53. C.-F. Chien, M. M. Kopecni, and R. J. Laub, *Anal. Chem.* **52,** 1407 (1980).
54. J. F. Parcher, J. R. Hansbrough, and A. M. Koury, *J. Chromatogr. Sci.* **16,** 183 (1978).
55. J. R. Mann and S. T. Preston, Jr., *J. Chromatogr. Sci.* **11,** 216 (1973).
56. D. F. Lynch, F. A. Palocsay, and J. J. Leary, *J. Chromatogr. Sci.* **13,** 533 (1975).
57. R. J. Laub, J. H. Purnell, and P. S. Williams, *Anal. Chim. Acta* **95,** 135 (1977).
58. G. P. Hildebrand and C. N. Reilley, *Anal. Chem.* **36,** 47 (1964).
59. R. J. Laub, J. H. Purnell, D. M. Summers, and P. S. Williams, *J. Chromatogr.* **155,** 1 (1978).
60. A. R. Cooper, C. W. P. Crowne, and P. G. Farrell, *Trans. Faraday Soc.* **62,** 2725 (1966); **63,** 447 (1967).
61. W. K. Al-Thamir, R. J. Laub, and J. H. Purnell, *J. Chromatogr.* **142,** 3 (1977).
62. D. F. Cadogan and D. T. Sawyer, *Anal. Chem.* **43,** 941 (1971).
63. V. Pretorius, *J. High Resolut. Chromatogr. Chromatogr. Commun.* **1,** 199 (1978).
64. A. Goldup, G. R. Luckhurst, and W. T. Swanton, *Nature (London)* **193,** 333 (1962).
65. R. P. W. Scott, *in* "Gas Chromatography 1958" (D. H. Desty, ed.), p. 189. Butterworths, London, 1958.
66. R. J. Laub and J. H. Purnell, *J. Chromatogr.* **161,** 49 (1978).
67. M. Rogozinski and I. Kaufman, *J. Gas Chromatogr.* **4,** 413 (1966).
68. D. T. Sawyer and D. J. Brookman, *Anal. Chem.* **40,** 1847 (1968).
69. R. J. Laub and J. H. Purnell, *Anal. Chem.* **48,** 1720 (1976).
70. R. J. Laub and J. H. Purnell, *J. Chromatogr.* **161,** 59 (1978).
71. R. J. Laub, J. H. Purnell, and D. M. Summers, *J. Chem. Soc. Faraday Trans. 1* **76,** 362 (1980).
72. J. F. Parcher and T. N. Westlake, *J. Phys. Chem.* **81,** 307 (1977).
73. J. Klein and H. Widdecke, *J. Chromatogr.* **147,** 384 (1978).
74. W. E. Acree, Jr., *J. Phys. Chem.* **86,** 1461 (1982).
75. G. M. Janini, J. W. King, and D. E. Martire, *J. Am. Chem. Soc.* **96,** 5368 (1974).
76. D. E. Martire, J. P. Sheridan, J. W. King, and S. E. O'Donnell, *J. Am. Chem. Soc.* **98,** 3101 (1976).
77. M. M. Kopecni, R. J. Laub, and Dj. M. Petkovic, *J. Phys. Chem.* **85,** 1595 (1981).
78. M. M. Kopecni, R. J. Laub, Dj. M. Petkovic, and C. A. Smith, *J. Phys. Chem.* **86,** 1008 (1982).
79. C.-F. Chien, M. M. Kopecni, R. J. Laub, and C. A. Smith, *J. Phys. Chem.* **85,** 1864 (1981).

80. L. R. Snyder and H. Poppe, *J. Chromatogr.* **184,** 363 (1980).
81. R. J. Laub, *Adv. Instrum.* **35**(2), 11 (1980).
82. R. J. Laub, *Am. Lab.* **13**(3), 47 (1981).
83. R. J. Laub, *Trends Anal. Chem.* **1,** 74 (1981).
84. R. J. Laub and J. H. Purnell, *J. High Resolut. Chromatogr. Chromatogr. Commun.* **3,** 195 (1980).
85. R. J. Laub and C. A. Smith, unpublished work.
86. C. Laurgeau, B. Espiau, and F. Barras, *Rev. Inst. Fr. Petr.* **34,** 669 (1979).
87. D. H. Desty, A. Goldup, and B. H. F. Whyman, *J. Inst. Petrol.* **45,** 287 (1959).
88. D. H. Desty and A. Goldup, *in* "Gas Chromatography 1960" (D. H. Desty, ed.), p. 162. Butterworths, London, 1960.
89. D. H. Desty, J. N. Haresnape, and B. H. F. Whyman, *Anal. Chem.* **32,** 302 (1960).
90. R. J. Laub, W. L. Roberts, and C. A. Smith, *J. High Resolut. Chromatogr. Chromatogr. Commun.* **3,** 355 (1980).
91. R. Dandeneau, P. Bente, T. Rooney, and R. Hiskes, *Am. Lab.* **11**(9), 61 (1976).
92. D. F. Cadogan and D. T. Sawyer, *Anal. Chem.* **43,** 941 (1971).
93. M. M. Kopecni, S. K. Milonjic, and R. J. Laub, *Anal. Chem.* **52,** 1032 (1980).
94. C.-F. Chien, M. M. Kopecni, and R. J. Laub, *J. High Resolut. Chromatogr. Chromatogr. Commun.* **4,** 539 (1981).
95. R. J. Laub and W. L. Roberts, *in* "Polynuclear Aromatic Hydrocarbons: Chemistry and Biological Effects" (A. Bjørseth and A. J. Dennis, eds.), p. 25. Battelle Press, Columbus, Ohio, 1980.
96. R. J. Laub, W. L. Roberts, and C. A. Smith, *in* "Polynuclear Aromatic Hydrocarbons: Chemical Analysis and Biological Fate," (M. W. Cooke and A. J. Dennis, eds.), p. 287. Battelle Press, Columbus, Ohio, 1981.
97. R. J. Laub, *in* "Chromatography, Equilibria, and Kinetics" (D. A. Young, ed.), p. 179. Royal Society of Chemistry, London, 1981.
98. H. Finkelmann, R. J. Laub, W. L. Roberts, and C. A. Smith, *in* "Polynuclear Aromatic Hydrocarbons: Physical and Biological Chemistry" (M. W. Cooke and A. J. Dennis, eds.), p. 275. Battelle Press, Columbus, Ohio, 1982.
99. W. K. Al-Thamir, J. H. Purnell, and R. J. Laub, *J. Chromatogr.* **176,** 232 (1979).
100. W. K. Al-Thamir, J. H. Purnell, and R. J. Laub, *J. Chromatogr.* **188,** 79 (1980).
101. R. J. Laub, *Anal. Chem.* **52,** 1219 (1980).
102. R. L. Pecsok and J. Apffel, *Anal. Chem.* **51,** 594 (1979).
103. E. sz. Kovats, *Helv. Chim. Acta* **41,** 1915 (1958).
104. J. Krupcik, J. Mocak, A. Simova, J. Garaj, and G. Guiochon, *J. Chromatogr.* **238,** 1 (1982).
105. P. Jones and C. A. Wellington, *J. Chromatogr.* **213,** 357 (1981).
106. W. K. Al-Thamir, R. J. Laub, J. H. Purnell, and C. A. Wellington, *J. Chromatogr.* **173,** 388 (1979).
107. S. N. Deming and M. L. H. Turoff, *Anal. Chem.* **50,** 546 (1978).
108. B. Sachok, R. C. Kong, and S. N. Deming, *J. Chromatogr.* **199,** 317 (1980).
109. J. L. Glach, J. J. Kirkland, K. M. Squire, and J. M. Minor, *J. Chromatogr.* **199,** 57 (1980).
110. W. J. Moore, "Physical Chemistry," 3rd ed., pp. 155–156. Prentice-Hall, Englewood Cliffs, New Jersey, 1962.
111. H. J. Issaq, J. R. Klose, K. L. McNitt, J. E. Haky, and G. M. Muschik, *J. Liq. Chromatogr.* **4,** 2091 (1981); H. J. Issaq and K. L. McNitt, *J. Liq. Chromatogr.* **5**(9) (1982), in press.
112. R. Lehrer, *Am. Lab.* **13**(10), 113 (1981).
113. L. B. Anderson and R. J. Laub, *J. Electroanal. Chem.* **122,** 359 (1981).
114. R. J. Laub, A. Pelter, and J. H. Purnell, *Anal. Chem.* **51,** 1878 (1979).

Index